Advancements in Human-Agent Teaming Research Infrastructure

As modern military operations become more complex, the need for rapid, coordinated decision-making across multiple domains—air, land, maritime, space, and cyberspace—has never been greater. The U.S. Department of War has identified artificial intelligence (AI) as a critical tool for overcoming these challenges, enabling faster and more efficient responses in high-stakes environments. However, for AI to be effective, it must function well within human teams and organizational structures. This book explores the evolving role of AI in military and defense operations, with a focus on how human-AI teams can be optimized for success.

This title offers a practical guide to the next generation of human-AI teaming. Built on the findings of a major research initiative led by the National Academies Board on Human-Systems Integration, this book compiles key insights into the infrastructure, metrics, and conceptual frameworks essential for advancing human-AI collaboration. Readers will gain an understanding of the latest testbed research, validated performance metrics, and critical concepts for evaluating human-AI teams. This book provides a foundation for improving decision-making and coordination in complex operational settings.

Advancements in Human-Agent Teaming Research Infrastructure: Testbeds, Metrics, and Concepts is an essential resource for researchers, engineers, and policymakers working in AI development, human factors, and defense strategy. It will also benefit military professionals and industry leaders seeking to understand how AI can enhance operational effectiveness while maintaining human oversight and trust.

Advancements in Human-Agent Teaming Research Infrastructure

Testbeds, Metrics, and Concepts

Edited by
Erin K. Chiou, Jason H. Wong,
Douglas S. Lange, and Julie Marble

CRC Press
Taylor & Francis Group
Boca Raton London New York

CRC Press is an imprint of the
Taylor & Francis Group, an **informa** business

Designed cover image: Dr. Laura Canil (www.canilvisuals.com)

First edition published 2026
by CRC Press
2385 NW Executive Center Drive, Suite 320, Boca Raton FL 33431

and by CRC Press
4 Park Square, Milton Park, Abingdon, Oxon, OX14 4RN

CRC Press is an imprint of Taylor & Francis Group, LLC

ISBN: 9781041096290 (hbk)
ISBN: 9781041106043 (pbk)
ISBN: 9781003655923 (ebk)

DOI: 10.1201/9781003655923

Typeset in Times
by codeMantra

Contents

SECTION 1 Testbeds

SECTION 2 Metrics

SECTION 3 Concepts

Foreword

> The challenge of AI is ensuring it remains our servant, not our master.
>
> *Norbert Wiener*

Artificial intelligence (AI) has become heavily integrated into a wide range of human activities at a rapid pace, resulting in both successes and failures that are rooted in the very complexities of cognition. AI has been created to take over or assist humans performance and decision-making, but it will almost never act completely alone. To be successful, we must face the urgent need to carefully integrate human and AI in order to claim the potential benefits of AI and avoid its many pitfalls.

On the downside, misalignments in human and AI performance can result in the inability of people to detect and correct AI errors in a timely manner, unintended alterations of human attention and decision-making due to biasing people toward AI guidance that may not be correct for the situation, both mistrust and overtrust of the AI, and confusion in understanding what the AI is doing and will do. Even with well-designed AI, unexpected problems can arise where the AI may do things that were not anticipated by its designers, or people may exhibit unforeseen changes in their behavior in response to the presence of the AI.

Detecting and addressing these issues requires new testbeds that will allow researchers to explore human-AI interaction and coordination in a controlled environment. While many AI systems, are being launched directly into operational systems (e.g., in automobiles, healthcare technologies, and warfare applications), this creates significant safety risks resulting from latent failures in the system. Human-AI testbeds are needed that provide a safe venue for the discovery of problems and the development of strategies and solutions for successful human-AI teaming.

Human-AI testbeds are envisioned as a critical support system for addressing important questions in the design of AI. They must enable the assessment of detailed aspects of cognition, including human situation awareness and shared situation awareness with the AI system, that underlies successful team performance. Measures of attention, decision biasing, mental models, decision-making, and relevant team processes, such as coordination, communication, and joint adaptation, should be afforded. These human-AI testbeds can provide critical information on the intertwined performance loop of people and AI working together, where the outputs of each form the inputs to the other, following a process of adaptation and change.

Unlike in operational environments, human-AI testbeds are ideal for exploring how people and AI interact and for finding methods to facilitate successful collaboration. Relevant scenarios (both typical and atypical problems) can be developed that provide the controllable, repeatable experiences necessary for scientific exploration and progress. Further, these same testbeds can be utilized as joint training environments, allowing both people and AI to learn what the other is capable of and how to perform effectively as a cohesive unit.

This book provides an important step forward in developing the human-AI testbeds that are so critical to this field. By detailing the capabilities that such testbeds need to incorporate and how to use them effectively, it provides a significant scaffold for building progress in this area. While rapidly developing AI technology holds much promise, it also engenders considerable peril. The challenge for this generation is in navigating this dichotomy.

Mica R. Endsley

SA Technologies

Preface

ORIGINS OF THIS VOLUME

The exponential growth of artificial intelligence (AI)/machine learning (ML) systems is driven by a rapid expansion in the volume and variety of data available for informed decision-making. While this growth represents a transformational shift across public and private sectors, the history of technology adoption offers a cautionary note. Many promising systems have failed to achieve sustained use due to unsolved technical challenges, low usability, or minimal contribution to mission effectiveness. Human-AI teaming has emerged as a response to this challenge, emphasizing the design of systems that can be integrated more effectively into human work contexts.

In 2020, the National Academies Board on Human-Systems Integration was tasked by the Air Force Research Laboratory 711th Human Performance Wing to identify research objectives for realizing effective human-AI teams in complex operational environments. The resulting consensus study (NASEM, 2022), developed by an 11-member committee and downloaded globally over 9,800 times as of this writing, identified 57 research objectives for human-AI teaming spanning technical, cognitive, and organizational dimensions.

Following the publication of that report, a group of 23 U.S.-based researchers in defense operations, machine learning, and human factors engineering convened in San Diego, CA, in August 2023 for a one-day workshop led by researchers from the Naval Information Warfare Center Pacific and Arizona State University. The purpose of this workshop was to prioritize a subset of these research objectives that could be addressed in the near, mid, and long term by the participants and their broader research networks. These priorities were later presented during a panel session held at the Annual Meeting of the Human Factors and Ergonomics Society (Wong et al., 2024). This volume is a direct continuation of that effort.

WHY HUMAN-AI TEAMING NOW

Since the 2023 workshop, U.S. policy has increasingly emphasized the need for rapid development, evaluation, and deployment of human-AI systems. In July 2025, America's AI Action Plan prioritized investment in "AI Interpretability, Control, and Robustness" and called for building an "AI Evaluations Ecosystem" (Executive Office of the President, 2025). Shortly thereafter, in January 2026, the U.S. Department of War released an AI Strategy highlighting experimentation in "AI-enabled battle management and decision support" (U.S. Department of War, 2026). Complementing these efforts, the U.S. National Science Foundation announced $2 million in August 2025 to support AI-ready testbeds for applications in transportation, agriculture, and disaster response (National Science Foundation, 2025). Across these initiatives, a common theme emerges: the need for environments, methods, and metrics that enable meaningful evaluation of human-AI systems under realistic conditions. Despite rapid advances in AI capabilities, the ability to rigorously test and evaluate human-AI teaming performance has not progressed at the same pace.

SCOPE AND CONTRIBUTIONS OF THIS BOOK

This edited volume brings together contributions focused on the design, development, implementation, and evaluation of human-agent teaming (HAT) testbeds, metrics, and concepts. We use the term HAT to emphasize that many of the systems studied extend beyond contemporary AI to include a broader class of automated information systems, including autonomous, semi-autonomous, and rule-based systems. Contributors to this volume include researchers and practitioners from academia, industry, and government, ensuring that perspectives span theoretical development, technological implementation, and operational application. Efforts were also made to include representation from a range of national and international contexts, aiming to broaden the diversity of perspectives across regions and domains. Submissions were solicited through professional networks, as well as through online forums affiliated with the Human Factors and Ergonomics Society and the International Ergonomics Association. Each chapter underwent a partially anonymous peer-review process involving the editors and contributing authors.

Contributors were asked to articulate the generalizability of their concept or use case, as well as the reasoning and tradeoffs underlying design decisions. For example, potential contributors were asked to make explicit why a given testbed, metric, or concept was needed, what alternatives were considered or might be related, and what limitations remained. For testbeds, this included reflecting on the balance between experimental control and ecological fidelity and the extent to which the platforms could be made available to others or modified to address broader research needs. For metrics, this included discussing how to balance scientific rigor (i.e., the frequency and quality of collected data) and unobtrusiveness (e.g., not constantly interrupting the task and introducing artificialities). For concepts, authors were asked to define key constructs, position them relative to related concepts, and discuss how those constructs shape assumptions about human-AI systems. Across all areas, contributors were encouraged to specify what their work clarifies, what remains unresolved, and how it may inform future research. While testbeds, metrics, and concepts are deeply intertwined in practice, chapters are organized into one of the three sections based on the editors' view of the primary contribution of the chapter. Therefore, some chapters listed in the metrics or concepts sections may also describe specific testbeds.

TESTBEDS, METRICS, AND CONCEPTS

The structure of this volume reflects three interrelated pillars of HAT research:

Testbeds provide the environments in which human-agent interactions can be studied and evaluated. Testbeds can range from controlled laboratory test environments to high-fidelity simulations and operationally relevant platforms. Although this volume does not cover the full range of different types and levels of testbed fidelity, it continues in the tradition of Brehmer and Dörner (1993) and Gray (2002), which demonstrated the value of computer-simulated microworlds and synthetic task environments for studying complex, dynamic systems. Several chapters discuss the continued importance of adequately representing the decision and behavioral contexts of complex work domains, such as aviation, cybersecurity, and disaster response.

Across the contributions in this volume, a consistent theme is that existing HAT testbeds remain limited in flexibility, extensibility, and standardization and that contemporary testbeds should increasingly accommodate adaptive, learning-enabled agents, introducing new challenges in experimental design, control, and evaluation. Purpose-built research platforms often support only a narrow set of experimental manipulations or scenario variations, while repurposed environments, such as commercial or game-based systems, can constrain customization and data collection. Several chapters highlight a push for more modular, configurable, and interoperable testbed architectures that can support a wider range of research questions. This includes the ability to vary team structures, task interdependencies, authority allocation, and communication patterns, as well as to capture multimodal data and human and agent behavior. Emerging approaches emphasize open-source and extensible designs that allow researchers to modify scenarios, integrate new agents, and instrument additional measures, supporting both replication and cumulative progress across research groups.

Together, these contributions suggest that future progress in HAT will depend not only on advances in AI capabilities but also on the development of shared, flexible, and research-oriented testbed infrastructures that can accommodate the diversity of teaming paradigms while supporting more scalable evaluation.

Metrics provide the means to evaluate performance and other outcomes of interest, such as trust and team coordination, within these environments. However, compared to advances in AI capabilities and testbed development, progress in metrics has been slower. While individual studies propose measures of human-AI interaction quality or team performance, there remains limited convergence on a single set of standardized, validated approaches. This lack of convergence may reflect a deeper challenge. HAT encompasses a wide range of interaction paradigms, from tightly coupled human-AI decision points, where reliance may be a key metric of interest, to more loosely integrated but interdependent coordination structures, where constructs like joint activity or shared mental models may be more relevant. As such, the notion of a single, universal metric of team effectiveness that could cover this wide range of interaction and teaming contexts may be fundamentally elusive.

At the same time, the contributions in this volume illustrate a few emerging directions in metric development. Metrics are increasingly being designed to capture temporal dynamics, including how trust is established, violated, and repaired over time, or how performance evolves in response to changing task conditions and system behaviors (Chapter 7). For example, *reliance* in the context of HAT can be viewed as a multidimensional phenomenon that reflects learning, calibration, and appropriate use of automation, rather than as a mere binary indicator of human use (Chapter 8). This highlights the importance of distinguishing between in-the-moment use and longer-term patterns of effective interaction. Further metrics like adaptive performance highlight the importance of measuring not only outcomes but also the processes by which human-agent teams recalibrate and recover following disruptions (Chapter 9). These contributions suggest that progress in HAT metrics may depend less on identifying a single standardized measure or set of measures and more on developing context-sensitive, theoretically grounded, and methodologically transparent approaches that can capture the dynamic and multidimensional nature of HAT.

Concepts provide theoretical grounding to frame problems, interpret results, and guide system design. The contributions in this volume illustrate how conceptual frameworks shape how human-agent systems are designed, trained, and evaluated. Several contributions point toward the need for richer models of human-AI interaction. For example, goal-based control reframes human-agent interaction by enabling operators to specify high-level goals and constraints that autonomous systems translate into coordinated actions, drawing on multiple goal pursuit theory and carrying implications for workload, trust, and scalable multi-agent collaboration (Chapter 10). Concepts like shared leadership or shifting authority challenge traditional human-centered hierarchies and require new ways of thinking about collaboration and control (Chapter 11). Similarly, joint activity theory focuses attention on the coupled performance of interacting agents under varying levels and types of challenge (Chapter 12), and machine theory of mind and cyber-social frameworks suggest that effective teaming may depend on the ability of both human and AI agents to represent and respond to one another's states and behaviors in a coordinated manner (Chapter 13). Emerging models that link operator reasoning and system autonomy to training outcomes support the need to understand how people learn to interact with increasingly capable agents (Chapter 14). Finally, work on ambient intelligent environments expands the notion of teaming beyond discrete human-agent interactions to include distributed, embedded systems operating across physical and digital spaces (Chapter 15).

These approaches move beyond viewing AI as an inert tool or subordinate assistant that relies completely on human input to function, instead framing it as a participant in shared cognitive and social processes. The concepts presented in this volume show that situating and evaluating human-agent teams involves a collection of concepts that influence how systems are built, how interactions are structured, and how success is defined and measured (e.g., from task performance to operator learning). Together, testbeds, metrics, and concepts form an interconnected foundation that enables the experimentation, evaluation, and interpretation of HAT, and ultimately its effective realization.

GAPS, TENSIONS, AND OPEN QUESTIONS

Several themes emerge across the chapters in this volume. First, operationalized human-agent teams vary, reflecting broader efforts to define HAT as adaptive, interdependent, and goal-directed interaction, which can differ meaningfully along dimensions such as authority structure, role differentiation, and task interdependence (Chung et al., 2025). Second, the policy emphasis on testing raises an important question of whether the field would benefit from more coordinated, dedicated efforts toward metric development and validation. Is the answer to collect even more data (e.g., at closer intervals and over longer time scales) for established metrics, or to pursue a more integrated approach? Finally, this volume reflects only a subset of ongoing work; it is a snapshot of current practice and thinking in the field and does not aim to be comprehensive. Therefore, important perspectives and approaches are likely to be absent.

HOW TO USE THIS VOLUME

This book is intended to serve multiple purposes. For researchers, it provides a reference for how testbeds, metrics, and related concepts are currently being developed and applied across domains. For practitioners, it offers insights into the practical challenges of designing and evaluating AI intended to be integrated into human work systems. For policymakers and program managers, it underscores progress and gaps in the infrastructure needed to support the ambitious vision of AI integration and effective HAT. More broadly, this volume is meant to be used as a point of departure, capturing a slice of how the research community is currently approaching the evaluation of human-agent systems, with the expectation that these approaches will continue to evolve. By engaging broadly and documenting the successes and unresolved challenges in this area, we hope to support more coordinated and effective efforts moving forward.

Erin K. Chiou, *Arizona State University*

Jason H. Wong

Douglas S. Lange

Julie Marble, *Applied Research Laboratory for Intelligence & Security*

REFERENCES

Brehmer, B. & Dörner, D. (1993). Experiments with computer-simulated microworlds: Escaping both the narrow straits of the laboratory and the deep blue sea of the field study. *Computers in Human Behavior*, *9*(2–3), 171–184.

Chung, H., Holder, T., Shah, J. A., & Yang, X. J. (2025). A systematic review and taxonomy of Human-Agent teaming testbeds. *Human Factors*, *68*(2), 197–238. https://doi.org/10.1177/00187208251376898

Executive Office of the President. (2025, July). *America's AI action plan*. The White House. https://www.whitehouse.gov/wp-content/uploads/2025/07/Americas-AI-Action-Plan.pdf

Gray, W. D. (2002). Simulated task environments: The role of high-fidelity simulations, scaled worlds, synthetic environments, and laboratory tasks in basic and applied cognitive research. *Cognitive Science Quarterly*, 2(2), 205–227.

National Academies of Sciences, Engineering, and Medicine. (2022). *Human-AI teaming: State-of-the-art and research needs*. The National Academies Press. https://doi.org/10.17226/26355

National Science Foundation. (2025, August). *NSF invests in AI-ready test-beds to accelerate U.S. leadership*. https://www.nsf.gov/news/nsf-invests-ai-ready-test-beds-accelerate-us-leadership

U.S. Department of War. (2026, January). *Artificial intelligence strategy for the Department of War.* https://media.defense.gov/2026/Jan/12/2003855671/-1/-1/0/ARTIFICIAL-INTELLIGENCE-STRATEGY-FOR-THE-DEPARTMENT-OF-WAR.PDF

Wong, J. H., Chiou, E. K., Gutzwiller, R. S., Cook, M. B., & Fallon, C. K. (2024). Human-artificial intelligence teaming for the U.S. Navy: Developing a holistic research roadmap. *Proceedings of the Human Factors and Ergonomics Society Annual Meeting*, *68*(1), 380–385. https://doi.org/10.1177/10711813241260352

About the Editors

Erin K. Chiou is Associate Professor of Human Systems Engineering at Arizona State University, USA, with a Ph.D. in Industrial and Systems Engineering from the University of Wisconsin-Madison, USA. Her research leverages social and psychological science to improve human-automation work system design. Recent projects have focused on AI-enabled decision-support systems with applications in complex and safety-critical settings. She directs the Automation Design Advancing People and Technology Laboratory.

Jason H. Wong is a senior cognitive scientist with a Ph.D. in Human Factors and Applied Cognitive Psychology from George Mason University, USA. His work leverages cognitive science to optimize how people use emerging technologies. Currently focused on human-AI teaming in defense, his expertise also includes human-centered design for augmented reality and aligning training systems with human performance and learning principles.

Douglas S. Lange is a senior scientist in Machine Learning and Artificial Intelligence with a Ph.D. in Computer Software Engineering from the Naval Postgraduate School, USA. He has extensive experience in research, development, and technology insertion for current and future command, control, communications, computers, intelligence, surveillance, and reconnaissance (C4ISR) systems.

Julie Marble directs research in human-autonomy teaming at the Applied Research Laboratory for Intelligence & Security and holds a Ph.D. in Human Factors and Cognitive Psychology from Purdue University, USA. She has more than 25 years of experience in human-machine teaming, human-robot interaction, and human reliability analysis, and she is a member of the Board on Human-System Integration at the National Academies of Sciences, Engineering, and Medicine.

Contributors

Richard Agbeyibor
School of Aerospace Engineering
Georgia Institute of Technology
Atlanta, Georgia

Momo Akib
College of Psychology
Florida Institute of Technology
Melbourne, Florida

Liana Algarín
Core4ce
Arlington, Virginia

Lillian K.E. Asiala
Sonalysts, Inc.
Waterford, Connecticut

Michael A. Bailey
USA

Jeffrey Bennett
Naval Information Warfare Center (NIWC) Pacific
San Diego, California

Stephanie A. Boettcher
University of South Florida
Tampa, Florida

Max Brunell
University of South Florida
Tampa, Florida

Mimi Cai
Department of Integrated Systems Engineering
The Ohio State University
Columbus, Ohio

Kyle Canady
College of Psychology
Florida Institute of Technology
Melbourne, Florida

Kendall Carmody
College of Aeronautics
Florida Institute of Technology
Melbourne, Florida

Meredith Carroll
College of Aeronautics
Florida Institute of Technology
Melbourne, Florida

Joshua Charvetto
Defence Science and Technology Group
Fairbairn, Canberra, Australia

Isabelle Clement
Human Computer Interaction
Iowa State University
Ames, Iowa

Samuel Coogan
School of Electrical and Computer Engineering
Georgia Institute of Technology
Atlanta, Georgia

Carmen Jimenez Cortes
College of Engineering and Engineering Technology
Kennesaw State University
Kennesaw, Georgia

Cassandra L. Crone
School of Psychological Sciences
Macquarie University
Sydney, New South Wales, Australia

Mary L. Cummings
Mason Autonomy and Robotics Center
George Mason University
George Mason, Virginia

Sanya Doda
School of Aerospace Engineering
Georgia Institute of Technology
Atlanta, Georgia

Michael C. Dorneich
Department of Industrial and Manufacturing Systems Engineering
Iowa State University
Ames, Iowa

Mark Dras
Performance and Expertise Research Centre
Macquarie University
Sydney, New South Wales, Australia
and
School of Computing
Macquarie University
Sydney, New South Wales, Australia

Yinuo Du
University of Texas at El Paso
El Paso, Texas

Thomas Fahy
Defence Science and Technology Group
Fairbairn, Canberra, Australia

Karen M. Feigh
School of Aerospace Engineering
Georgia Institute of Technology
Atlanta, Georgia

Maria José Ferreira
Carnegie Mellon University
Pittsburgh, Pennsylvania

Kingsley Fletcher
Defence Science and Technology Group
Fairbairn, Canberra, Australia

Stephen B. Gilbert
Department of Industrial and Manufacturing Systems Engineering
Iowa State University
Ames, Iowa

Cleotilde Gonzalez
Carnegie Mellon University
Pittsburgh, Pennsylvania

Eva Gößwein
Department of General Psychology: Cognition
University of Duisburg-Essen
Duisburg, Germany
and
Department of Mechatronics
University of Duisburg-Essen
Duisburg, Germany
and
Chair for Ergonomics and Innovation Management
Chemnitz University of Technology
Chemnitz, Germany

Daniel Griffin
University of South Florida
Tampa, Florida

Cassandra Heffernan
Defence Science and Technology Group
Fairbairn, Canberra, Australia
and
University of Duisburg-Essen
Duisburg, Germany
and
Chemnitz University of Technology
Chemnitz, Germany

Simon Hosking
Defence Science and Technology Group
Australian Department of Defence
Fairbairn, Canberra, Australia

Lixiao Huang
Center for Human, Artificial Intelligence, and Robot Teaming
Arizona State University
Tempe, Arizona

Mohammadreza Jalaeian
Department of Integrated Systems Engineering
The Ohio State University
Columbus, Ohio

Rachel W. Kallen
School of Psychological Sciences
Macquarie University
Sydney, New South Wales, Australia
and
Performance and Expertise Research Centre
Macquarie University
Sydney, New South Wales, Australia

Jack Kolb
School of Aerospace Engineering
Georgia Institute of Technology
Atlanta, Georgia

Nathan Lau
Grado Department of Industrial and Systems Engineering
Virginia Tech
Blacksburg, Virginia

Magnus Liebherr
Department of General Psychology: Cognition
University of Duisburg-Essen
Duisburg, Germany
and
Department of Mechatronics
University of Duisburg-Essen
Duisburg, Germany

Brian Lu
Naval Information Warfare Center (NIWC) Pacific
San Diego, California

Tailia Malloy
University of Luxembourg
Esch-sur-Alzette, Luxembourg

James E. McCarthy
Sonalysts, Inc.
Waterford, Connecticut

Volodymyr Miloserdov
Carnegie Mellon University
Pittsburgh, Pennsylvania

Anirudh More
Department of Health Systems and Implementation Science
Virginia Tech
Roanoke, Virginia

Dane A. Morey
Department of Integrated Systems Engineering
The Ohio State University
Columbus, Ohio

Patrick Nalepka
School of Psychological Sciences
Macquarie University
Sydney, New South Wales, Australia
and
Performance and Expertise Research Centre
Macquarie University
Sydney, New South Wales, Australia

Ranjani Narayanan
School of Aerospace Engineering
Georgia Institute of Technology
Atlanta, Georgia

Amanda K. Newendorp
Department of Industrial and Manufacturing Systems Engineering
Iowa State University
Ames, Iowa

Quintin Oliver
College of Psychology
Gaming Research Integration for Learning Laboratory
Kettering, Ohio

Gaurav Patil
School of Psychological Sciences
Macquarie University
Sydney, New South Wales, Australia
and
Performance and Expertise Research Centre
Macquarie University
Sydney, New South Wales, Australia

Joshua Pearman
Michigan State University
East Lansing, Michigan

Baptiste Prébot
Carnegie Mellon University
France

Mukhunth Raghavan
University of South Florida
Tampa, Florida

Michael F. Rayo
Department of Integrated Systems Engineering
The Ohio State University
Columbus, Ohio

Michael J. Richardson
School of Psychological Sciences
Macquarie University
Sydney, New South Wales, Australia
and
Performance and Expertise Research Centre
Macquarie University
Sydney, New South Wales, Australia

Amy Ryder
PatchPlus Consulting
Pittsburgh, Pennsylvania

James Simpson
Department of Mathematics and Computer Science
Philipps-Universität Marburg
Marburg, Germany

Amanda Thayer
College of Psychology
Florida Institute of Technology
Melbourne, Florida

Jaquelin Tsui
Naval Information Warfare Center (NIWC) Pacific
San Diego, California

Adan Vela
Department of Industrial Engineering and Management Systems
University of Central Florida
Orlando, Florida

Tianzi Wang
General Motors
Detroit, Michigan

Jessica Wildman
College of Psychology
Florida Institute of Technology
Melbourne, Florida

Jennifer Winner
Air Force Research Laboratory
Wright-Patterson Air Force Base, Ohio

Section 1

Testbeds

1 Testbed Requirements to Support Future Research Directions in Human-Agent Teaming

Amanda K. Newendorp, Isabelle Clement, Michael C. Dorneich, and Stephen B. Gilbert

1.1 INTRODUCTION

Human-agent teaming (HAT) is an active field of research, and the rapidly advancing capabilities of artificial intelligence (AI) raise new research questions and challenges for HAT (Endsley, 2023; NASEM, 2021). Researchers utilize simulated, experimental testbeds for HAT to study the collaborative performance of humans and AI agents working together on complex, interdependent tasks. These simulated environments allow researchers to manipulate and measure team dynamics, such as shared cognition, trust, and communication (Chung et al., 2025). Flexible testbeds based on realistic use cases are needed to advance research in HAT (NASEM, 2021). However, much HAT research has been conducted using purpose-built testbeds (Chung et al., 2025). Because purpose-built testbeds vary widely across labs and disciplines, it is difficult to standardize methods, compare results, and reuse testbeds. These testbeds fall into two broad categories: existing applications and games adapted for studying HAT, and testbeds explicitly designed for studying HAT. Existing applications and games are advantageous because of their low software development costs but are often limited in usefulness by licensing restrictions, insufficient data logging, and limited customization capabilities (Asiala et al., 2023). Purpose-built HAT testbeds can address these limitations but often contain only the features and attributes needed for the creators' own research questions (Chung et al., 2025). The goal of this chapter is to identify HAT testbed requirements that researchers should consider when designing new purpose-built testbeds that are flexible enough to address current and future HAT research needs.

Previous reviews have identified common characteristics of empirical HAT studies (Hidalgo et al., 2023; O'Neill et al., 2022), attributes of existing HAT testbeds (Chung et al., 2025), and future research directions in HAT (Lyons et al., 2021; NASEM, 2021). This chapter consolidates these reviews to identify the requirements for testbeds that support future HAT research. These requirements will enable researchers to

DOI: 10.1201/9781003655923-2

develop more flexible HAT testbeds that can be used by interdisciplinary researchers to address diverse research questions. Researchers could adapt these testbeds rather than designing their own software, thereby reducing startup costs and increasing standardization across labs. Additionally, using the same testbeds for multiple studies will facilitate meta-analyses comparing HAT findings across similar experimental contexts.

To identify these testbed requirements, a systematic literature review was conducted to categorize trends in HAT research, including both big-picture research questions and core human factors issues. The big-picture questions are based on trends in research questions. These include emergent topics (e.g., advances in AI) and understudied topics (e.g., larger team configurations). The core human factors issues are based on established human factors topics, such as trust and transparency, that affect HAT interactions and outcomes. The questions and issues that emerged were then explored in detail to generate implications for HAT testbed requirements.

1.2 METHODS

A combination of systematic literature review, affinity diagramming, and topic review was conducted to develop testbed requirements. The systematic literature review was guided by the Preferred Reporting Items for Systematic Reviews and Meta-Analyses (PRISMA) guidelines (Page et al., 2021) to minimize biases during the review process (Lame, 2019). To develop testbed requirements for future HAT studies, the process focused on HAT literature review papers published from 2018 to 2025. This time period has seen a proliferation of literature reviews, providing a ready source of comprehensive assessments of the current state of HAT research. The resulting corpus of HAT literature reviews was analyzed for recurring research questions, gaps, future work in HAT, and human factors issues that have been studied. An affinity diagramming process was used to group closely related topics, resulting in: (1) big-picture questions and (2) core human factors issues in HAT. Each question and topic was then explored in greater detail to generate implications for HAT testbed requirements. Figure 1.1 illustrates the process for identifying, screening, including, and analyzing the literature to develop testbed requirements.

1.2.1 Paper Selection and Keywords

To identify papers, the following search terms were used: ("scoping review" OR "research trends" OR "systematic review" OR "overview" OR "review" OR "literature review") AND ("human-autonomy teaming" OR "human-automation teaming" OR "human-machine teaming" OR "human-robot teaming" OR "human-agent teaming" OR "human-AI teaming"). The Dimensions database (www.dimensions.ai) was searched for publications between 2018 and 2025. This database offers access to a diverse range of research areas, providing a representative sample of papers in HAT. The search resulted in an initial corpus of 5,000 papers.

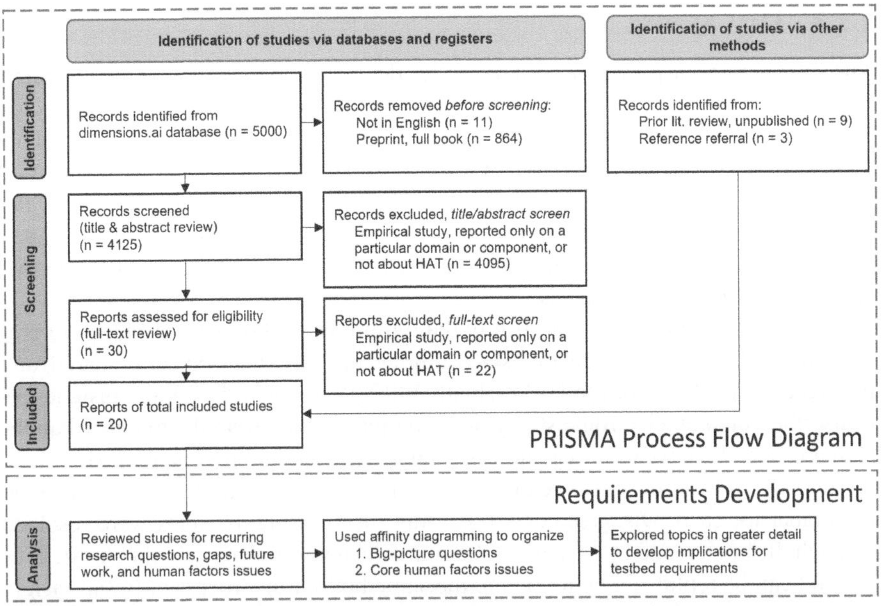

FIGURE 1.1 Search process based on the PRISMA method and analysis method.

1.2.2 Paper Screening and Eligibility Assessment

Based on a review of titles and abstracts, two authors independently assessed the eligibility of the papers. Papers were included if they described a general HAT literature scoping review or overview. The exclusion criteria were:

- Paper was not in English
- Paper was a preprint or a monograph (full book)
- Paper focused on an empirical study with human participants
- Paper focused only on specific components of HAT (e.g., a review focused on trust or transparency in HAT)
- Paper focused on a particular domain (e.g., a review specific to medicine, gaming, etc.).

Empirical studies were excluded to align with the goal of identifying broader research questions, gaps, and future directions in HAT, rather than summarizing findings from individual studies. Including empirical studies could have overrepresented themes that have already been well studied. Screening of titles and abstracts eliminated 4,095 papers, leaving 30 papers. A further 22 papers were removed during full-text screening based on the exclusion criteria, resulting in a corpus of eight papers. Additional papers were added based on two other methods. Nine papers were added from a prior unpublished literature review. Three additional papers were identified by reviewing the references of the remaining papers. This process yielded 20 papers.

1.2.3 Data Analysis

The final corpus of papers was analyzed to identify insights based on key research trends, questions, and gaps. Unlike many systematic reviews, this review extracted key information from each paper without a predetermined coding scheme. This appraoach allowed insights to be determined bottom-up from the data rather than top-down from pre-determined catgeoities. Two authors independently coded the same three papers and then met to compare findings. After confirming that they had identified the same insights, the remaining 17 papers were divided between the two authors for independent analysis.

To organize the findings, an affinity diagramming method was employed to create representations of the relationships among insights that could be shared and interpreted by others (Beyer & Holtzblatt, 1997; Brassard, 1996). Each insight was transcribed onto an electronic sticky note and placed on a shared whiteboard application using Miro, where three authors silently grouped and regrouped notes. Once the notes were in groups, the team collectively discussed each cluster and decided on a label for each. Disagreements regarding the grouping of notes were resolved through discussion and by reviewing the source material for additional context. This method can be replicated by following the same bottom-up process of initial insight extraction and shared agreement on affinity diagram-derived themes. This process was used to encompass a broad range of research areas in HAT. The outcomes are intended to apply to interdisciplinary audiences (e.g., social scientists, systems engineers, and human factors researchers).

After identifying emergent HAT questions and topics through affinity diagramming, a set of papers was collected for each. These additional papers were identified using dimensions.ai, with a modified version of the original search query for each HAT question and topic. For example, HAT papers about training were identified with the following query: (agent OR automation OR autonomy OR autonomous OR robot) AND (training). These topics were then reviewed to develop implications for testbed requirements. The overarching goal was to identify aspects of current and future HAT research that should be considered in the design of a research testbed.

1.3 RESULTS

The affinity diagramming process yielded seven key questions in HAT and four core HAT human factors issues (see Figure 1.2). The questions and issues were mapped to the final corpus of 20 papers, as shown in Table 1.1. A paper was mapped to one of the questions or topics if it included a substantive discussion of the topic (i.e., merely mentioning a topic did not warrant mapping).

1.3.1 Big-Picture Questions in HAT

The big-picture questions in HAT are based on broad trends in HAT and AI. They consider how rapidly developing AI capabilities and human perceptions of increasingly autonomous agents influence future HAT systems. These questions ask what research is needed to inform the design of real-world HAT solutions, including training,

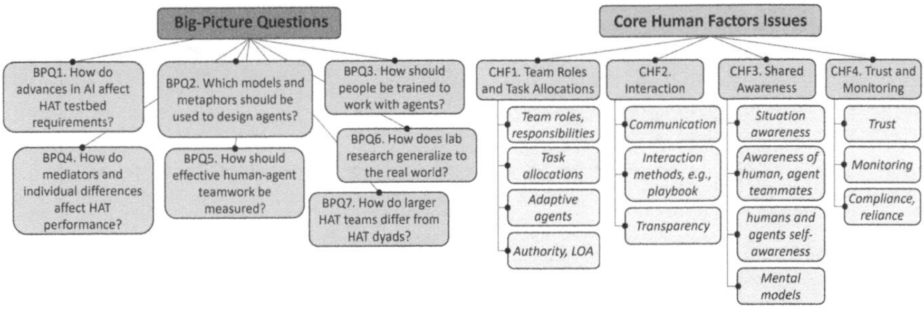

FIGURE 1.2 Affinity diagram from analysis of the corpus of papers to identify big-picture questions and core human factors issues.

consideration of individual differences, performance and teamwork measures, and generalizability. This chapter does not offer answers to all of these questions; rather, it describes current thinking on these topics and presents implications for HAT testbed design requirements at the end of each subsection.

1.3.1.1 How Do Changes in AI Technology Affect HAT Testbed Requirements? (BPQ1)

AI systems are increasingly capable of collaborating on complex cognitive tasks, which creates a greater need to help the system and the user understand each other (Berretta et al., 2023; Patel et al., 2025). Interacting with these systems requires social and teamwork constructs, such as communication, cohesion, and shared situation awareness (Lyons et al., 2021). Whereas humans primarily calibrate their trust in automation by observing its abilities, their trust in autonomous agent teammates is additionally predicted by overall team communication and coordination (Chiou & Lee, 2023). These factors can be examined through interdisciplinary lenses (NASEM, 2021; Schmutz et al., 2024), such as sociotechnical theory (Ang et al., 2025), which offer unique insights (e.g., social scientists bring expertise on human behavior, while computer scientists bring knowledge of how agents interpret that human behavior). Agents increasingly play the role of teammates, and this shift introduces more social influences and implications to HAT (Lyons et al., 2021; NASEM, 2021; Wynne & Lyons, 2018). HAT testbeds should support this shift by enabling researchers to manipulate agent capabilities (e.g., autonomy level, social intelligence).

Current limitations of AI systems have implications for testbed requirements (Endsley, 2023; NASEM, 2021). First, while naturalistic language (e.g., in large language models [LLMs]) is easy to understand, it can be challenging to trust without sufficient references and rationale (communication and explanation style). Second, there is little visibility into the AI system's internal logic and rules, making it difficult to form accurate mental models. Finally, the AI system's ability to adapt over time makes it challenging to maintain a current, accurate mental model (bi-directional information exchange).

AI technology has introduced more non-deterministic agents to HAT (e.g., those built on machine learning or LLMs). Their probabilistic nature and dependence on

TABLE 1.1
Mapping of Papers to the Big-Picture Questions (BPQ) and Core Human Factors (CHF) Issues

Paper	BPQ1	BPQ2	BPQ3	BPQ4	BPQ5	BPQ6	BPQ7	CHF1	CHF2	CHF3	CHF4
Ang et al. (2025)	X					X		X	X	X	X
Berretta et al. (2023)	X	X						X	X	X	X
Capel and Brereton (2023)	X							X	X		X
Chen et al. (2014)			X	X				X	X	X	X
Cooke and Lawless (2021)		X	X		X	X		X	X	X	
Endsley (2023)	X		X			X		X	X	X	X
Holder et al. (2023)	X		X	X	X	X		X	X	X	X
Lyons et al. (2021)	X	X	X	X		X		X	X	X	X
McNeese et al. (2023)	X	X	X			X	X	X	X	X	
Mouloua et al. (2019)						X		X			X
Myers et al. (2024)	X	X	X		X			X	X		X
NASEM (2021)	X	X	X	X	X	X	X	X	X	X	X
O'Neill et al. (2022)	X	X	X	X	X	X		X	X	X	X
Patel et al. (2025)	X						X		X	X	X
Schelble et al. (2020)	X		X	X		X		X	X	X	X
Schmutz et al. (2024)	X		X	X		X	X		X	X	X
Shively et al. (2018)	X							X	X	X	X
Steinberg (2021)		X		X	X	X		X	X	X	
Tummala et al. (2025)	X		X				X	X	X	X	X
Wynne and Lyons (2018)		X		X	X			X	X	X	X

training data raise novel HAT requirements (NASEM, 2021). This non-deterministic nature suggests a need for "uncertainty-aware" agents that can convey uncertainty to users in an interpretable way (NASEM, 2021; Tomsett et al., 2020). HAT testbeds should include scenarios in which varying levels of agent uncertainty affect HAT outcomes. Furthermore, biased training data can produce biased agents, and humans may hold biases toward AI systems (NASEM, 2021; Tummala et al., 2025). These biases may interact with one another, and they can also be exploited in adversarial attacks. Future HAT systems are subject to ethical considerations, including the incorporation of human values and data privacy (Ang et al., 2025; Capel & Brereton, 2023). These ethical considerations could be incorporated into testbed scenarios (e.g., agents that can fire on military targets more quickly than a human but may occasionally make mistakes).

Ongoing developments in AI, such as large-scale foundation models and agentic LLM systems, are increasingly integral to HAT research. For example, LLMs have been used to generate courses of action in military research (Goecks & Waytowich, 2024) and to generate conversational agents across multiple domains (Rheu et al., 2021). As new AI technologies are developed, HAT testbeds should allow researchers to interface with AI agents external to the testbed. This would enable researchers not only to study new types of agents but also to compare similar agents or training datasets in different experimental contexts. An application programming interface (API) that defines input-output protocols will enable the testbed to communicate and interact with external agent software, allowing researchers to study new technology without redesigning the testbed.

1.3.1.1.1 Implications for HAT Testbed Requirements

- The testbed should provide a way to manipulate AI-related abilities, such as level of autonomy, social intelligence, communication style, and explanation style.
- The testbed should include scenarios that vary in uncertainty, enabling assessment of how different levels of uncertainty affect decision-making outcomes and performance metrics.
- The testbed should include scenarios in which participants are asked to contend with conflicting value systems or biased training data.
- The testbed should define an API that enables it to interface with external agents or widgets, ensuring compatibility with future AI systems.

1.3.1.2 Which Models and Metaphors Should Be Used to Design Agents? (BPQ2)

Researchers in HAT debate which models, metaphors, and research disciplines should be leveraged in the design of agents (Berretta et al., 2023; McNeese et al., 2023; Myers et al., 2024; NASEM, 2021; O'Neill et al., 2022). One approach to integrating more capable agents is to leverage existing relationship paradigms, such as a butler or assistant (Desai & Twidale, 2023; Newendorp et al., 2024), which can reduce cognitive load by helping users build a mental model of the agent. Since humans are accustomed to working with complex, dynamic human teammates, a human-like teammate paradigm may be effective for describing agents. Agents can be modeled along two dimensions: humanness (the ability to connect with humans) and autonomy (the ability to function without human oversight) (de Visser et al., 2018). High humanness may afford high-bandwidth communication for conveying complexity with minimal demand on human resources (e.g., gestures or conversational tone carry more meaning than words alone). Highly autonomous agents may benefit from more human-like behaviors, such as strategies to repair trust when agents make errors. On the other hand, agents with low autonomy, narrow functionality, and limited communication may not require human-likeness. HAT testbeds should enable researchers to vary an agent's level of automation (autonomy) and social intelligence, appearance, or communication modes (humanness).

Alternatively, agent interactions can be built on humans' history of collaborating with animals (Billings et al., 2012; Steinberg, 2021). Human-animal teaming utilizes animals' unique strengths (e.g., dogs' sense of smell) and different communication modalities (verbal commands from humans to dogs and physical actions from dogs to humans). Rather than trying to build an AI agent that can reason at the level of a human teammate, in this paradigm the agent has limited, specific skills that mesh well with those of its human teammates. An HAT testbed should allow researchers to study a range of paradigms by varying an agent's role (e.g., optimizing for agent and human strengths) and interaction style (e.g., verbal vs. non-verbal communication).

Finally, other researchers have suggested that agents should be considered only sophisticated tools rather than teammates (Sarkar, 2023; Shneiderman, 2020). Automated agents lack ethical or legal responsibility, possess unique capabilities, and require some level of human supervision (Shneiderman, 2020). Although AI agents emulate human behavior, they do not possess true intelligence or deep understanding. Ultimately, agents are products of human design, and their capabilities reflect underlying human programming (Sarkar, 2023). Although many HAT researchers instead support the agent-as-teammate perspective (NASEM, 2021), an HAT testbed should support the entire range of agent metaphors, from agent-as-a-tool to agent-as-an-animal.

1.3.1.2.1 Implications for HAT Testbed Requirements

- To support a range of metaphors, the HAT testbed should enable researchers to vary agent characteristics, such as: level of automation, authority, communication style, teamwork skills, emotional intelligence, interaction style (e.g., verbal vs. non-verbal communication).

1.3.1.3 How Should People Be Trained to Work with Autonomous Agents? (BPQ3)

While the previous question focused on how to design the agent teammate, this question focuses on how to "design" the human teammate in HAT through training. Although human-human team training has been extensively studied (e.g., Aguinis & Kraiger, 2009), human-agent team training is still in its early stages. Because autonomous agents are unfamiliar to many people and are being integrated into workplaces, there is a need to develop effective training methods (Endsley, 2023; Lyons et al., 2021; McNeese et al., 2023; NASEM, 2021; O'Neill et al., 2022). One aspect of HAT training is helping users build an accurate mental model of the agent and set team expectations (NASEM, 2021). This has sometimes been accomplished outside the testbed through pre-task training protocols (Farah et al., 2026; Johnson et al., 2023). Other training approaches (such as co-training, cross-training, and agent-led training) can also be studied, as agent behavior has been shown to influence team behavior (McNeese et al., 2017).

Co-training involves team members rehearsing a task together. The testbed should enable teams to work together for an extended period, such as through multiple trials. An ideal HAT testbed would enable researchers to generate scenarios with adjustable properties (e.g., task difficulty, length), allowing human and agent teammates to practice their teamwork and task skills. Cross-training involves team members training in each other's roles to form a more flexible and resilient team (Lancaster et al., 2025). Flexibility in a team is the ability to adapt to changing circumstances, embrace new roles, share skills with teammates, and be proactive when adapting to unexpected challenges (Iftikhar et al., 2023). Just as all-human teams benefit from learning each other's roles (Salas et al., 2008), humans can be trained to perform agent tasks when needed (Cooke & Lawless, 2021). This type of cross-training is an important strategy among competitive e-gamers who regularly engage in HAT-like interactions with in-game AI agents (Lancaster et al., 2025). Training should ensure that humans know when it is appropriate to provide backup to a teammate (Cooke & Lawless, 2021). Testbeds can support this type of training by allowing human and agent teammates to change tasks and roles during the experiment.

An agent teammate may also take on the role of a coach, helping human teammates improve their performance (NASEM, 2021). Underlying all of these training approaches is a testbed requirement to provide robust methods for measuring both individual and team performance so that researchers can objectively assess the effectiveness of the training (see Section 1.3.1.5). The testbed should automatically capture as many metrics as possible to lower the burden of post-experiment analysis. Task performance measures can be captured using objective metrics, such as time spent on the task, error counts, or mission success.

1.3.1.3.1 Implications for HAT Testbed Requirements

- The testbed should enable researchers to include an agent in multiple roles, either as a teammate to complete tasks or as a trainer/coach.
- The testbed should include measures of individual tasks, team tasks, as well as team performance to evaluate the effectiveness of training methods.

1.3.1.4 How Do Mediators and Individual Differences Influence HAT Performance? (BPQ4)

This question is based on recommendations to study the effects of individual differences and emergent team behaviors on HAT outcomes (Chen & Barnes, 2014; NASEM, 2021; Wynne & Lyons, 2018). The Input-Mediator-Output-Input (IMOI) model is frequently used to study teamwork (Salas et al., 2008) and has been recommended for use in future HAT studies (Hidalgo et al., 2023; O'Neill et al., 2022). This model describes the relationship among inputs (individual differences), mediators (emergent teamwork states and processes), and outputs (outcomes related to the team's goals and performance). A feedback loop represents carryover effects from previous experiences. When the feedback loop is included, knowledge of mediators and outputs is carried into the next session as input. Humans learn more about their agent teammates over time, and sophisticated agents can learn their human teammates' behaviors and preferences over time (de Visser et al., 2018). This feedback can lead to decreased trust in an autonomous agent if performance and behavior are mismatched (de Visser et al., 2016) or increased implicit coordination with a sophisticated autonomous agent (McNeese et al., 2017).

An HAT testbed can support these research questions by enabling the study of individual differences in both agents and humans, which affect HAT performance and require further study (Chen & Barnes, 2014; Wynne & Lyons, 2018). Individual differences are an important consideration when integrating autonomous systems into applied settings (Schelble et al., 2020) and when designing HAT training protocols (Schmutz et al., 2024). An agent's individual differences can be manipulated directly by varying its reliability, training data, communication style, or level of automation. While researchers have limited control over their participants' characteristics, the testbed can be designed to facilitate the study of specific individual differences. For example, the testbed scenario could be set in a domain where it is easy to recruit both novices and experts (e.g., pilots vs. non-pilots). Alternatively, the testbed could assist researchers in generating multiple trials of increasing difficulty, allowing a subset of participants to develop expertise through repeated trials. An HAT testbed should also enable researchers to study the interactions among inputs, mediators, and outputs by allowing them to manipulate multiple variables, such as task allocations, team compositions, or situational factors.

1.3.1.4.1 Implications for HAT Testbed Requirements

- The testbed should enable researchers to manipulate individual agent differences, such as reliability, training data bias, communication style, and automation level.
- The testbed should enable researchers to create multiple scenarios in which these variables can be manipulated.
- The testbed should enable researchers to manipulate multiple variables (e.g., task allocations, team composition, time pressure) to study interaction effects.

1.3.1.5 How Should Effective Human-Agent Teamwork Be Measured? (BPQ5)

Although the IMOI model provides a valuable framework for defining experiments about teamwork and classifying teaming constructs (Hidalgo et al., 2023; O'Neill et al., 2023), there is still significant ongoing work to define and validate teaming constructs and measurable behaviors (Cooke et al., 2000; Goodwin et al., 2018; Myers et al., 2024; NASEM, 2021; Roberts et al., 2022). Team constructs are often measured indirectly through subjective assessments (Sottilare et al., 2017; Wiese et al., 2015). There are few validated objective measures of team skills (O'Neill et al., 2022, 2023). Team skill metrics are more challenging to capture, as they often involve a combination of behaviors, including teammate monitoring, backup behavior, communication, and collaboration. Behavioral markers (Sottilare et al., 2018) offer examples of countable behaviors that serve as proxies for team skill constructs, such as "frequency with which team members remind each other to follow through on their tasks" (p. 241) as an indicator of low trust. An initial framework for operationalizing team metrics based on observable behaviors (Dorneich et al., 2023) was proposed, grounded in a teaming action loop (Rousseau et al., 2006) to situate team constructs, behaviors, and metrics. Team metrics were organized into four stages of team execution (preparation, execution, evaluation, and adjustment) and emergent team chemistry categories. Such a framework would enable researchers to assess teamwork in a reproducible manner, correlate teamwork and taskwork, and assess the efficacy of different design approaches for supporting team performance. This is an active area of research, with work needed to establish a fully validated framework with metrics that are relevant, credible, replicable, diagnostic, and verifiable. When possible, a testbed should classify and collect behaviors to support the evaluation of team functioning. Future work would identify which behaviors could be feasibly assessed via machine learning classifiers that observe actions in the testbed and interpret dialogue between team members using natural language processing. The testbed might also suggest pre-populated options for validated (or commonly used) measures of teamwork, allowing comparison of results across studies.

HAT testbed design can leverage research on which task design features elicit desirable teaming behaviors. Collaborative video games, such as Overcooked 2, have been analyzed to identify game features (e.g., information asymmetry, shared challenges) that induce specific teamwork competencies (Farah et al., 2023; Tokadlı et al., 2018). An HAT testbed could include these task features to elicit desirable teaming behaviors.

1.3.1.5.1 Implications for HAT Testbed Requirements

- The testbed should include a mechanism for researchers to select which events and behaviors to record.
- The testbed should track and count measurable teamwork behaviors, potentially including pre-populated options for validated teamwork measures.
- The testbed should include tasks and scenarios that are designed to elicit teamwork (e.g., task interdependence, information asymmetry, shared challenges).

1.3.1.6 How Does Lab Research Generalize to the Real World? (BPQ6)

As automation becomes more integral to modern workplaces, HAT researchers are calling for more longitudinal or field studies to understand how lab-based results extend to the real world (Cooke & Lawless, 2021; Endsley, 2023; Goodwin et al., 2018; Holder et al., 2023; O'Neill et al., 2022; Schaefer et al., 2021; Steinberg, 2021). Lab research often involves small teams unfamiliar with each other working together over a relatively short period. This can exclude important team tasks such as team selection, training, planning, rehearsal, evaluation, and maintenance (Steinberg, 2021). Some experiments have incorporated a feedback loop by asking participants to perform multiple trials over an extended period and examining learning effects (McNeese et al., 2017), but it is rare to find published field studies in which the feedback loop occurs naturally.

However, recognizing a need for longitudinal and field testing does not negate the need for experimental testbeds. Longitudinal tests can be performed in the lab by bringing the same participants back for multiple sessions, and a high degree of ecological fidelity (realism) can be ensured by basing studies on realistic use cases (Schmutz et al., 2024) and including subject matter experts when designing an HAT testbed (Mallick et al., 2023). Multiple sessions can help humans learn about an agent, influencing their actions in later trials (Wynne & Lyons, 2018) and the formation of trust (de Visser et al., 2020). HAT testbeds can support these studies by allowing researchers to manipulate situational factors that reflect real-world characteristics, such as time pressure, attentional demands, task load, task complexity, task criticality, and uncertainty.

For the experimenter, designing multiple scenarios for repeated trials can be burdensome. The challenge lies in keeping the control variables constant while changing the details enough to eliminate learning effects. For instance, if a team of soldiers needs to clear a building, repeated trials using the same building layout will likely cause performance to be determined by familiarity with the layout as much as by an increase in teaming skills. An experimenter would need to create a new building layout for each trial. A testbed that can automatically generate isomorphic scenarios based on a scenario's key features would facilitate the rapid development of trials for use in a longitudinal experiment.

1.3.1.6.1 Implications for HAT Testbed Requirements

- The testbed should allow researchers to manipulate situational factors, including time pressure, attentional demands, task load, task complexity, task criticality, and uncertainty.
- The testbed should enable researchers to include sufficient realism to represent real-world scenarios.
- The testbed should allow researchers to build custom scenarios that represent real-world use cases.
- The testbed should enable researchers to quickly develop isomorphic scenarios to support longitudinal testing while accounting for learning effects between trials.

1.3.1.7 How Do Larger HAT Teams Differ from HAT Dyads? (BPQ7)

This question is based on recommendations to study larger human-agent teams (Chung et al., 2025; NASEM, 2021; Patel et al., 2025; Tummala et al., 2025). Future teams may include multiple agents and/or multiple humans performing complex tasks, such as in a command-and-control setting (Patel et al., 2025). Larger teams may require different training protocols and exhibit different teamwork dynamics (NASEM, 2021; Schmutz et al., 2024). However, the majority of HAT testbeds support teams of one human and one agent (the most common team composition) or one human and multiple agents (such as swarms of agents, the second most common composition) (Chung et al., 2025). One notable exception is the CERTT testbed, which supports two humans and one agent and has been used in more than 30 published studies as of 2025 (Chung et al., 2025).

Larger team compositions can lead to different agent roles and communication styles. While many HAT studies include subordinate agents that provide support, agents with a higher level of automation may oversee a network of humans or agents (Chung et al., 2025). This would have implications for communication, such as information flow and asymmetry. A supervisory agent might be situated at the center of a hub-and-spoke communication network and disseminate information to other team members (Chung et al., 2025). To support communication analysis, HAT testbeds should log the communication streams of all teammates using a common timestamp.

1.3.1.7.1 Implications for HAT Testbed Requirements

- The testbed should allow researchers to vary the team composition, including teams with multiple humans and/or multiple agents.
- The testbed should include agent roles unique to larger teams, such as an agent overseeing a team of humans and/or agents.
- The testbed should log the actions of each human and agent using a common timestamp.

1.3.2 Core Human Factors Issues in HAT

In addition to the big-picture questions in HAT, core human factors issues are key topics related to HAT performance. In the IMOI model, these issues are classified as inputs (e.g., team composition) and mediators (e.g., emergent properties such as team trust). When conducting experiments in a testbed, these core issues often serve as independent variables in a study. In this review, closely related topics are discussed together. For example, agent transparency, communication style, and interaction methods are discussed as aspects of team interaction in Section 1.3.2.2. Each topic is then related to testbed requirements for independent variables.

1.3.2.1 Team Roles and Task Allocations (CHF1)

Team roles and responsibilities include task allocation, authority, and levels of automation. Allocating responsibilities among team members begins with understanding the unique strengths of each human and agent and building the right teams (Chen & Barnes, 2014). In human teams, each person brings unique knowledge, skills, and abilities that may qualify them for a specific role (Cooke & Lawless, 2021; Steinberg, 2021). Similarly, there may be value in creating unique agents rather than a pool of identical agents (Steinberg, 2021). This could lead to unique groupings of teammates and responsibilities. This heterogeneity of roles and responsibilities, along with interdependence among teammates, is a defining feature of teams, as opposed to groups (NASEM, 2021). An HAT testbed can provide multiple roles for both humans and agents, such as piloting a drone, providing recommendations, or supervising a team.

Traditionally, automation roles and function allocation are determined during the design phase, where functions are assigned principally in terms of the capabilities of the automation and humans (Dekker & Woods, 2002; Fitts, 1951). However, teams must respond appropriately to changing conditions, which can involve dynamically reassigning tasks during operations (Scerbo, 1996). This raises a need to consider levels of agent autonomy and whether agents have the authority to reassign tasks. When a system adapts during task execution, performance can be improved and workload reduced, but often at the cost of situation awareness and skill degradation (Dorneich et al., 2017; Volz & Dorneich, 2020).

A framework for characterizing adaptive systems describes both the types of adaptations and their triggers (Feigh et al., 2012). Types of adaptations include modifications to function allocation (who performs each task), task scheduling (duration and priority of task timing), interactions (how the system interacts with users, such as communication modality and display layout), and content (the information provided to the user). These adaptations may be triggered by the operator, system, environment, task or mission, and spatiotemporal triggers. Testbeds can support these studies by enabling researchers to use event triggers when scripting experiments.

Other challenges with adaptive task allocation include team members' need to clearly understand their responsibilities, avoid role confusion, and know when and how to back each other up (Cooke & Lawless, 2021). There must be effective mechanisms to facilitate a smooth transfer of tasks and roles (Endsley, 2023). Future work to address these challenges could include developing user interface (UI) elements to facilitate shared awareness of each team member's status, quickly briefing team members on newly assigned tasks, and developing training approaches to ensure team members can back each other up (discussed in Sections 1.3.1.3 and 1.3.2.3).

1.3.2.1.1 Implications for HAT Testbed Requirements

HAT testbeds can be used to study various team configurations and task allocations. This could include testing different combinations of humans and agents filling the available roles, transferring control of assets among different teams, or reallocating tasks in response to changing conditions. Agent support might include performing a suitable set of tasks and appropriately adjusting and reassigning tasks when conditions change. Independent variables may include:

- The number of humans and agents on the team
- Which tasks, and how many tasks, are assigned to each human or agent
- The method of allocating tasks among humans and agents
- Which humans or agents have the authority to reassign tasks and at what level of agent autonomy
- Which events trigger the reassignment of tasks
- The number of agents supervised by each human.

1.3.2.2 Team Interaction (CHF2)

Team interactions include topics such as communication, interaction, and explanation styles. Agent transparency and communication are consistently listed among the key issues of interest in HAT (Chen & Barnes, 2014; Cooke & Lawless, 2021; Endsley, 2023; Lyons et al., 2021; NASEM, 2021; Schaefer et al., 2021; Shively et al., 2018) and have been explored in explainable AI (XAI) research (Holder et al., 2023; NASEM, 2021). Humans and agents must be "able to explain and to be explicable to each other" (Cooke & Lawless, 2021, p. 74). Agent communication can refer to how an agent explains its own behavior and how it participates in team coordination and collaboration. Transparency refers to how an automated agent explains its own behavior. In either case, both the communication style and modality are of interest.

Agent transparency can be modeled using the Situation Awareness-Based Transparency (SAT) model, which describes three levels of transparency in an agent: the agent's perception of the environment, the agent's comprehension of the environment, and the agent's projection of the future (Chen et al., 2014, 2018). This framework is often employed in empirical studies in which agents communicate with varying levels of transparency (e.g., Loft et al., 2023; Mercado et al., 2016). High transparency is often preferred and associated with higher performance, and it does not always come at the cost of increased workload (van de Merwe et al., 2024).

Communication style also includes an agent's vocabulary, syntax, and etiquette. Human-like expressions such as feedback and apologies are associated with higher trust in agents (de Visser et al., 2016, 2018). Rigid, simplistic communication protocols have been associated with lower team performance (Demir et al., 2017) and less efficient communication (more repeated requests and fewer status updates) (McNeese et al., 2017). However, more structured and restricted communication (e.g., controlled natural language) may be an effective way for humans and autonomous agents to share information and understand each other during a military mission (Patel et al., 2013). Automated agents can also be programmed to adapt their communication style to better interact with humans. Studies on automation etiquette have shown that an intelligent tutoring agent that systematically adjusts its interaction style can improve user performance, motivation, confidence, and satisfaction (Yang & Dorneich, 2018a, 2018b).

The increasing availability of conversational agents provides a growing area of research to consider. A systematic review of conversational agent research identified five themes of trust-enhancing factors: the agent's social intelligence, voice characteristics and communication style, the agent's anthropomorphic appearance, non-verbal communication, and performance quality (Rheu et al., 2021). However, researchers have also suggested that naturalistic verbal communication may not always be appropriate in HAT. As agents' algorithms and source material become increasingly complex, natural language is neither necessary (Chen & Barnes, 2014) nor effective for providing transparency (Endsley, 2023).

A multi-modal approach may be appropriate for agent communication. Just as soldiers communicate through text or non-verbal cues in certain situations, teammate agents may also require context-dependent, non-verbal communication methods (Cooke & Lawless, 2021). This communication may also occur through visual elements on an interface, such as written text or visualizations (Wohleber et al., 2023). Communication modality may be symmetric (e.g., all humans and agents communicate verbally) or asymmetric (e.g., humans communicate verbally while agents communicate visually). This communication sometimes involves conveying high-level goals or using a "playbook" of preset behaviors, and the necessary granularity of these instructions may be lower for newer AI systems (NASEM, 2021).

1.3.2.2.1 Implications for HAT Testbed Requirements

Agent communication in HAT may include explanations of its own behavior, recommendations to the team, and other contributions to team coordination and collaboration. This communication may take different forms and include varying levels of detail. Independent variables may include:

- Amount of transparency in agent recommendations and explanations
- Agent communication style (e.g., verbosity, formality, tone, naturalness)
- Agent communication modality (e.g., verbal, graphical, behavioral, visual, auditory)
- Human-agent communication symmetry (matched or unmatched communication modalities)
- Visualization style of agent recommendations
- Granularity of control of the agent (e.g., playbook vs. high-level goals).

1.3.2.3 Shared Team Awareness (CHF3)

This topic includes awareness of the environment (situation awareness), of each other (mental models and awareness of teammates), and of oneself (self-awareness). Situation awareness is defined as "the perception of the elements in the environment within a volume of time and space, the comprehension of their meaning, and the projection of their status in the near future" (Endsley, 1995, p. 36). This definition was used to develop the SAT model for agent transparency (Chen et al., 2014, 2018) and can inform the design of HAT interfaces (Endsley & Jones, 2016). Shared situation awareness extends this definition to include the degree to which each team member has sufficient situation awareness (Endsley, 1995). This concept is closely related to other concepts, such as interactive team cognition (Cooke et al., 2013) and shared mental models (Cannon-Bowers et al., 1993; Mathieu et al., 2000). It has been identified in HAT reviews as a key issue for future study (Endsley, 2023; Lyons et al., 2021; NASEM, 2021; Patel et al., 2025; Schaefer et al., 2021; Shively et al., 2018; Steinberg, 2021; Tummala et al., 2025).

HAT testbeds can support shared situation awareness by providing UI elements, such as common operating pictures (COPs), which offer a common visualization of strategic, tactical, and operational information to facilitate shared awareness (McNeese et al., 2006). Context- and role-dependent information displays, which selectively provide situationally appropriate information, are associated with higher situation awareness and team performance (Michailovs et al., 2023) and lower workload (Simonson et al., 2022). HAT testbeds should allow researchers to customize UI elements that facilitate situation awareness.

Shared situation awareness also includes mental models of each teammate's capabilities, goals, effectiveness, limitations, and projected actions (NASEM, 2021). Future work could explore the type and amount of information to share, how a teammate might use this information to help the team, and when to transfer control of tasks to another teammate (NASEM, 2021). While theory of mind has been used to understand human cognition, an analogous theory of behavior has been proposed to explain the actions of agents without attributing them to human-like cognition (McNeese et al., 2023; NASEM, 2021). An HAT testbed should allow researchers to manipulate the amount of information about each human and agent that is provided to each teammate.

1.3.2.3.1 Implications for HAT Testbed Requirements

HAT testbeds should include a mechanism to support shared situation awareness. This might include a COP or role-specific information views. The testbed should also allow researchers to influence the mental models team members form of each other by manipulating the flow of information. Independent variables might include:

- Which UI elements are included to support situation awareness
- What information is provided to each teammate about the environment and each other.

1.3.2.4 Team Trust, Reliance, and Compliance (CHF4)

Unless an agent is fully autonomous and perfectly reliable, humans will be required to monitor it. The degree to which humans monitor an agent is often discussed in the context of trust and reliance on automation (Chen et al., 2018; Endsley, 2023; Hauptman & McNeese, 2022; Mouloua et al., 2019; Parasuraman & Riley, 1997; Shively et al., 2018). At one extreme, low trust and reliance are associated with increased monitoring of agents (automation underuse) and the potential neglect of other tasks (Parasuraman & Riley, 1997). At the other extreme, high trust and reliance are associated with insufficient monitoring of agents (automation misuse); the more a human trusts the agent, the slower they are to detect and react to automation failures (Parasuraman & Riley, 1997). Automation misuse may become a more significant challenge as humans place greater trust and reliance in increasingly capable agents (Endsley, 2023).

The monitoring of agents can be influenced by operator workload (Parasuraman & Riley, 1997). When performing concurrent tasks, humans are less likely to reject bad agent recommendations and are more likely to exhibit a bias toward trusting automation (Tatasciore et al., 2023). Future work should explore design solutions to keep human operators up to date when multitasking or monitoring multiple systems (Chen & Barnes, 2014). Testbeds can support this research by allowing researchers to manipulate workload through task difficulty or secondary tasks.

Trust is a key topic in HAT (Lyons et al., 2021; NASEM, 2021; Shively et al., 2018), human-robot interaction (Hancock et al., 2011), and conversational agent interfaces (Rheu et al., 2021). An agent's or robot's characteristics (e.g., performance reliability, communication style) predict trust formation (Hancock et al., 2011; Lee & See, 2004; Rheu et al., 2021). Although trust in and reliance on automation are sometimes grouped together, trust is not the sole predictor of reliance; other factors (e.g., emotional responses to technology) also influence reliance on automation (Lee & See, 2004). Similarly, transparency alone does not lead to trust (Chen et al., 2014). Rather, a combination of transparency and reliability leads to appropriately calibrated trust (e.g., high transparency and low reliability result in appropriately low trust).

1.3.2.4.1 Implications for HAT Testbed Requirements

Testbeds should support studies on trust and monitoring and allow researchers to manipulate the flow of information, enabling human operators to stay up to date while multitasking or monitoring multiple systems (Chen & Barnes, 2014). HAT testbeds can test design affordances that facilitate appropriately calibrated trust and reliance in an agent, such as transparency, reliability, and communication style. HAT testbeds can be used to study the relationship between task load and human trust and reliance on agents. Independent variables might include:

- Agent characteristics, such as transparency, reliability, and communication style
- Number of tasks assigned to each human and agent
- Number of agents supervised by each human
- Task difficulty and attentional demands (e.g., secondary tasks).

1.4 CONCLUSION

This chapter identified testbed requirements based on seven big-picture research questions in HAT that are expected to form the basis of much future research, and on four ongoing core human factors topics that are expected to inform the selection of independent variables in future research. These requirements were synthesized from the current literature to help researchers develop more flexible HAT testbeds that can be used to study both current and future research questions across disciplines. Widespread adoption of testbeds that meet these requirements would lead to lower development costs for running HAT studies, a greater ability to compare findings across studies, and more substantial support for future research directions.

The list of requirements in this chapter is not exhaustive. The requirements do not provide guidance on scripting experimental scenarios, as scenario requirements are expected to be driven by the research question. There are also requirements that do not relate to these big-picture research questions or core human factors issues, such as researcher pain points, software infrastructure requirements, the need to support researchers with varying coding resources, or requirements related to supporting in-person and remote experiments (NASEM, 2021). Finally, while HAT topics such as XAI are important considerations in experimental design, this chapter does not offer a comprehensive review of all HAT topics. Rather, the focus is on HAT research trends that generate testbed requirements.

REFERENCES

Aguinis, H., & Kraiger, K. (2009). Benefits of training and development for individuals and teams, organizations, and society. *Annual Review of Psychology*, *60*, 451–474. https://doi.org/10.1146/annurev.psych.60.110707.163505

Ang, K. C. S., Sankaran, S., & Liu, D. (2025). Advancing sociotechnical systems theory: New principles for human-robot team design and development. *Applied Ergonomics*, *129*, 104604. https://doi.org/10.1016/J.APERGO.2025.104604

Asiala, L., McCarthy, J. E., & Huang, L. (2023). *Improving the state of the art for training Human-AI teams: Technical report #3--Analysis of testbed alternatives*. https://doi.org/10.48550/arXiv.2309.03213

Berretta, S., Tausch, A., Ontrup, G., Gilles, B., Peifer, C., & Kluge, A. (2023). Defining human-AI teaming the human-centered way: A scoping review and network analysis. *Frontiers in Artificial Intelligence*, *6*, 1250725. https://doi.org/10.3389/frai.2023.1250725

Beyer, H., & Holtzblatt, K. (1997). *Contextual design: Defining customer-centered systems* (1st ed., Number 2). Morgan Kaufmann.

Billings, D. R., Schaefer, K. E., Chen, J. Y. C., Kocsis, V., Barrera, M., Cook, J., Ferrer, M., & Hancock, P. A. (2012). *Human-Animal trust as an analog for Human-Robot trust: A review of current evidence*. https://apps.dtic.mil/sti/html/tr/ADA559369/

Brassard, M. (1996). *The memory jogger plus+: Featuring the seven management and planning tools*. Goal QPC.

Cannon-Bowers, J. A., Salas, E., & Converse, S. (1993). Shared mental models in expert team decision making. In N. J. Castellan (Ed.), *Individual and group decision making* (1st ed., p. 341). L. Erlbaum Associates.

Capel, T., & Brereton, M. (2023). What is human-centered about human-centered AI? A map of the research landscape. *Conference on Human Factors in Computing Systems – Proceedings, CHI'*23. https://doi.org/10.1145/3544548.3580959

Chen, J. Y. C., & Barnes, M. J. (2014). Human – Agent teaming for multirobot control: A review of human factors issues. *IEEE Transactions on Human-Machine Systems, 44*(1), 13–29. https://doi.org/10.1109/THMS.2013.2293535

Chen, J. Y. C., Lakhmani, S. G., Stowers, K., Selkowitz, A. R., Wright, J. L., & Barnes, M. (2018). Situation awareness-based agent transparency and human-autonomy teaming effectiveness. *Theoretical Issues in Ergonomics Science, 19*(3), 259–282. https://doi.org/10.1080/1463922X.2017.1315750

Chen, J. Y. C., Procci, K., Boyce, M. W., Wright, J. L., Garcia, A., & Barnes, M. (2014). *Situation awareness-based agent transparency* [US Army Research Laboratory]. https://apps.dtic.mil/sti/html/trecms/AD1143367/

Chiou, E. K., & Lee, J. D. (2023). Trusting automation: Designing for responsivity and resilience. *Human Factors, 65*(1), 137–165. https://doi.org/10.1177/00187208211009995

Chung, H., Holder, T., Shah, J. A., & Yang, X. J. (2025). A systematic review and taxonomy of human–agent teaming testbeds. *Human Factors*. https://doi.org/10.1177/00187208251376898

Cooke, N. J., Gorman, J. C., Myers, C. W., & Duran, J. L. (2013). Interactive team cognition. *Cognitive Science, 37*(2), 255–285. https://doi.org/10.1111/COGS.12009

Cooke, N. J., & Lawless, W. F. (2021). Effective human-artificial intelligence teaming. In W. F. Lawless, R. Mittu, D. A. Sofge, T. Shortell, & T. A. McDermott (Eds.), *Systems engineering and artificial intelligence* (pp. 61–75). Springer International Publishing. https://doi.org/10.1007/978-3-030-77283-3_4

Cooke, N. J., Salas, E., Cannon-Bowers, J. A., & Stout, R. J. (2000). Measuring team knowledge. *Human Factors, 42*(1), 151–173. https://doi.org/10.1518/001872000779656561

de Visser, E. J., Monfort, S. S., McKendrick, R., Smith, M. A. B., McKnight, P. E., Krueger, F., & Parasuraman, R. (2016). Almost human: Anthropomorphism increases trust resilience in cognitive agents. *Journal of Experimental Psychology: Applied, 22*(3), 331–349. https://doi.org/10.1037/XAP0000092

de Visser, E. J., Pak, R., & Shaw, T. H. (2018). From 'automation' to 'autonomy': The importance of trust repair in human–machine interaction. *Ergonomics, 61*(10), 1409–1427. https://doi.org/10.1080/00140139.2018.1457725

de Visser, E. J., Peeters, M. M. M., Jung, M. F., Kohn, S., Shaw, T. H., Pak, R., & Neerincx, M. A. (2020). Towards a theory of longitudinal trust calibration in human–robot teams. *International Journal of Social Robotics, 12*(2), 459–478. https://doi.org/10.1007/s12369-019-00596-x

Dekker, S. W. A., & Woods, D. D. (2002). MABA-MABA or Abracadabra? Progress on human–automation coordination. *Cognition, Technology & Work, 4*(4), 240–244. https://doi.org/10.1007/S101110200022

Demir, M., Amazeen, P. G., McNeese, N. J., Likens, A., & Cooke, N. J. (2017). Team coordination dynamics in human-autonomy teaming. 236. https://doi.org/10.1177/1541931213601542

Desai, S., & Twidale, M. (2023). Metaphors in voice user interfaces: A slippery fish. *ACM Transactions on Computer-Human Interaction*, *30*(6). https://doi.org/10.1145/3609326

Dorneich, M. C., Dudley, R., Letsu-Dake, E., Rogers, W., Whitlow, S. D., Dillard, M. C., & Nelson, E. (2017). Interaction of automation visibility and information quality in flight deck information automation. *IEEE Transactions on Human-Machine Systems*, *47*(6), 915–926. https://doi.org/10.1109/THMS.2017.2717939

Dorneich, M.C., Gilbert, S., Francis, R., Talyat, M., Driskell, J., & Didic, E. (2023). Team Skill Metrics that Span Human-Human and Human-Agent Teams: An Initial Assessment, *Proceedings of the Human Factors and Ergonomics Society Annual Meeting*. Washington DC, October 23–27. *67*(1), 123–130, http://doi.org/10.1177/21695067231192892.

Endsley, M. R. (1995). Toward a theory of situation awareness in dynamic systems. *Human Factors*, *37*(1), 32–64. https://doi.org/10.1518/001872095779049543

Endsley, M. R. (2023). Ironies of artificial intelligence. *Ergonomics*. https://doi.org/10.1080/00140139.2023.2243404

Endsley, M. R., & Jones, D. G. (2016). *Designing for situation awareness: An approach to user-centered design* (2nd ed.). CRC Press. https://doi.org/10.1080/10686967.2017.11918512

Farah, Y. A., Banuelos-Moriel, A., & Dorneich, M. C. (2023). Evaluating the consistency of cooperative video games in inducing teamwork behaviors. *Proceedings of the Human Factors and Ergonomics Society Annual Meeting*. https://doi.org/10.1177/21695067231196239

Farah, Y. A., Kim, J., Newendorp, A. K., Shah Abadi, G., Dorneich, M. C., & Gilbert, S. B. (2026). Teaming up with an AI agent: Training humans to develop human-agent teamwork skills. *Journal of Cognitive Engineering and Decision Making*. https://doi.org/10.1177/15553434261430062

Feigh, K. M., Dorneich, M. C., & Hayes, C. C. (2012). Toward a characterization of adaptive systems: A framework for researchers and system designers. *Human Factors*, *54*(6), 1008–1024. https://doi.org/10.1177/0018720812443983

Fitts, P. M. (1951). *Human engineering for an effective air-navigation and traffic-control system* (P. M. Fitts, Ed.). National Research Council. https://psycnet.apa.org/record/1952-01751-000

Goecks, V. G., & Waytowich, N. (2024). COA-GPT: Generative pre-trained transformers for accelerated course of action development in military operations. *2024 International Conference on Military Communication and Information Systems, ICMCIS 2024*. https://doi.org/10.1109/ICMCIS61231.2024.10540749

Goodwin, G. F., Blacksmith, N., & Coats, M. R. (2018). The science of teams in the military: Contributions from over 60 years of research. *The American Psychologist*, *73*(4), 322–333. https://doi.org/10.1037/AMP0000259

Hancock, P. A., Billings, D. R., Schaefer, K. E., Chen, J. Y. C., De Visser, E. J., & Parasuraman, R. (2011). A meta-analysis of factors affecting trust in human-robot interaction. *Human Factors*, *53*(5), 517–527. https://doi.org/10.1177/0018720811417254

Hauptman, A. I., & McNeese, N. J. (2022). Overcoming the lumberjack effect through adaptive autonomy. *Proceedings of the Human Factors and Ergonomics Society Annual Meeting*, *66*(1), 1075–1079. https://doi.org/10.1177/1071181322661372

Hidalgo, M., Rebensky, S., Nguyen, D., Cohen, M., Temple, L., & Fegley, B. (2023). Measurement and manipulation in human-agent teams: A review. *Human Factors and Simulation*, *83*(83). https://doi.org/10.54941/AHFE1003559

Holder, E., Chen, J. Y. C., Liggett, K., Bridgham, P., Briscombe, N., Eskridge, T., Carvalho, M., & Burski, L. (2023). Human interactions. In A. Kott (Ed.), *Autonomous Intelligent Cyber Defense Agent (AICA)* (pp. 253–274). Springer. https://doi.org/10.1007/978-3-031-29269-9_12

Iftikhar, R., Chiu, Y. T., Khan, M. S., & Caudwell, C. (2023). Human–agent team dynamics: A review and future research opportunities. *IEEE Transactions on Engineering Management, 71*, 10139–10154.

Johnson, C. J., Demir, M., McNeese, N. J., Gorman, J. C., Wolff, A. T., & Cooke, N. J. (2023). The impact of training on human–autonomy team communications and trust calibration. *Human Factors*. https://doi.org/10.1177/00187208211047323

Lame, G. (2019). Systematic Literature Reviews: An Introduction. Proceedings of the Design Society: International Conference on Engineering Design, 1(1), 1633–1642. https://doi.org/10.1017/DSI.2019.169

Lancaster, C. M., Duan, W., Mallick, R., & McNeese, N. J. (2025). Human-centered team training for human-AI teams: From training with AI tools to training for AI teammates. *Proceedings of the ACM on Human-Computer Interaction, 9*(2). https://doi.org/10.1145/3710998

Lee, J. D., & See, K. A. (2004). Trust in automation: Designing for appropriate reliance. *Human Factors, 46*(1), 50–80. https://doi.org/10.1518/hfes.46.1.50_30392

Loft, S., Bhaskara, A., Lock, B. A., Skinner, M., Brooks, J., Li, R., & Bell, J. (2023). The impact of transparency and decision risk on human–automation teaming outcomes. *Human Factors, 65*(5), 846–861. https://doi.org/10.1177/00187208211033445

Lyons, J. B., Sycara, K., Lewis, M., & Capiola, A. (2021). Human-autonomy teaming: Definitions, debates, and directions. *Frontiers in Psychology, 12*. https://doi.org/10.3389/FPSYG.2021.589585

Mallick, R., Sawant, S., Brady, C., McNeese, N. J., Madathil, K. C., & Bertrand, J. (2023). Can we build it? Yes, we can! Development procedure of high-fidelity simulation environments for human-agent teams. *Proceedings of the Human Factors and Ergonomics Society Annual Meeting, 67*(1), 1617–1622. https://doi.org/10.1177/21695067231192225

Mathieu, J. E., Goodwin, G. F., Heffner, T. S., Salas, E., & Cannon-Bowers, J. A. (2000). The influence of shared mental models on team process and performance. *Journal of Applied Psychology, 85*(2), 273–283. https://doi.org/10.1037/0021-9010.85.2.273

McNeese, M. D., Pfaff, M. S., Connors, E. S., Obieta, J. F., Terrell, I. S., & Friedenberg, M. A. (2006). Multiple vantage points of the common operational picture: Supporting international teamwork. *Proceedings of the Human Factors and Ergonomics Society Annual Meeting, 50*(3), 467–471. https://doi.org/10.1177/154193120605000354

McNeese, N. J., Demir, M., Cooke, N. J., & Myers, C. (2017). Teaming with a synthetic teammate: Insights into human-autonomy teaming. *Human Factors: The Journal of the Human Factors and Ergonomics Society, 60*(2), 262–273. https://doi.org/10.1177/0018720817743223

McNeese, N. J., Flathmann, C., O'Neill, T. A., & Salas, E. (2023). Stepping out of the shadow of human-human teaming: Crafting a unique identity for human-autonomy teams. *Computers in Human Behavior, 148*, 107874. https://doi.org/10.1016/J.CHB.2023.107874

Mercado, J. E., Rupp, M. A., Chen, J. Y. C., Barnes, M. J., Barber, D., & Procci, K. (2016). Intelligent agent transparency in human-agent teaming for multi-UxV management. *Human Factors, 58*(3), 401–415. https:// doi.org/10.1177/0018720815621206

Michailovs, S., Pond, S., Schmitt, M., Irons, J., Stoker, M., Visser, T. A. W., Huf, S., & Loft, S. (2023). The impact of information integration in a simulation of future submarine command and control. *Human Factors, 65*(7). https://doi.org/10.1177/00187208211045872

Mouloua, M., Ferraro, J. C., Parasuraman, R., Molloy, R., & Hilburn, B. (2019). Human monitoring of automated systems. In M. Mouloua, & P. A. Hancock (Eds.), *Human performance in automated and autonomous systems* (1st ed., pp. 1–26). CRC Press. https://doi.org/10.1201/9780429458323-1

Myers, C. W., Cooke, N. J., Gorman, J. C., & McNeese, N. J. (2024). Introduction to the emerging cognitive science of distributed human-autonomy teams. *Topics in Cognitive Science, 16*(3), 377–390. https://doi.org/10.1111/TOPS.12744

NASEM. (2021). *Human-AI teaming: State-of-the-art and research needs* (pp. 1–126). https://doi.org/10.17226/26355

Newendorp, A. K., Sanaei, M., Perron, A. J., Sabouni, H., Javadpour, N., Sells, M., Nelson, K., Dorneich, M. C., & Gilbert, S. B. (2024, May). Apple's knowledge navigator: Why doesn't that conversational agent exist yet? *Proceedings of the 2024 CHI Conference on Human Factors in Computing Systems*. https://doi.org/10.1145/3613904.3642739

O'Neill, T. A., Flathmann, C., McNeese, N. J., & Salas, E. (2023). Human-autonomy Teaming: Need for a guiding team-based framework? *Computers in Human Behavior, 146*, 107762. https://doi.org/10.1016/J.CHB.2023.107762

O'Neill, T., McNeese, N. J., Barron, A., & Schelble, B. (2022). Human-autonomy teaming: A review and analysis of the empirical literature. *Human Factors, 64*(5), 904–938. https://doi.org/10.1177/0018720820960865

Page, M. J., McKenzie, J. E., Bossuyt, P. M., Boutron, I., Hoffmann, T. C., Mulrow, C. D., Shamseer, L., Tetzlaff, J. M., Akl, E. A., Brennan, S. E., Chou, R., Glanville, J., Grimshaw, J. M., Hróbjartsson, A., Lalu, M. M., Li, T., Loder, E. W., Mayo-Wilson, E., McDonald, S., … Moher, D. (2021). The PRISMA 2020 statement: An updated guideline for reporting systematic reviews. *BMJ, 372*. https://doi.org/10.1136/BMJ.N71

Parasuraman, R., & Riley, V. (1997). Humans and automation: Use, misuse, disuse, abuse. *Human Factors, 39*(2), 230–253. https://doi.org/10.1518/001872097778543886

Patel, J., Boardman, M., Files, B., Gregory, F., Lamb, S., Sarkadi, S., Tešić, M., & Yeung, N. (2025). Give us a hand, mate! A holistic review of research on human-machine teaming. *BMJ Mil Health, 171*(5), 392–396. https://doi.org/10.1136/MILITARY-2024-002737

Patel, J., Dorneich, M. C., Mott, D., Bahrami, A., & Giammanco, C. (2013). Improving coalition planning by making plans alive. *IEEE Intelligent Systems, 28*(1), 17–25. https://doi.org/10.1109/MIS.2012.88

Rheu, M., Shin, J. Y., Peng, W., & Huh-Yoo, J. (2021). Systematic review: Trust-building factors and implications for conversational agent design. *International Journal of Human-Computer Interaction, 37*(1), 81–96. https://doi.org/10.1080/10447318.2020.1807710

Roberts, A. P. J., Webster, L. V., Salmon, P. M., Flin, R., Salas, E., Cooke, N. J., Read, G. J. M., & Stanton, N. A. (2022). State of science: Models and methods for understanding and enhancing teams and teamwork in complex sociotechnical systems. *Ergonomics, 65*(2), 161–187. https://doi.org/10.1080/00140139.2021.2000043

Rousseau, V., Aubé, C., & Savoie, A. (2006). Teamwork behaviors: A review and an integration of frameworks. *Small Group Research, 37*(5), 540–570. https://doi.org/10.1177/1046496406293125

Salas, E., Cooke, N. J., & Rosen, M. A. (2008). On teams, teamwork, and team performance: Discoveries and developments. *Human Factors, 50*(3), 540–547. https://doi.org/10.1518/001872008X288457

Sarkar, A. (2023). Enough with "human-AI collaboration." *Conference on Human Factors in Computing Systems – Proceedings*. https://doi.org/10.1145/3544549.3582735

Scerbo, M. W. (1996). Theoretical perspectives on adaptive automation. In R. Parasuraman & M. Mouloua (Eds.), *Automation and human performance* (1st ed., pp. 37–63). CRC Press. https://www.taylorfrancis.com/chapters/edit/10.1201/9781315137957-3/

Schaefer, K. E., Perelman, B., Rexwinkle, J., Canady, J., Neubauer, C., Waytowich, N., Larkin, G., Cox, K., Geuss, M., Gremillion, G., Metcalfe, J. S., DeCostanza, A., & Marathe, A. (2021). Human-autonomy teaming for the tactical edge: The importance of humans in artificial intelligence research and development. In W. F. Lawless,

R. Mittu, D. A. Sofge, T. Shortell, & T. A. McDermott (Eds.), *Systems engineering and artificial intelligence* (pp. 115–148). Springer International Publishing. https://doi.org/10.1007/978-3-030-77283-3_7

Schelble, B. G., Flathmann, C., & McNeese, N. (2020). Towards meaningfully integrating human-autonomy teaming in applied settings. *HAI 2020- Proceedings of the 8th International Conference on Human-Agent Interaction* (pp. 149–156). https://doi.org/10.1145/3406499.3415077

Schmutz, J. B., Outland, N., Kerstan, S., Georganta, E., & Ulfert, A. S. (2024). AI-teaming: Redefining collaboration in the digital era. *Current Opinion in Psychology*, *58*, 101837. https://doi.org/10.1016/J.COPSYC.2024.101837

Shively, R. J., Lachter, J., Brandt, S. L., Matessa, M., Battiste, V., & Johnson, W. W. (2018). Why human-autonomy teaming? In C. Baldwin (Ed.), *Advances in intelligent systems and computing* (Vol. 586, pp. 3–11). Springer. https://doi.org/10.1007/978-3-319-60642-2_1

Shneiderman, B. (2020). Design lessons from AI's two grand goals: Human emulation and useful applications. *IEEE Transactions on Technology and Society*, *1*(2), 73–82. https://doi.org/10.1109/TTS.2020.2992669

Simonson, R. J., Keebler, J. R., Blickensderfer, E. L., & Besuijen, R. (2022). Impact of alarm management and automation on abnormal operations: A human-in-the-loop simulation study. *Applied Ergonomics*, *100*, 103670. https://doi.org/10.1016/J.APERGO.2021.103670

Sottilare, R. A., Shawn Burke, C., Salas, E., Sinatra, A. M., Johnston, J. H., & Gilbert, S. B. (2017). Designing adaptive instruction for teams: A meta-analysis. *International Journal of Artificial Intelligence in Education*, *28*(2), 225–264. https://doi.org/10.1007/S40593-017-0146-Z

Sottilare, R. A., Shawn Burke, C., Salas, E., Sinatra, A. M., Johnston, J. H., & Gilbert, S. B. (2018). Designing adaptive instruction for teams: A meta-analysis. *International Journal of Artificial Intelligence in Education*, *28*(2), 225–264. https://doi.org/10.1007/S40593-017-0146-Z

Steinberg, M. (2021). Toward system theoretical foundations for human-autonomy teams. In W. F. Lawless, R. Mittu, D. A. Sofge, T. Shortell, & T. A. McDermott (Eds.), *Systems engineering and artificial intelligence* (pp. 77–92). Springer International Publishing. https://doi.org/10.1007/978-3-030-77283-3_5

Tatasciore, M., Bowden, V., & Loft, S. (2023). Do concurrent task demands impact the benefit of automation transparency? *Applied Ergonomics*, *110*. https://doi.org/10.1016/J.APERGO.2023.104022

Tokadlı, G., Ouverson, K., Meusel, C., Garcia, A., Gilbert, S. B., & Dorneich, M. C. (2018). An analysis of video games using the dimensions of human-agent interaction. *Proceedings of the Human Factors and Ergonomics Society Annual Meeting*, *62*(1), 716–720. https://doi.org/10.1177/1541931218621163

Tomsett, R., Preece, A., Braines, D., Cerutti, F., Chakraborty, S., Srivastava, M., Pearson, G., & Kaplan, L. (2020). Rapid trust calibration through interpretable and uncertainty-aware AI. *Patterns*, *1*(4). https://doi.org/10.1016/j.patter.2020.100049

Tummala, V. S., Burris-Melville, T. S., & Eskridge, T. C. (2025). AI as a team member: Redefining collaboration. *Journal of Leadership Studies*, *18*(4), 67–80. https://doi.org/10.1002/JLS.70003

van de Merwe, K., Mallam, S., & Nazir, S. (2024). Agent transparency, situation awareness, mental workload, and operator performance: A systematic literature review. *Human Factors*, *66*(1), 180–208. https://doi.org/https://doi.org/10.1177/00187208221077804

Volz, K. M., & Dorneich, M. C. (2020). Evaluation of cognitive skill degradation in flight planning. *Journal of Cognitive Engineering and Decision Making*, *14*(4), 263–287. https://doi.org/10.1177/1555343420962897

Wiese, C. W., Shuffler, M. L., & Salas, E. (2015). Teamwork and team performance measurement. In *International Encyclopedia of the Social & Behavioral Sciences* (2nd ed., pp. 96–103). https://doi.org/10.1016/B978-0-08-097086-8.22017-5

Wohleber, R. W., Stowers, K., Barnes, M., & Chen, J. Y. C. (2023). Agent transparency in mixed-initiative multi-UxV control: How should intelligent agent collaborators speak their minds? *Computers in Human Behavior*, *148*, 107866. https://doi.org/10.1016/J.CHB.2023.107866

Wynne, K. T., & Lyons, J. B. (2018). An integrative model of autonomous agent teammate-likeness. *Theoretical Issues in Ergonomics Science*, *19*(3), 353–374. https://doi.org/10.1080/1463922X.2016.1260181

Yang, E., & Dorneich, M. C. (2018a). Affect-aware adaptive tutoring based on human–automation etiquette strategies. *Human Factors*, *60*(4), 510–526. https://doi.org/10.1177/0018720818765266

Yang, E., & Dorneich, M. C. (2018b). Evaluating human–automation etiquette strategies to mitigate user frustration and improve learning in affect-aware tutoring. *Applied Sciences*, *8*(6), 895. https://doi.org/10.3390/APP8060895

2 Buy vs. Build
A Structured Assessment of Testbed Alternatives

Lillian K.E. Asiala, James E. McCarthy, and Lixiao Huang

2.1 INTRODUCTION

Across many domains, there is growing interest in the use of AI to increase the speed at which decisions can be made and actions can be taken (NASEM, 2021). Much of this work focuses on the use of AI systems as tools (cf. Sarkar, 2023; Shneiderman, 2021; Shneiderman & Endsley, 2022). However, our work is based on the premise that autonomous agents will eventually be able to work with humans as teammates, not merely as tools (*cf.* Beans, 2018; Chiou & Lee, 2021; Gervits et al., 2020; Phillips et al., 2011, 2013; Seeber et al., 2020; Warta et al., 2016). In our view, this distinguishes automation from autonomy. We define automation as technology that requires human supervision and control, providing a tool that humans can use to improve their performance (McNeese et al., 2018; Shneiderman, 2020). Alternatively, autonomous systems are quasi-cognitive systems that use AI or similar technologies to make decisions and/or take actions within a specified domain (McNeese et al., 2018; USAF, 2013). These systems can independently respond to novel task-related performance challenges. If the autonomous system is also capable of coordination and cooperation, we can begin to view the system as an autonomous teammate that is capable of both independent and interdependent action (Chiou & Lee, 2021).

We observed that most efforts employed idiosyncratic testbeds for their studies. That is, no commonly accepted testbed unified findings across the field. On the one hand, the use of numerous separate testbeds could be viewed positively. If multiple laboratories using very different testbeds reported similar findings, the external validity of those findings would be enhanced. On the other hand, this pattern also made it extremely difficult to replicate findings. Since replication is an issue that continues to befuddle social science research, one contribution we could make would be to create an open-source testbed that various labs could adopt to facilitate their research. We hoped that such an environment would advance research in the field while facilitating replication efforts.

Before investing in the creation of a novel testbed, we wanted to assess whether an existing system was likely to meet the community's needs and, failing that, which features and capabilities the community felt would be important to include in a testbed. In this report, we present the process that we used and the results that were

DOI: 10.1201/9781003655923-3

produced. In Part 1, we describe the stakeholder engagement process and the process for using the outcome of that engagement to develop and apply evaluation criteria. In Part 2, we describe the application of this process to evaluate existing testbeds and report the results of that evaluation.

2.2 PART 1: TESTBED EVALUATION FRAMEWORK

The framework that we used to evaluate existing testbeds to support human-autonomy teaming consisted of four phases. First, we engaged with stakeholders to identify critical testbed attributes. Second, using the results of that process, we defined evaluation criteria. Third, we selected candidate testbeds for review. Fourth, we used a system of two raters to apply the evaluation criteria to the selected testbeds.

2.2.1 Theoretical Background

In our initial examination of the human-AI teaming literature, we engaged in a number of conversations with researchers in the field. This led to the creation of a testbed taxonomy consisting of three dimensions: Interdependency Level, Task Relevancy, and Level of Agent Sophistication. This taxonomy informed our conceptualization of testbeds and provided a theoretical framework for formalizing our evaluation.

The first dimension, *level of interdependency*, is likely the most important. The interdependency dimension reflects the extent to which team members are dependent on others to complete the collective task. Fortunately, earlier work allows us to operationalize this dimension. First, consider the work of Saavedra et al. (1993). In the business-focused "group work" domain, these researchers called upon the earlier work of Thompson (1967) and Van de Ven et al. (1976) to discuss the range of interdependence shown in Figure 2.1. At one extreme (*pooled workflow*), there is little to no interdependence; individuals complete the assigned task independently, and the "team product" is the sum of the individual effort (*e.g.*, the individual workers in a given call center). In the *sequential workflow*, individuals are responsible for producing their work product and passing it to the next team member for further processing (*i.e.*, a classic "assembly line" process). In this configuration, there is a one-way flow of information, resources, work product, *etc.* A greater level of interdependence is present in the remaining two cases. In the *reciprocal model*, there is a two-way flow between "adjacent" workers. This "give and take" requires greater levels of coordination and cooperation than the one-way flow that is present in the sequential model. Finally, in the *team model*, there is multi-way sharing of information, resources, work product, *etc.* Each member of the team can participate in a give and take with any other member of the team (see also, Singh et al., 2017).

Later, Arthur et al. (2005) defined two factors that influence interdependency. The first factor, "workflow," was adapted from the work of Saavedra et al. (1993), Tesluk et al. (1997), Thompson (1967), and Van de Ven et al. (1976). Using a self-report tool, team participants indicated the workflow pattern (similar to those in Figure 2.1) that most closely resembled the one that occurred in their team. To complement the workflow measure, Arthur et al. (2005) developed a measure of "Team Relatedness." The relatedness scale asked respondents to assess the degree to which an individual can

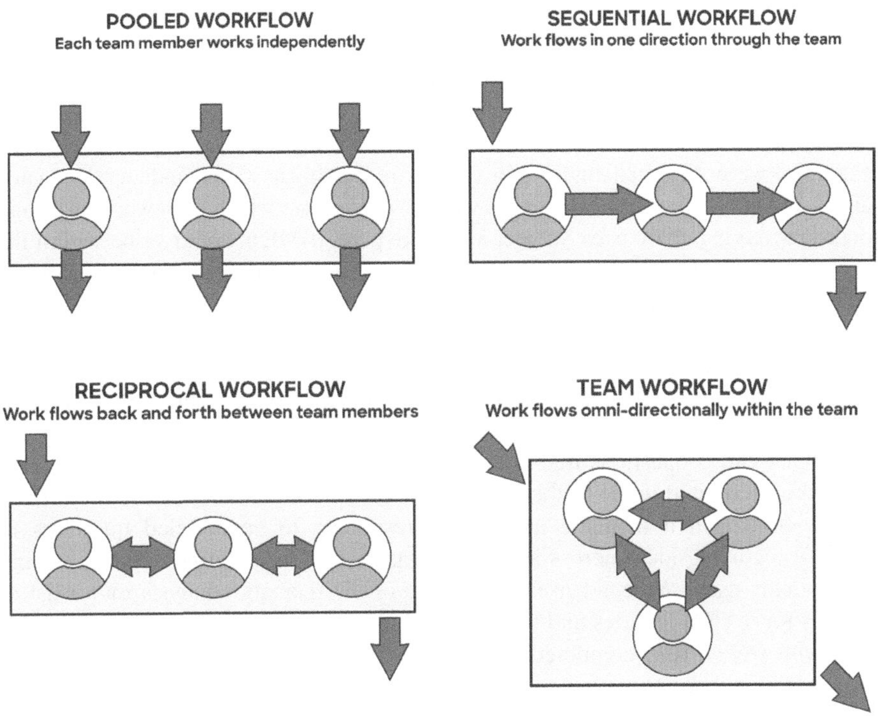

FIGURE 2.1 The Workflow Hierarchy defined by Thompson (1967) and Van de Ven et al. (1976) – (adapted from Saavedra et al., 1993) illustrates the flow of work between teammates in pooled, sequential, reciprocal, and team workflow.

perform a given task. Within this measure, five levels of "relatedness" are defined for raters, who must then assign one of the five values to each task that needs to be performed. Both instruments demonstrated high levels of interrater reliability, ranging from 0.80 to 0.96. Critically, tasks designed by Arthur et al. (2005) to have a greater level of interdependence were rated higher on both scales than those designed to represent individual taskwork. Interestingly, the two instruments were only moderately correlated with each other, indicating that they assess different aspects of the task.

The second taxonomy dimension, *sophistication*, refers to the capabilities of the agents that are operational within the testbed. This factor refers to the extent to which the testbed limits the agents that researchers can integrate into the task environment, or the capabilities that those agents can bring to bear. Agent sophistication is largely a qualitative dimension of this taxonomy; in our research, we have not discovered robust operational definitions of sophistication.

The third taxonomy dimension, which we labeled *relevancy*, refers to how directly the testbed task transfers to the application environment. Again, while a given research environment is likely to be more or less closely related to a given transfer environment, this distinction is largely qualitative and has not been operationalized. We will revisit this taxonomy in Part 2, where we discuss the application of the framework for our specific evaluation purposes.

2.2.2 Identifying Testbed Attributes of Interest

Guided by our theoretical background and our interactions with human-autonomy teaming researchers, our team had several preexisting notions about which testbed attributes would be important to stakeholders in autonomy-teaming research. For example, we anticipated interest in certain research questions, independent variables, and performance measures, as well as ideas about which teamwork tasks and decision-making challenges Subject Matter Experts (SMEs) would value within the testbed. We used these anticipated values to guide our engagement with research stakeholders. We developed surveys to validate these notions for both researchers (McCarthy et al., 2023b) and military SMEs with hands-on experience in the application of human-AI teaming and its challenges (McCarthy et al., 2023a). These surveys included Likert-type items that asked participants to rate their agreement with the importance of these testbed features. We complemented these "validation" items with open-ended questions that allowed respondents to suggest other features that might be useful within the testbed.

We conducted a thematic analysis of responses to open-ended questions to identify recurring ideas across answers, regardless of the terminology or phrasing respondents used. We condensed the results of the thematic analysis into a list of themes for testbed features and then mapped individual responses to entries on that list. Items were further combined thematically until the list yielded a suitable number of features to serve as criteria to guide evaluation. Next, we established the relative value of each feature. If all features are weighted equally, then the value of each testbed is the sum of those scores. However, some features are typically more important than others when selecting a testbed. Feature weights were aligned with priorities for this analysis. Next, we developed a set of testbed scoring standards. Using these standards, evaluators rated the features of each testbed on a scale from 1 to 10 based on the extent to which it met the criteria. Since this process has an element of subjectivity, the research team employed two independent raters to assess the testbeds. After both evaluators completed their assessments, they met to review the scores and resolve any noticeable discrepancies between them.

After the evaluation, we calculated a score for each solution to provide a quantitative measure of suitability. For a sample use of the evaluation tool in practice, consider the hypothetical "Solution 1" and "Solution 2" shown in Table 2.1. Solution 1 meets all the scoring criteria and achieves a perfect score, whereas Solution 2 meets only some of the criteria.

The methods guiding these processes and the results of the analyses are provided in more detail in Part 2.

2.3 PART 2: TESTBED EVALUATION CASE STUDY

Using the testbed taxonomy as an organizational framework, we applied the results of the survey to the development of criteria for evaluating existing human-autonomy teaming testbeds. The goal of the evaluation was to determine whether an existing testbed possessed the features that would encourage researchers from multiple labs to adopt it. If not, we would create a new open-source testbed with that goal in mind. In this section, we describe how we combined survey results with our own goals and

TABLE 2.1
Sample Evaluation Tool

Evaluation Criteria	Weighting Factor	Solution 1		Solution 2	
		Rate	Score	Rate	Score
Data collection and performance measures	3	10	30	10	30
Implementation factors	3	10	30	5	15
Teaming factors	2	10	20	10	20
Task features	2	10	20	0	0
Scenario authoring	2	10	20	0	0
Data processing	1	10	10	6	6
System architecture	1	10	10	3	3
Agents	1	10	10	10	10
Totals			**150**		**84**

priorities to develop testbed evaluation criteria, how we identified testbeds for evaluation, and the outcome of the evaluation process and subsequent decision.

2.3.1 Development of Evaluation Criteria

Evaluation criteria are necessary to conduct an assessment of available solutions to a given problem, as they provide defined standards or factors against which the suitability of a solution is assessed. We used the results of our engagement with researchers and subject matter experts in hybrid teams as the basis for developing our evaluation criteria.

Our work began with a thematic analysis of survey results. We created a master list of testbed features by compiling themes from open-ended questions and hypothesized features presented as Likert-style items. This process yielded 289 individual entries. We then sorted those entries into superordinate feature descriptions to account for feature overlap among responses to open-ended items and validation items. For example, one Likert-style item asked participants to rate their level of agreement with the statement, "The [testbed] should be implemented as an 'open-source' tool." In addition to general agreement with this item, six respondents mentioned that the testbed should have an "open-source architecture" in their responses to other questions in the survey. We categorized these entries within a superordinate feature category: "The [testbed] should be 'open source.'"

The superordinate feature-categorization process yielded more than one hundred superordinate categories. To create a more reasonable feature set, we dropped categories with fewer than three responses. We also combined categories with a high degree of conceptual overlap, narrowing the list to 21 features. Since 21 is still an unwieldy number of individual criteria for evaluating testbeds, we further consolidated this list into seven broad criteria that subsume these 21 features. In addition to considering the stakeholder survey results and the testbed taxonomy described in Part 1, we considered our own goals for the testbed. For us, it was important that the testbed function as a data collection and analysis tool. We also aimed to develop the testbed as an open-source tool that multiple labs could utilize and contribute to. Given those goals, we added two features to the list (bringing the total to 23): "implementation cost" and "speed of implementation," under a testbed criterion we called "implementation factors." This factor accounted for the challenges (or lack thereof) of using an existing testbed as a starting point for achieving those goals. Table 2.2 summarizes the full list of testbed criteria and considerations for each. Refer to Table 2.3 for a comprehensive list of features organized by these criteria.

Other researchers have produced similar lists of features (sans implementation features) in their research on human-autonomy teaming. For example, Cavanah and her colleague (2020) frequently discuss the value of testbeds with tasks that are easy to learn and that have a scalable level of task difficulty (both of which are mentioned in the criteria above). They also often discuss the importance of establishing multiple interdependent roles that humans or agents can fill. They note that there should be multiple ways to complete the tasks and that the environment should enable several team members to be active in parallel.

TABLE 2.2
Driving Factors in the Selection of a Human-AI Teaming Testbed for Sonalysts' Research Initiative

Testbed Criteria	Considerations
Data collection and performance measurement	The testbed must store outcome and process assessments on individuals and teams.
Implementation factors	As an internally funded effort, we will want to ensure that our budget is directed to data collection and analysis to the greatest degree possible, rather than siphoned away to the development of tools that may not be widely employed.
Teaming factors	It will be important to have teams comprising at least four members, and we will want a high degree of flexibility in assigning specific roles to humans or agents. Additionally, researchers must be able to assess communication among team members effectively.
Task features	It will be important for the testbed to provide tasks with a high degree of psychological fidelity and the ability to modulate several task factors, such as difficulty, uncertainty, time pressure, information density, signal-to-noise ratio, and data reliability.
Scenario authoring	The testbed should support customizing scenarios that meet their research objectives and designing scenario events for testing competencies and measuring the performance of those competencies.
Data processing	The testbed should support ease of data export for researchers. The environment's inclusion of "onboard" data analysis tools can also be considered.
Testbed architecture	A variety of system architectures and approaches are likely to be acceptable. However, the testbed's architecture should enable researchers to make changes to the testbed as their research progresses.
Agents	In line with our taxonomy discussion, we aim to assess the ease with which researchers can integrate agents of varying sophistication.

TABLE 2.3
Testbed Evaluation Criteria, Criteria Weights, and Features

Feature	Likert Strength Score	Themes Overlap Score	Internal Priority Ranking	Evaluation Criteria	Evaluation Criteria Z-Score	Final Weight
Can be instrumented to collect a variety of data, including individual and team status.	2.11	3	6	Data collection/ performance measurement	1.71	3
Collect process measures.	2.54	3	9			
Collect outcome measures.	0	3	17			
Implementation cost	0	2	1	Implementation factors	−0.808	3
Speed of implementation	0	1	2			
Support teams with multiple defined roles (4–12).	2.24	3	7	Teaming factors	0.742	2
Allow each role to be filled by a human or an agent.	1.76	3	8			
Distinguish between the source and intended recipient for each communication act.	0	2	23			
Ease the creation of unique scenarios aligned with research objectives.	1.93	3	11	Scenario authoring	0.406	2
Facilitate the definition of linkages among competencies, scenario events, and assessment routines.	1.95	2	19			

(Continued)

TABLE 2.3 (*Continued*)
Testbed Evaluation Criteria, Criteria Weights, and Features

Feature	Likert Strength Score	Themes Overlap Score	Internal Priority Ranking	Evaluation Criteria	Evaluation Criteria Z-Score	Final Weight
The task should include adjustable levels of information density, signal-to-noise ratio, and data reliability.	0	1	3	Task features	0.333	2
Tasks should impose conflicts among individual goals and team goals.	0	1	4			
Task demands should replicate or impose information processing demands analogous to real-world tasks.	1.92	3	5			
Tasks should require communication among teammates (*e.g.*, spoken, text, chat).	1.90	3	10			
The performance task should be easy to learn but potentially challenging.	1.45	3	12			
Tasks should have adjustable difficulty levels, uncertainty, time pressure, *etc*.	2.15	3	13			
Tasks should require significant coordination and interdependency.	2.15	2	22			
Include "quick look" visualizations of individual variables and variable pairs.	1.22	2	14	Data processing	−0.37	1
Facilitate data export	2.33	2	18			
Modular/flexible	0	2	16	System architecture	−0.53	1
Open source	2.13	2	20			
Permits collaboration across labs.	2.19	2	21			
Facilitate integration of agents with differing capabilities.	1.86	2	15	Agents	−1.48	1

2.3.2 Creation of a Rating Tool

To conduct a quantitative assessment of testbed alternatives, we created an evaluation tool that used the testbed criteria presented in Table 2.2. The next step was to establish the relative value of each criterion.

We used three factors to determine criterion weights: (1) Likert-style item strength, (2) overlap between researcher and SME survey themes, and (3) internal research priorities. The relative contributions of each of these factors are summarized in Table 2.3. In the stakeholder surveys, respondents reported their agreement with the inclusion of various features in the testbed. We mapped each of those Likert-style items to the feature list and calculated a mean agreement score among respondents for those features. Only 16 of the 23 total features were represented by Likert-style items. The remaining seven features included implementation factors associated with our specific goals or were derived from themes identified in responses to the open-response survey items. Therefore, their "item strength" score was "0."

Second, both the researcher and SME surveys included open-response questions that prompted respondents to list ideal testbed features. To quantify the analysis of these open-ended responses, we mapped themes from the responses to the testbed features described in Section 2.3.1. Note that many of these qualities were derived from the survey results provided by stakeholders designated as researchers, so more research themes were mapped to qualities than SME themes. We created a 1–3 scale to reflect the degree of overlap between each testbed quality and the researcher and SME thematic analyses, where "1" represented no overlap with either, "2" represented overlap with responses from one of the stakeholder groups, and "3" represented overlap with responses from both stakeholder groups.

Third, we used our internal research objectives to further prioritize the qualities. First, given our internal priorities, we ranked the features by relative importance. Items were divided into three sections: the top seven items, the bottom seven items, and the remaining items in the middle. If an item appeared in the top seven, we added 1/3 of a point to its score. If an item appeared in the bottom seven, we subtracted 1/3 of a point from its score. No changes were made to the other nine mid-ranked qualities. We averaged these three factors and produced Z-scores based on those averages for each criterion. The Z-scores were used to rank the importance of each criterion and assign weights of 1, 2, or 3.

This process was effective in producing weights for seven of the eight evaluation criteria. The only criterion with a weight inconsistent with our project's needs was "Implementation Factors." This is to be expected, as it is directly related to our specific priorities, and we did not anticipate that it would be a consideration for the research community as a whole. Therefore, we did not create Likert items about implementation features, and our respondents did not spontaneously list them in their responses to open-ended questions. As a consequence, this criterion received a very low score. However, cost and schedule are always important. Therefore, we overrode the scoring formula to reflect the practical significance of implementation factors and changed the weight of that criterion from 1 to 3. The final set of weights is shown in Table 2.3.

Next, the team developed testbed scoring standards (shown in Table 2.4). Using these standards, evaluators rated each attribute on a scale from 1 to 10. The rating was

TABLE 2.4
Scoring Standards for Testbed Evaluation Criteria

Evaluation Criteria	Description	Scoring Standard 0–3	4–6	7–10
Data collection/ performance measures	The testbed is instrumented to collect data on individuals and teams for both outcomes and processes.	The testbed collects a few fixed data points.	The testbed collects a variety of outcome and process measures.	The testbed includes rich instrumentation and likely supports the ability to define unique measures of interest.
Implementation factors	The cost of implementation is low, and the speed of implementation is fast. The testbed is primarily designed for data collection and analysis, rather than the development of specific tools.	The software architecture and/or problem space are complex. This will lead to lengthy and costly development and/or modification timelines.	The architecture is reasonable, but various factors suggest that a significant amount of software engineering and/or subject-matter expertise will be required to develop or modify specific scenarios.	The system features a well-considered modular architecture. The problem space and authoring tools are simple enough to enable speedy and cost-effective development and modification.
Teaming factors	The testbed supports teams of 4–12 members, with flexible assignments of roles to humans or agents, and measures communication between teammates.	The team size is small and fixed. Roles are not differentiated, and autonomous agents are not incorporated. Communication between teammates is not recorded.	Team size is appropriate (*e.g.*, 4–6), and autonomous and human teammates can be assigned to multiple roles. Team communication is recorded.	Team size and configuration (autonomous/human teammates) are very flexible, and roles are (or can be) clearly defined. All modes of communication are recorded for evaluation.
Scenario authoring	The testbed should support the ability to create scenarios that meet our research objectives and make connections between competencies, scenario events, and performance measures.	Scenarios are fixed or only slightly adjustable.	The scenarios have moderate authorability, and the process seems reasonably straightforward.	The testbed includes scenario authoring tools that enable a high degree of expressiveness and tailoring.

(*Continued*)

TABLE 2.4 (*Continued*)
Scoring Standards for Testbed Evaluation Criteria

Evaluation Criteria	Description	Scoring Standard 0–3	4–6	7–10
Task features	The testbed should provide tasks with a high degree of psychological fidelity and the ability to modulate several task factors, such as difficulty, uncertainty, time pressure, information density, signal-to-noise ratio, and data reliability.	The testbed tasks are fixed and artificial, and do not require a high level of information processing.	The tasks require at least some information processing, and a level of customization is possible to modulate basic difficulty factors.	The testbed includes tasks that require a significant amount of information processing. These tasks feature many variables, making it reasonable to modulate important complexity dimensions.
Data processing	The testbed enables data export and visualization.	Data export is cumbersome and/or not possible. Data visualization is not possible.	Data can be exported in conventional formats (*e.g.*, Excel or CSV). Analysis and visualization tools are not available.	There is significant support for exporting data in various formats. Statistical analysis tools are provided. It is possible to generate data visualizations.
System architecture	The testbed needs to facilitate flexible changes as research progresses, be open-source, and allow for collaboration among distributed teams and labs.	The architecture is fixed and monolithic.	The system architecture is reasonable and should not present insurmountable impediments to system modification.	The testbed employs a well-structured, modular architecture. SDKs and/or other tools are available to support system maintenance and modification.
Agents	The testbed should make it relatively easy to "plug in" agents with various capabilities.	Agents cannot be incorporated into the system. The only option is a "Wizard of Oz" emulation of agents.	Agents can be developed and/or interfaced with the testbed. The interface may require significant software engineering skills.	Well-defined APIs exist to facilitate the integration of agents.

based on a combination of the proportion of features within each criterion met by the testbed and the rater's general assessment that the criterion was met. For example, the "System Architecture" criterion has three key features: modularity and flexibility, an open-source framework, and the ability to permit collaboration across academic and industry-based labs. If a testbed has two of those three features, one might assign it a score of 6 or 7 to reflect the proportion of features met. Whether the testbed received a rating of 6 or 7 depended on how well the rater felt other architectural features of the testbed would benefit the community. After two independent raters had assessed the testbeds according to these scoring standards, they met to review their scores and resolve discrepancies.

2.3.3 Identifying Candidate Testbeds

To conduct an analysis of existing testbeds, we selected testbed environments demanding a high level of interdependency, as this would ensure that our tasks required the type of teamwork most representative of the domains in which we typically work and would likely be the most challenging for agents to support smoothly. We reviewed the human-autonomy teaming literature and identified three categories of such testbeds: (1) In-House Teaming Testbeds (for testbeds developed from scratch for human-AI teaming research); (2) Commercial Game Platforms (for testbeds built on proprietary/off-the-shelf games); and (3) Open-Source Game Platforms (for testbeds based on modifiable, open-source games). Table 2.5 shows the items reviewed within each category. More details about the testbeds can be found in Asiala et al. (2023).

2.3.3.1 In-House Teaming Testbeds

Researchers have specifically developed the first category of testbeds to assess team performance, particularly in the context of human-AI teaming. These environments are of significant interest because they are likely to include the necessary instrumentation for robust data collection.

2.3.3.2 Commercial Game Platforms

An alternative to using custom-developed testbeds is to employ commercial games. While the results of our surveys indicated a strong preference for open-source testbeds, our review revealed that some researchers successfully used commercial games.

Commercial games are better capitalized than other testbeds, which makes for a richer and more compelling exercise environment. They are also more likely to closely resemble a relevant task environment, which can improve performance and immersion. By using commercial games, researchers can benefit from the significant skill and investment the developers put into creating the environment. Further, commercial games often attract a large online user community with significant technical skills. The associated user community can use those skills to create extensions to the gaming environment that benefit researchers (Leung et al., 2005). In other words, a large, technically skilled user base is more capable of extending commercial games—something the custom testbeds do not benefit from.

However, there are complications that researchers must consider when using commercial games as the foundation of a research testbed. One is the cost of the game

TABLE 2.5
Categories of Testbeds and the Selected Instances

Categories	Explanations	Selected Instances
In-house teaming testbeds	Testbeds developed from scratch for human–AI teaming research	1. Agile-Teams Testbed (AT2B) 2. Blocks World for Teams (BW4T) 3. MazeWorld 4. Multi-Agent Computing Environment (MACE) 5. Michigan Intelligent Coordination Experiment (MICE) 6. Reconfigurable Testbed for Teams (RT4T)
Commercial game platforms	Testbeds built on proprietary/off-the-shelf games	1. Overcooked! 2 (Bishop et al., 2020) 2. Overcooked!-based "Onion Soup" testbed variation (Carroll et al., 2019) 3. Overcooked!-based "Apple Juice" testbed variation (Gao et al., 2020) 4. Overcooked!-based "Salad" testbed variation (Wu et al., 2021) 5. Overcooked!-based "Gather Ingredients" testbed Variation (Song et al., 2019) 6. Overcooked!- based "Sushi" testbed variation (Buehler et al., 2021) 7. Situation Authorable Behavior Research Environment (SABRE; Leung et al., 2005) 8. Artificial Social Intelligence for Successful Teams (ASIST) Study 3 Testbed (ASIST Saturn+; Huang et al., 2022a) 9. ASIST Study 4 Testbed (ASIST Dragon; +; Huang et al., 2022a) 10. Black Horizon
Open-source game platforms	Testbeds based on modifiable, open-source games	1. Globulation 2. Unvanquished 3. Netrek 4. Crossfire

and any licensing restrictions that would limit their ability to use the game and report on the resultant findings. In most cases, the costs and other restrictions are not onerous; however, researchers must clearly understand their implications.

Researchers must consider the degree to which they can modify the game to satisfy the requirements of various experimental designs, as well as the relative ease with which they can author novel challenges designed to assess specific research hypotheses (Leung et al., 2005). Many gaming environments that tend to be used in research include scripting tools that allow users to expand the game in interesting ways. However, the research team must determine whether it possesses the necessary skills to employ these tools (Carbonaro et al., 2005). Often, the online community of users can help in this context by providing new challenges or game features or by answering questions that the research team may have as it works with the game (Carbonaro et al., 2005; Leung et al., 2005). For this reason, the vitality of the game's user community is another consideration when evaluating the suitability of a commercial game as the foundation for a testbed.

The last set of considerations pertains to experimental throughput. Some commercial games are so complex that they require a significant amount of time to develop basic competency. Until that level of competency in gameplay is achieved, any differences associated with the experimental intervention may be swamped by the noise generated by less-than-competent gameplay. In a related vein, the challenges presented by many games can be quite time-consuming to complete. While this may create a rich body of teamwork and taskwork performance data, it can also limit research throughput. In many cases, it is helpful to select games that are easier to master, with sufficient challenges (*i.e.*, scenarios, chapters, levels, etc.) but short in duration (Leung et al., 2005).

2.3.3.3 Open-Source Game Platforms

Open-source games may represent a compromise between custom-built researcher testbeds and commercial games. These games are generally developed by enthusiasts as a hobby and are maintained by the community. They often take longer to mature than commercial games and can lack the production values of many commercial games but generally exceed those of custom-built testbeds. They are published under an open-source license such as the MIT License, GNU General Public License, or BSD License. The combination of a sophisticated gaming experience and an extensible code base makes them an attractive option for human-AI teaming testbeds.

As an example of such a testbed, *Spring* (Bakkes et al., 2009) is an open-source real-time strategy game (see https://springrts.com/ and https://github.com/spring/spring). Like most games in this genre, players compete against one another by gathering resources, building structures, and developing military units to protect and capture land and resources. In *Spring*, the game ends when one player/team destroys the opponent's "Commander" unit.

In the environment in which we work, it is important to meet various industrial security standards, such as those imposed by the Cybersecurity Maturity Model Certification (CMMC). It can be difficult to ascertain whether a given program meets those standards when evaluating open-source tools. With open-source software, additions and edits can be proposed and accepted by maintainers, which potentially presents security risks. Research into the risks associated with open-source software is ongoing. Ultimately, these concerns led us to eliminate open-source games from further consideration. However, researchers in other environments may not have this constraint and thus may be more open to this approach.

2.3.4 Case Study Results and Discussion

The research team applied the scoring tool to the identified testbeds. Table 2.6 summarizes the final scores for each testbed.

Using our criteria and weights, the highest-scoring testbeds were *Neverwinter Nights*, the two ASIST testbeds (Saturn+ and Dragon), BW4T, and Black Horizon.

ASIST Dragon, ASIST Saturn+, BW4T, Black Horizon, and *Neverwinter Nights* scored significantly higher than the other testbeds evaluated for implementation as a human-AI teaming testbed. Based on this, the team gave additional consideration to these options.

TABLE 2.6
Consolidated Evaluation Scores

Testbed	Evaluator 1 Score	Evaluator 2 Score
ASIST Dragon	101	101
Neverwinter Nights	94	106
ASIST Saturn+	87	95
Blocks World for Teams (Johnson et al., 2009)	85	78
Black Horizon	75	85
Overcooked!-based "Salad" testbed variation (Wu et al., 2021)	62	66
Overcooked!-based "Onion Soup" testbed variation (Carroll et al., 2019)	59	59
Overcooked!-based "Apple Juice" testbed variation (Gao et al., 2020)	51	52
Overcooked!-based "Sushi" testbed variation (Buehler et al., 2021)	48	35
Overcooked!–based "Gather Ingredients" testbed variation (Song et al., 2019)	35	36
Overcooked! 2	36	27

First, the research team opted to deprioritize Black Horizon from further consideration. While Black Horizon supports multi-player operations, those players compete individually, and the game does not support team-based operations.

Second, we determined that *Neverwinter Nights* did not meet our testbed needs. A significant concern was that *Neverwinter Nights* is a commercial game. The researchers who responded to our survey clearly preferred an open-source solution. Third, a similar concern led us to deprioritize the ASIST testbeds. Like SABRE, which was based on *Neverwinter Nights*, the ASIST testbeds are built around Microsoft's *Minecraft* product. As a result, they might violate the open-source preference expressed by our researcher respondents.

The fourth tool we investigated was BW4T. While BW4T is open source, it has drawbacks that give us pause. Notably, a significant portion of the development of BW4T occurred outside the United States, and its primary developers have since ceased supporting the environment. Both factors may make BW4T incompatible with CMMC certification—a critical feature for our work in the defense industry.

The evaluation process led us to believe that no existing testbed had the features that would lead it to be widely adopted within the research community. As a result, we decided to develop a novel testbed and drew inspiration from the reviewed programs, particularly BW4T and the ASIST testbeds. Called the Team Coordination Testbed (TCT), our novel testbed presents a task similar to those used in the BW4T and ASIST testbeds (McCarthy, Asiala, & Bailey, in press). The design decisions for this testbed were based on the criteria developed for the evaluation of the existing testbeds. The team developed a Concept of Operations for the TCT to meet the requirements specified by researchers and SMEs in their survey responses and to enable it to score well within the systematic assessment tool described in this chapter. For example, the TCT supposes a team with four defined roles whose members must work together to achieve a common goal. Each role has defined responsibilities, and the players must work as a team and share information in order to achieve that goal. These design features are derived from the teaming factors criterion. Furthermore, the TCT design goal is to enable humans and AI agents to fill any team role. In addition, because the TCT is being developed "from scratch," it can be released as an open-source environment that researchers from across the domain can tailor and enhance. To facilitate these community contributions, we will adopt a curated approach to the testbed release. Labs that receive the software will be granted a license free of charge and, in exchange, will be responsible for informing Sonalysts of any modifications to the testbed software. Knowledge of these additions allows Sonalysts to confirm that they do not introduce security vulnerabilities. More details on the TCT are available in "The Team Coordination Testbed: An Open-Source Testbed for Assessing Highly Interdependent Teams" (McCarthy, Asiala, & Bailey, in press).

REFERENCES

Arthur, W. Jr., Edwards, B. D., Bell, S. T., Villado, A. J., & Bennett, W. Jr. (2005). Team task analysis: Identifying tasks and jobs that are team-based. *Human Factors*, *47*(3), 654–669. https://doi.org/10.1518/001872005774860087

Asiala, L. K. E., McCarthy, J. E., & Huang, L. (2023). *MUM-T initiative technical report #3: Analysis of testbed alternatives*. https://doi.org/10.48550/arXiv.2309.03213

Bakkes, S., Spronck, P., & Van den Herik, J. (2009). Rapid and reliable adaptation of video game AI. *IEEE Transactions on Computational Intelligence and AI in Games*, *1*(2), 93–104. https://doi.org/10.1109/TCIAIG.2009.2029084

Beans, C. (2018). Can robots make good teammates? *Proceedings of the National Academy of Sciences*, *115*(44), 11106–11108. https://doi.org/10.1073/pnas.1814453115

Buehler, M. C., Adamy, J., & Weisswange, T. H. (2021). Theory of Mind Based Assistive Communication in Complex Human Robot Cooperation. https://doi.org/10.48550/arXiv.2109.01355

Carbonaro, M., Cutumisu, M., McNaughton, M., Onuczko, C., Roy, T., Schaeffer, J., Szafron, D., Gillis, S. & Kratchmer, S. (2005). Interactive story writing in the classroom: Using computer games. *Proceedings of DiGRA 2005 Conference: Changing Views: Worlds in Play*. https://doi.org/10.26503/dl.v2005i1.152

Carroll, M., Shah, R., Ho, M. K., Griffiths, T., Seshia, S., Abbeel, P., & Dragan, A. (2019). On the utility of learning about humans for human-AI coordination. *Advances in Neural Information Processing Systems*, 32 (1–12).

Cavanah, E., Ford, Z., Jasper, A., Stonewall, J., Gilbert, S. B., & Dorneich, M. (2020). Creating metrics for human-agent teams. *Proceedings of the Human Factors and Ergonomics Society Annual Meeting*, *64*(1), 349–353. https://doi.org/10.1177/1071181320641079

Chiou, E. K., & Lee, J. D. (2021). Trusting automation: Designing for responsivity and resilience. *Human Factors*, *65*(1), 137–165. https://doi.org/10.1177/00187208211009995

Gao, X., Gong, R., Zhao, Y., Wang, S., Shu, T., & Zhu, S. C. (2020). Joint mind modeling for explanation generation in complex human-robot collaborative tasks. *2020 29th IEEE International Conference on Robot and Human Interactive Communication (RO-MAN)* (pp. 1119–1126). https://doi.org/10.1109/RO-MAN47096.2020.9223595

Gervits, F., Thurston, D., Thielstrom, R., Fong, T., Pham, Q., & Scheutz, M. (2020). Toward genuine robot teammates: Improving human-robot team performance using robot shared mental models. In *Proceedings of the 19th international conference on autonomous agents and MultiAgent systems*, Richland, SC; (429–437).

Johnson, M., Jonker, C. M., van Riemsdijk, M. B., Feltovich, P. J., & Bradshaw, J. M. (2009). Joint activity testbed: Blocks World for Teams (BW4T). *ESAW*, *9*, 254–256. https://doi.org/10.1007/978-3-642-10203-5_26

Leung, A., Diller, D., & Ferguson, W. (2005). SABRE: A game-based testbed for studying team behavior. *Proceedings of the Fall Simulation Interoperability Workshop (SISO)* Orlando, FL (pp. 18–23).

McCarthy, J. E., Asiala, L. K. E., Maryeski, L., & Warren, N. (2023a). *Improving the state of the art for training Human-AI teams: Technical report #1-- Results of subject-matter expert knowledge elicitation survey*. https://doi.org/10.48550/arXiv.2309.03211

McCarthy, J. E., Asiala, L. K. E., Maryeski, L., & Warren, N. (2023b). *Improving the state of the art for training Human-AI teams: Technical report #2-- Results of researcher knowledge elicitation survey*. https://doi.org/10.48550/arXiv.2309.03212

McCarthy, J. E., Asiala, L. K. E., Bailey, M. A. (in press). The Team Coordination Testbed: An open-source testbed for assessing highly interdependent teams. In E. K. C. Chiou, D. S. Lange, J. H. Wong, & J. Marble (Eds.), *Advancements in human agent teaming research infrastructure: Testbeds, metrics, and concepts*. CRC Press.

McNeese, N. J., Demir, M., Cooke, N. J., & Myers, C. (2018). Teaming with a synthetic teammate: Insights into human-autonomy teaming. *Human Factors*, *60*(2), 262–273. https://doi.org/10.1177/0018720817743223

National Academies of Sciences, Engineering, and Medicine. (2021). *Human-AI teaming: State-of-the-art and research needs*. The National Academies Press.

Phillips, E., Ososky, S., Grove, J., & Jentsch, F. (2011). From tools to teammates: Toward the development of appropriate mental models for intelligent robots. *Proceedings of the Human Factors and Ergonomics Society Annual Meeting*, *55*(1), 1491–1495. https://doi.org/10.1177/1071181311551310

Phillips, E., Rivera, J., & Jentsch, F. (2013). Developing a tactical language for future robotic teammates. *Proceedings of the Human Factors and Ergonomics Society Annual Meeting*, *57*(1), 1283–1287. https://doi.org/10.1177/1541931213571284

Sarkar, A. (2023). Enough with "human-AI collaboration". *Extended Abstracts of the 2023 CHI Conference on Human Factors in Computing Systems* (pp. 1–8). https://doi.org/10.1145/3544549

Saavedra, R., Earley, P. C., & Van Dyne, L. (1993). Complex interdependence in task-performing groups. *Journal of Applied Psychology*, *78*(1), 61.

Seeber, I., Bittner, E., Briggs, R. O., De Vreede, T., De Vreede, G. J., Elkins, A., Maier, R., Merz, A. B., Oeste-Reiß, S., Randrup, N., Schwabe, G., & Söllner, M. (2020). Machines as teammates: A research agenda on AI in team collaboration. *Information & Management*, *57*(2), 103174. https://doi.org/10.1016/j.im.2019.103174

Shneiderman, B. (2020). Human-centered artificial intelligence: Reliable, safe & trustworthy. *International Journal of Human–Computer Interaction*, *36*(6), 495–504. https://doi.org/10.1080/10447318.2020.1741118

Shneiderman, B. (2021, July 21). *19th NOTE: Human-Centered AI google group*. https://groups.google.com/g/human-centered-ai/c/syqiC1juHO.c

Shneiderman, B., & Endsley, M. (2022, June 29). *65th NOTE: Human-Centered AI google group*. https://groups.google.com/g/human-centered-ai/c/chqfvaI3AkQ

Singh, R., Sonenberg, L., & Miller, T. (2017). Communication and shared mental models for teams performing interdependent tasks. *International Workshop on Coordination, Organizations, Institutions, and Norms in Agent Systems* (pp. 81–97). https://doi.org/10.1007/978-3-319-46882-2_10

Song, Y., Wang, J., Lukasiewicz, T., Xu, Z., & Xu, M. (2019). Diversity-driven extensible hierarchical reinforcement learning. *Proceedings of the AAAI Conference on Artificial Intelligence*, *33*(1), 4992–4999. https://doi.org/10.1609/aaai.v33i01.33014992

Tesluk, P., Mathieu, J. E., Zaccaro, S. J., & Marks, M. (1997). Task and aggregation issues in the analysis and assessment of team performance. In M. T. Brannick, E. Salas, & C. W. Prince (Eds.) *Team performance assessment and measurement*. Psychology Press (pp. 209–236).

Thompson, J. D. (1967). *Organizations in action*. McGraw-Hill.

USAF. (2013). *Autonomy science and technology strategy*. Dayton, OH: Airforce Research Laboratory.

Van de Ven, A. H., Delbecq, A. L., & Koenig Jr, R. (1976). Determinants of coordination modes within organizations. *American Sociological Review*, *41*, 322–338.

Warta, S. F., Kapalo, K. A., Best, A., & Fiore, S. M. (2016). Similarity, complementarity, and agency in HRI: Theoretical issues in shifting the perception of robots from tools to teammates. *Proceedings of the Human Factors and Ergonomics Society Annual Meeting*, *60*(1), 1230–1234. https://doi.org/10.1177/1541931213601287

Wu, S. A., Wang, R. E., Evans, J. A., Tenenbaum, J. B., Parkes, D. C., & Kleiman-Weiner, M. (2021). Too many cooks: Bayesian inference for coordinating multi-agent collaboration. *Topics in Cognitive Science*, *13*(2), 414–432. https://doi.org/10.1111/tops.12525

3 The Team Coordination Testbed

An Open-Source Testbed for Assessing Highly Interdependent Teams

James E. McCarthy, Lillian K.E. Asiala, and Michael A. Bailey

3.1 INTRODUCTION

A consensus report produced for the Air Force Research Laboratory by the National Academies of Sciences, Engineering, and Medicine documented a widespread and increasing interest in supporting human-artificial intelligence (AI) teaming (NASEM, 2021). Using this report as a springboard, our team initiated a research program to explore ways to optimize the performance of human-AI teams. To provide a foundation for our research, we are exploring options for a synthetic task environment (STE) that can serve as a testbed for our studies.

Building on the results of recent surveys (McCarthy & Asiala, 2023a) of subject-matter experts (SMEs) and researchers within the human-AI teaming domain (McCarthy & Asiala, 2023b), the research team identified the testbed qualities on which we will focus, identified potential STEs, and conducted a quantitative comparison of the candidates (Asiala et al., in press). Three strong candidates emerged from that work. One was the Blocks World for Teams (BW4T) testbed, developed by researchers at Delft University of Technology in the Netherlands and the Florida Institute for Human and Machine Cognition (Johnson et al., 2009). The other two systems stemmed from the Defense Advanced Research Projects Agency (DARPA) Artificial Social Intelligence (ASI) for Successful Teams (ASIST) program (Elliot, 2021), which aimed to study ASI as an advisor in human-AI teaming. These two testbeds were ASIST Saturn+ (Study 3) and ASIST Dragon (Study 4). BW4T and the ASIST testbeds are described in more detail in Section 3.2.

While these three testbeds scored well in our analysis, they each had issues that raised concerns as we considered adopting them for our work. For example, the ASIST testbeds are built around Microsoft's *Minecraft* product. As a result, they might violate the open-source preference expressed by our researcher respondents. The DARPA testbeds were also designed to support research in a different domain—the development

DOI: 10.1201/9781003655923-4

of agents that employ "social intelligence" to maximize teamwork within simulated search-and-rescue scenarios. While BW4T is open-source, much of its development occurred outside the United States, and its developers have ceased supporting this environment. The foreign origin and lack of ongoing support may be incompatible with the Cybersecurity Maturity Model Certification process established by the United States Government. While important to us, these factors may be irrelevant to other researchers. As a result, we felt there might be value in creating a new testbed inspired by BW4T, the ASIST testbeds, and other systems we reviewed. In this chapter, we describe the emerging testbed and how researchers can access it to support their work.

3.2 FOUNDATIONS

To place our work in context, we begin with a discussion of BW4T and the two DARPA ASIST testbeds.

BW4T asks users to locate colored blocks within a maze and return them to a designated drop location in a specific sequence. The goal is to find and drop the blocks in the specified order as quickly as possible. BW4T collects three types of performance data: time to complete the tasks; player and agent "idle time" (any time a human or agent player was not doing something task-related); and various types of error rates. Error rates pertain to the amount of time two players are holding the same color block, indicating redundancy in the task; the number of boxes lost during the task; and the number of times a player was blocked from entering a room by another player's presence, indicating a lack of team coordination (Johnson et al., 2015).

Each player controls a "bot" and has their own view of the display. The contents of a compartment are displayed when the bot enters and disappear when the bot leaves. Players complete tasks via pop-up menus. Other portions of the display provide "health" information, such as the battery life of the controlled bot, and a chat area for communication with other players. A given player can send messages to an individual player or all players. In the current version of the software, players can only send specific pre-defined messages.

The ASIST testbeds presented an urban search-and-rescue task scenario (Huang et al., 2022). In ASIST Saturn+, a team of three players is responsible for searching for, stabilizing, and transporting victims to safe zones within a collapsed building across two 17-minute missions. The victims suffered injuries at various levels of severity and required different levels of treatment. Players assume one of three roles within the team: Medic, Engineer, or Transporter. Each role possesses unique knowledge and capabilities while also sharing common skills (*e.g.*, transporting stabilized victims at various speeds). The team's goal is to maximize its point total by rescuing victims. The ASIST Dragon testbed is a dedicated environment designed for bomb-disposal scenarios. It is available 24/7 and admin-free, allowing participants to join a waiting room at any time. Every three participants form a team and automatically spin up *Minecraft* servers to launch a new mission. In Dragon, all members share the same bomb-disposal role, but differences in tool allocation determine each member's bomb-disposal ability and influence team coordination. Within both testbeds, ASI agents actively monitor players' states, predict their actions, and intervene to improve teamwork.

The ASIST testbeds include several features that developers can use to control scenario difficulty. Three perturbations were used to increase the complexity of a scenario: (1) a structural collapse that creates new rubble that the engineer must clear; (2) GPS-based map signal loss; and (3) a random bomb explosion that causes injury and fire. Additional adjustable factors include the number and location of victims, bombs, and blockages. The ASIST testbeds, such as those developed by Cooke and her colleagues, enabled the introduction of perturbations into ongoing scenarios. The testbeds also feature robust data collection and performance measurement capabilities, including structured data (*e.g.*, survey responses, JSON logs, researcher analyses, and annotator-coded metadata) and unstructured data (*e.g.*, screen recordings, communication transcripts, and study-level metadata). The ASIST Dragon testbed extends the capabilities of Saturn+ by enabling simultaneous performance measurement across multiple teams and supporting the management of geographically distributed teams. It also expands the modalities of human-AI communication (*e.g.*, by adding artifacts), enabling the study of team formation, resource allocation, risk avoidance, and teamwork behaviors.

3.3 PRIORITIES

Based on lessons learned from our review of existing testbeds (BW4T, ASIST Saturn+, ASIST Dragon, and others), our team developed a set of assumptions about which testbed attributes would be valued within the research community. We then examined those hypotheses through the lens provided by surveys of researchers and SMEs. Asiala et al. (in press) provide a more detailed description of this process. This process enabled us to create a weighted list of attributes an ideal testbed would possess. This list is shown in Table 3.1. We recognized that some of these attributes were likely unachievable during the first phase of the effort. However, we wanted to establish them as goals so that the overall architecture would facilitate, rather than hinder, their eventual achievement. We return to these attributes in Section 10 to summarize our ability to achieve each goal.

TABLE 3.1
Targeted Testbed Attributes

Testbed Attribute	Description
Data collection/performance measures	The testbed should be instrumented to collect data on individuals and teams for both outcomes and processes. Ideally, the testbed should include rich instrumentation and support the definition of custom measures of interest.
Implementation factors	Initial development and ongoing enhancement of the testbed should be relatively inexpensive. Therefore, the testbed should use a well-considered modular architecture. The problem space and authoring tools should be simple enough to enable speedy and cost-effective development and modification. The testbed should be primarily designed for data collection and analysis.
Teaming factors	The testbed should support teams of 4–12 members, with flexible assignments of roles to humans or agents. Team size and configuration (including autonomous and human teammates) should be highly flexible, with clearly defined roles. It should include one or more communication methods among teammates. All modes of communication should be recorded for evaluation.

(Continued)

TABLE 3.1 (*Continued*)
Targeted Testbed Attributes

Testbed Attribute	Description
Scenario authoring	The testbed should support the creation of scenarios that meet a given research team's objectives and facilitate connections among competencies, scenario events, and performance measures. The testbed should include scenario authoring tools that enable a high degree of expressiveness and tailoring.
Task features	The testbed should provide tasks with a high degree of psychological fidelity and the ability to modulate several task factors, such as difficulty, uncertainty, time pressure, information density, signal-to-noise ratio, and data reliability. The testbed should include tasks that require substantial information processing. These tasks should involve multiple variables, making it reasonable to modulate key complexity dimensions.
Data processing	The testbed should enable data export and visualization. It should include significant support for exporting data in various formats. Statistical analysis tools should be provided, including the ability to generate data visualizations.
System architecture	The testbed should facilitate flexible changes as research progresses and be provided as "open-source," allowing for collaboration among distributed teams and labs. The testbed should employ a well-structured, modular architecture. Software Development Kits (SDKs) and/or other tools should be available to support system maintenance and modification.
Agents	The STE should make it relatively easy to "plug in" agents with various capabilities. To facilitate this, well-defined APIs should be included.

3.4 OVERVIEW OF TESTBED AND GAME PLAY

The preceding work provided a foundation for developing an open-source Team Coordination Testbed (TCT). The TCT implements a basic "find stuff and do stuff" game mechanic similar to the tasks presented within BW4T and the ASIST testbeds. The current version of the testbed (*i.e.*, the "Galileo Skin") has team members searching an abandoned spacecraft in a decaying orbit to locate and extract cargo before the spacecraft reenters the atmosphere and is destroyed. Later versions of the game could be "reskinned" to support tasks with a similar structure, such as search-and-rescue, bomb localization and defusing, or various intelligence, surveillance, and reconnaissance tasks.

The TCT game environment is a rectangular maze of passageways leading to compartments representing the abandoned spacecraft (see Figure 3.1). The loading dock is a special compartment that must be included in each scenario. This is the space to which the team must deliver cargo.

The other compartments will be scattered throughout the maze. Each compartment will include zero or more pieces of cargo that the team must transport to the loading dock. Each compartment is open to a passageway along one of the bulkheads. Compartments may be generic or designed for a specific purpose. Generic compartments have no features that provide clues to their intended use or the type of cargo they may contain. Special-use compartments have features that provide clues to their intended use and the contents they may contain, as shown in Figure 3.1. While special-use compartments are

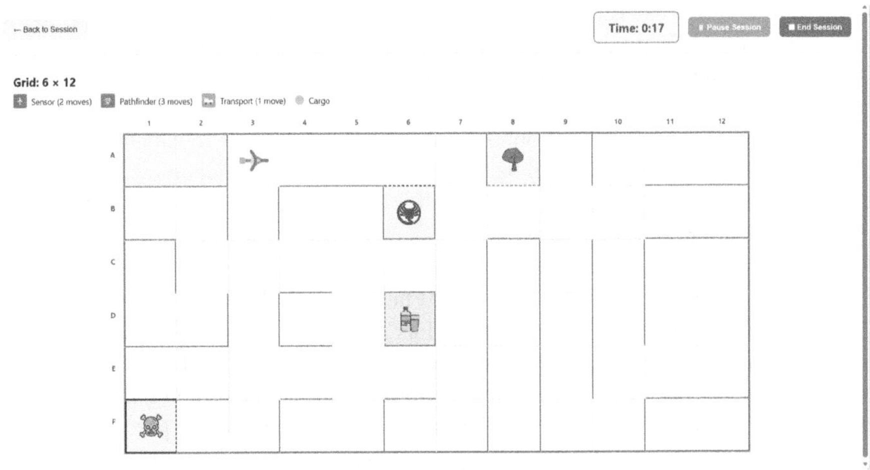

FIGURE 3.1 Notional TCT game space.

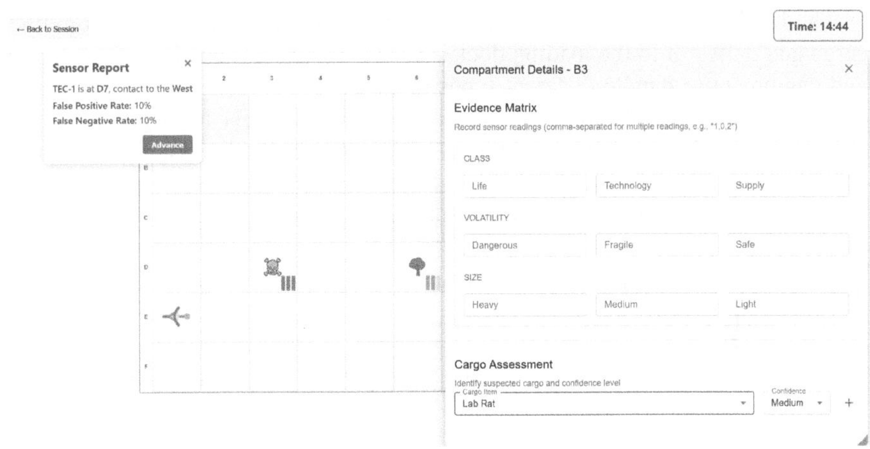

FIGURE 3.2 Sample TCT screen.

designed for specific classes of cargo, any cargo class may be placed in any compartment. In the current environment, cargo is characterized by three dimensions: Class (Life Form, Technology, or Supply), Volatility (Safe, Fragile, or Dangerous), and Size (Light, Medium, or Heavy), yielding up to 27 cargo types.

The task is completed interdependently by a mission team. The initial TCT defines four roles: a ship's engineer who develops a map of the ship; a sensor technician who uses incoming data to come to conclusions about the type of cargo in different compartments throughout the ship; a logistics expert who equips and, together with the ship's engineer, guides teams through the cargo recovery process; and a commander who is responsible for facilitating the exchange of information, coordinating extraction activities, and safely extracting crew members before time runs out. A sample screen is shown in Figure 3.2.

Team members use three types of tools to complete the task. The sensor technician's primary tools are sensors tuned to specific cargo attributes. For example, there are sensors designed to detect living organisms, fragile objects, or heavy items. Each sensor searches the space (semi-autonomously or under manual control). When a sensor encounters a space that contains cargo with the attribute to which it is tuned, it issues a report. Sensor speed can be traded off against sensor accuracy. Additionally, the false-positive and false-negative rates for each sensor class (Class, Volatility, or Size) can be adjusted independently.

The ship's engineer uses tools called pathfinders to map the space (semi-autonomously or under manual control). When a pathfinder encounters a bulkhead, it reports the location of the bulkhead and its attributes. Bulkhead attributes include material (armored or not), air system status (sealed or open), and lock type (present or absent). These three attributes combine to define eight compartment types. Pathfinders do not trade speed for accuracy; they are always 100 percent accurate. Instead, they trade speed for completeness. When set to "slow," pathfinders move at 45 seconds per grid square and report all attributes in all four directions. However, when set to "fast," pathfinders move at 15 seconds per grid square and limit their reports to new features encountered in all four directions each time they advance.

The Logistics Expert works with transport teams that are dispatched to retrieve cargo and deliver it to the loading dock. Various transport teams have the expertise and equipment to handle different types of cargo. Unlike sensors and pathfinders, transport teams do not move automatically; a team member must guide them. In addition, unlike sensors and pathfinders, transport teams may open doors and enter compartments. If a transport team is not correctly configured to transport a piece of cargo, the transfer may fail, and the transport team and/or the cargo may be damaged.

3.5 CONTROLLING UNCERTAINTY, DIFFICULTY, AND WORKLOAD

Among the top priorities for the testbed was the ability of researchers to modulate task difficulty, as driven by factors such as uncertainty, time pressure, information density, signal-to-noise ratio, and data reliability. Similarly, we wanted researchers to be able to define scenarios with varying workload levels. In this section, we review the "levers" that the testbed provides to modulate these factors.

Difficulty is a function of the search area, the number of objects to be retrieved, the required precision for the retrieval operation, and the time available to the crew to complete the challenge. The TCT allows authors to specify the absolute horizontal and vertical dimensions of the game space. A larger game space provides a larger area to search within the specified time, thereby making a given scenario more or less difficult. Similarly, the author can specify the number, location, and contents of compartments. Including only a few compartments produces a low "signal-to-noise" ratio; including more compartments increases that ratio. Including multiple objects within a space can add complexity to sensor reports, reduce certainty, and increase difficulty. Similarly, aligning stored cargo with the compartment type can reduce difficulty, whereas placing cargo in unexpected or "surprising" compartments can increase difficulty. The structure of the maze walls is related to both the size of the search space and the number and location of the compartments (*e.g.*, passageways in the current example). A simple maze will allow sensors, pathfinders, and transport teams to navigate more easily and with less involvement from team members. A more complex maze will increase navigation complexity and team member involvement while producing more pathfinder reports. The scenario designer sets the required level of retrieval precision. A designer may choose to accept cargo items in any order. Alternatively, the design may dictate a specific retrieval order. Doing so will increase the task's complexity. The team's task is completed against a running clock. The time a designer allocates to the team to complete its mission is a powerful source of performance stress and task difficulty.

Teams seldom work in contexts in which they have complete or entirely accurate information. Operational uncertainty (*e.g.*, the fog of war) is often present, and its level can have a pronounced effect on task difficulty, individual workload, and both individual and team performance. The TCT provides two ways for the author to modulate uncertainty. First, the author can trade off accuracy for speed. Suppose a researcher wants to increase the rate at which sensors move through the space. In that case, the "cost" is lower accuracy (expressed as higher rates of false-positive and false-negative findings). A slower speed setting is associated with more accurate sensor reports and less uncertainty. The second way a designer can control uncertainty is by independently setting the false-positive and false-negative rates for each sensor type (*i.e.*, class, volatility, or weight sensors). In this way, if needed in a specific setting, the designer can, for example, create fast and accurate volatility sensors.

In addition to difficulty and uncertainty, researchers are often interested in controlling the level of workload in teamwork scenarios. In addition to the features already discussed that affect workload (*e.g.*, maze complexity, signal-to-noise ratio, or uncertainty level), the TCT includes features that also affect workload. The first is the number of active sensors or pathfinders in the scenario. As the number of these tools increases, so does the number of reports they generate. While these reports are important sources of information, they also directly increase the workload for the sensor technician and ship's engineer, who must process them. Moreover, they have an indirect effect on the team members with whom the sensor technician and ship's engineer must interact—specifically, the logistics expert and commander. A similar effect can be achieved by increasing the rate at which the sensors and pathfinders traverse the space. Faster transit speeds are likely to result in increased direct and indirect workload. A significant source of additional workload (primarily for the sensor technician and ship's engineer) is created by shifting from automated to manual navigation. Manual navigation adds a significant "secondary task" for team members, which can interfere with their processing of reports and their judgments of cargo and compartment types.

3.6 ASSESSING INDIVIDUAL AND TEAM PERFORMANCE

Having discussed the levers available to researchers to modulate difficulty, uncertainty, workload, and related variables, we turn our attention to the other side of the equation. What performance attributes can the TCT assess?

The TCT is being designed and developed as a research tool. Therefore, we have adopted an intensive, data-driven approach to the architecture. Within this approach, essentially every sensor or pathfinder report, every operator action, and every communication between team members is recorded in a time-tagged database. Separate services can scan the database tables to assess performance in a manner meaningful to a particular research effort.

For the prototype version of the TCT, we have designed performance measures that are appropriate to the current cargo-retrieval "skin." These measures focus on the speed and accuracy with which each team member completes their respective tasks. We have also defined measures of team performance, primarily the team's ability to secure the cargo and safely evacuate the ship.

To complement these performance measures, our team also developed plans to assess Situation Awareness (SA). The SA probes are modeled on the Situation Awareness Global Assessment Technique (Endsley, 1995) and can be enabled or disabled within a session. Additionally, the testbed's open-source nature enables researchers to implement an alternative SA assessment approach if a given approach is more suitable for their work.

A third group of measurements will capture team coordination actions. Cannon-Bowers and Salas (1998) led a series of studies designed to identify the factors that distinguished high- and low-performing teams. The researchers identified four critical dimensions: Support, Leadership, Information Exchange, and Communication (SLIC). Research has shown that performance along the SLIC dimensions distinguishes experienced teams from inexperienced teams and is indicative of effective team performance.

Most, if not all, of the SLIC teamwork dimensions involve communicating with other teammates. The initial version of the testbed implements a highly structured communication mechanism. It is likely insufficiently expressive to allow meaningful differences among teams. However, as we move toward more natural language-based communication, meaningful assessment will become possible. This maturation will also enable us to transition from having skilled observers assign scores to various dimensions to employing AI-based agents for scoring. The use of natural language within the testbed, combined with automated scoring, will enhance our ability to apply consistent rubrics to both all-human and hybrid teams. As with other features of the testbed, our open-source approach will allow researchers to add alternative measures of team coordination as desired.

3.7 ASSESSING PERFORMANCE COVARIATES

The three types of in-session performance measures described in the preceding section will be complemented by covariate measures assessed through pre- and post-session surveys.

Two distinct categories of constructs have a demonstrable impact on team performance. The first is the individual trait group, which focuses on relatively stable characteristics arising from an individual's personality or from their developmental background within a particular environment. These traits can be described broadly in terms of temperament and ability. Traits related to temperament encompass those that fall within traditional models of personality, such as the "Big Five" model. They also encompass the person's general predisposition to individualism or collectivism, interest in working in groups, and locus of control. We also categorize general abilities, such as cognitive and interpersonal abilities (*e.g.*, emotional competence and intelligence), under this "trait-level" category. Generally, these qualities affect many aspects of an individual's experience within and beyond the team dynamic. These trait-level constructs can be measured once before a given session and applied to understand performance across all subsequent sessions.

The second group of constructs pertains to an individual's beliefs and attitudes regarding *their* team and is specific to working in a team environment. Teamwork-related beliefs and attitudes are more likely to be informed by an individual's experience operating in particular group settings and, therefore, could be more changeable over time. While these constructs are conceptualized as an individual's attitude toward a team, the measures typically ask respondents to report on their experience with a particular team at the time of measurement (rather than on their general attitude toward teams or their propensity to work well with others). An individual's experience or training may alter these constructs. Therefore, it is more appropriate to measure these states at the end of each session.

In assessing individual and team performance, as well as individual states and traits, the TCT maintains extensive data. To simplify data organization, the development team has provided an easy way to export the various data collections as Excel files.

3.8 ARCHITECTURAL OVERVIEW

The TCT was designed with a modular and data-driven architecture to facilitate easier modification and adoption of the software by individual research labs. Standard modifications could include updating the "skin" of the testbed (*e.g.*, changing from the space-based environment to an urban search-and-rescue environment); adding or modifying measures of performance (*e.g.*, creating a measure of SA or shared mental models); or enhancing system functionality (*e.g.*, transitioning from the highly structured approach to communication currently in place to one based on natural language). The TCT is built on a client-server model, allowing researchers to run experiments and view results in real time via a web browser.

To facilitate community development of TCT, we are adopting a "curated" approach to its open-source release. Under this approach, each lab that requests and receives a copy of the software will be granted a non-exclusive, royalty-free, and free-of-charge license to use it. If the recipient makes any changes to the software, they will be responsible for informing us and sharing their modifications with us. If we believe the modifications are safe, secure, and of general value to the community, we may incorporate the modification(s) into the central codebase and include them in subsequent distributions to other recipients. Recipients may also redistribute the code. However, the same condition applies: anyone who receives the code, from us or from another lab, must share it with us for possible inclusion in the codebase and redistribution to other requesting labs.

To request a copy of the open-source license and the code itself, please contact Jim McCarthy at mccarthy@sonalysts.com.

3.9 CONCLUSION

A literature review and analysis of alternatives led us to believe we could contribute to the field of hybrid teaming by creating a reusable, customizable testbed in which various performance measures could be used to assess a range of hybrid teaming interventions. Surveys of researchers in this domain, reviews of existing research environments, and consideration of our own research goals led us to establish a set of goals for what we labeled the TCT. From the outset, we recognized that the initial version of the testbed would not fully meet those goals, but we aimed to lay a foundation for our team and other researchers to extend as research needs dictate. Table 3.2 summarizes the status of the testbed.

TABLE 3.2
Status of Testbed Attributes

Testbed Attribute	Description
Data collection/ performance measures	The testbed was developed with a data-centric design that enhances visibility into system states and operator actions. The current version includes a variety of performance measures, and this data-centric approach should facilitate the definition of additional measures.
Implementation factors	The testbed features an open, modular architecture and a data-driven design. The architecture was established to ease scenario authoring, performance measurement, and software modification.
Teaming factors	The testbed skin supports teams of four highly interdependent players. This number can be reduced by combining roles or increased by subdividing roles (*e.g.*, sensor technicians could specialize in individual sensor types and report their results to a sensor coordinator). Inter-player communication is supported, but is more structured than is ideal.
Scenario authoring	The testbed includes an initial authoring tool for defining game boards. Future enhancements may include the ability to incorporate dynamic events or change the environment's "skin" (*e.g.*, from cargo retrieval to search-and-rescue).
Task features	The testbed requires significant information processing, and designers can modulate several task factors, including difficulty, uncertainty, time pressure, information density, signal-to-noise ratio, and data reliability.
Data processing	The testbed enables data export using various organizational schemes. Currently, data are exported as Microsoft Excel-compatible files. We have not yet incorporated analysis or visualization tools.
System architecture	The testbed architecture and licensing framework facilitate flexible changes as research progresses and collaboration among distributed teams and labs is established. The testbed employs a well-structured, modular architecture. We have not yet developed SDKs and/or other tools to support system maintenance and modification.
Agents	The data-driven nature of the testbed should enable researchers to "plug in" agents with various capabilities. However, the current version supports all-human teams, and our immediate goal is to generate baseline data that can serve as a point of comparison for agent-enhanced teams. We are actively seeking research partners with whom we can collaborate on this effort.

The next step in our research program will be to establish performance baselines with all-human teams. These baselines will serve as an important point of comparison as we introduce agents with different designs and assess their performance relative to each other and to all-human teams. We would welcome the opportunity to collaborate with researchers to establish these baselines.

We look forward to working with the community to enhance this testbed. Doing so will enhance our ability to replicate and compare results across the community.

REFERENCES

Asiala, L. K. E., McCarthy, J. E., Huang, L. (in press). Buy vs. Build: A structured assessment of testbed alternatives. In E. K. C. Chiou, D. S. Lange, J. H. Wong, & J. Marble (Eds.), *Advancements in human agent teaming research infrastructure: Testbeds, metrics, and concepts*. CRC Press.

Cannon-Bowers, J. A., & Salas, E. (1998). Individual and team decision making under stress: Theoretical underpinnings. In J. A. Cannon-Bowers & E. E. Salas (Eds.), *Making decisions under stress: Implications for individual and team training* (pp. 17–38). American Psychological Association.

Elliott, J. (2021). Artificial Social Intelligence for Successful Teams (ASIST). Arlington, Virginia, United States: Defense Advanced Research Projects Agency (DARPA).

Endsley, M. R. (1995). Measurement of situation awareness in dynamic systems. *Human Factors*, *37*(1), 65–84.

Huang, L., Freeman, J., Cooke, N., Colonna-Romano, J., Wood, M. D., Buchanan, V., & Caufman, S. J. (2022, April 11). *Exercises for artificial social intelligence in Minecraft search-and-rescue for teams*. https://doi.org/10.17605/OSF.IO/JWYVF

Johnson, M., Bradshaw, J., Duran, D., Vignati, M., Feltovich, P., Jonker, C., & van Riemsdijk, M. B. (2015). RT4T: A reconfigurable testbed for joint human-agent-robot teamwork. *Proceedings of the Tenth Annual ACM/IEEE International Conference on Human-Robot Interaction* (pp. 254–256).

Johnson, M., Jonker, C. M., van Riemsdijk, M. B., Feltovich, P. J., & Bradshaw, J. M. (2009, November). Joint activity testbed: Blocks World for Teams (BW4T). *ESAW*, (9), 254–256.

McCarthy, J.E., & Asiala, L.K.E. (2023a). *MUM-T initiative technical report #1: Results of subject-matter expert knowledge elicitation survey* [Unpublished manuscript].

McCarthy, J.E., & Asiala, L.K.E. (2023b). *MUM-T initiative technical report #2: Results of researcher knowledge elicitation survey* [Unpublished manuscript].

National Academies of Sciences, Engineering, and Medicine. (2021). *Human-AI teaming: State-of-the-art and research needs*. The National Academies Press.

4 Human Agent Teaming Testbeds

Crewed Autonomous Aircraft Flight Simulator

Richard Agbeyibor, Sanya Doda, Carmen Jimenez Cortes, Jack Kolb, Adan Vela, Samuel Coogan, and Karen M. Feigh

4.1 INTRODUCTION

Developments in AI and autonomy are reshaping the landscape across sectors. As defense and commercial investments in autonomous systems mature, autonomous aircraft may soon be able to competently "aviate, navigate, and communicate" on their own – fundamental aviation tasks that were previously performed only by trained pilots. Certain specialized aviation missions, however, involve complex tasks that cannot readily be automated or offboarded. For these missions, there may soon be a requirement for mission specialists, who are not trained pilots or AI engineers, to team with an agent aboard an autonomous aircraft.

The Crewed Autonomous Aircraft Flight Simulator (CAAFS) was developed to support experimental research exploring how non-pilot crew members interact with autonomous flight systems in specialized aviation missions. The simulator platform enables human-in-the-loop studies of human-autonomy teaming (HAT) in scenarios where operators collaborate with an automated or autonomous pilot (AP) while managing dynamic mission demands.

Our goal was to create a low-cost, "hi-fidelity testbed" (Gray, 2002) that could be built from commercial off-the-shelf and open-source components, making it accessible to academic researchers while still providing sufficient realism to elicit authentic operator behavior. Existing HAT simulators and testbeds in government and industry labs (Chen et al., 2018; Monk et al., 2025; USAARL, 2025) offer operational realism but are often inaccessible to researchers due to cost, schedule, or security constraints. We designed the CAAFS testbed to fill that gap and to enable joint experimentation by researchers in human factors, autonomy, robotics, and aerospace engineering.

The testbed supports a wide range of mission profiles, including intelligence, surveillance, and reconnaissance (ISR); aeromedical evacuation (MedEvac); and other specialized aviation missions that are well suited to integrating real-time collaboration between humans and APs. By combining a modular architecture, immersive

DOI: 10.1201/9781003655923-5

visualization, flexible authority structures, and multimodal sensing, the CAAFS testbed allows researchers to probe critical HAT questions.

Key design decisions were driven by the need to observe and evaluate teaming behaviors – not just performance metrics. The CAAFS testbed enables investigation of interaction breakdowns, trust calibration, workload shifts, and fluency in shared decision-making. The testbed uses realistic visuals and interfaces to surface cognitive biases and teaming behaviors in ecologically valid but repeatable settings. Team fluency, defined as the quality of coordinated interaction between human and agent teammates (Hoffman, 2019), is the core enabler and measure of teaming in our research on HAT in crewed autonomous aircraft. We operationalize fluency through six primary teaming metrics: workload, trust, situation awareness (SA), user interface (UI) interaction, hand movement and gaze behavior, and physiological state.

This chapter describes the testbed's architecture, capabilities, and example research applications. We explain the design tradeoffs, summarize lessons learned from its use, and share insights for researchers building similar testbeds. Our hope is that the CAAFS testbed, with its open-source components, can support broader adoption of controlled, realistic HAT research in emerging domains like advanced air mobility, crewed-uncrewed teaming, and autonomous emergency response.

4.2 RESEARCH OBJECTIVES

Our research aims to model effective teaming between non-pilot human crew members and APs in high-stakes missions. Unlike prior fluency work focused on turn-based robotics (Hoffman, 2019), we examine team fluency under varied authority, responsibility, and transparency constructs.

We focus on MedEvac and ISR scenarios, which involve adaptive collaboration and present distinct challenges to team fluency. These scenarios are representative of future naval aviation where autonomous systems support human crew in missions that can change unexpectedly due to both external forces (e.g., landing sites no longer secure upon arrival, enemy fire, denied environments) and internal forces (e.g., emergent criticality of patient care). This motivates three core research objectives:

1. **Characterize challenges to team fluency** created by human biases, cognitive limitations, and task characteristics – where human and autonomous agents have different capabilities, responsibilities, and authority levels.
2. **Quantify the impact of fluency** on mission effectiveness in HAT contexts, including MedEvac and ISR scenarios with varying levels of authority and responsibility delegation.
3. **Develop and evaluate mitigation strategies** such as adaptive autonomy, multimodal interfaces, and explainable agent behaviors to improve teaming performance.

4.3 SIMULATOR CAPABILITIES

The simulator provides a full-scale electric Vertical Takeoff and Landing (eVTOL) cabin with crew seating and seatbelts, touchscreen interfaces, simulated radios,

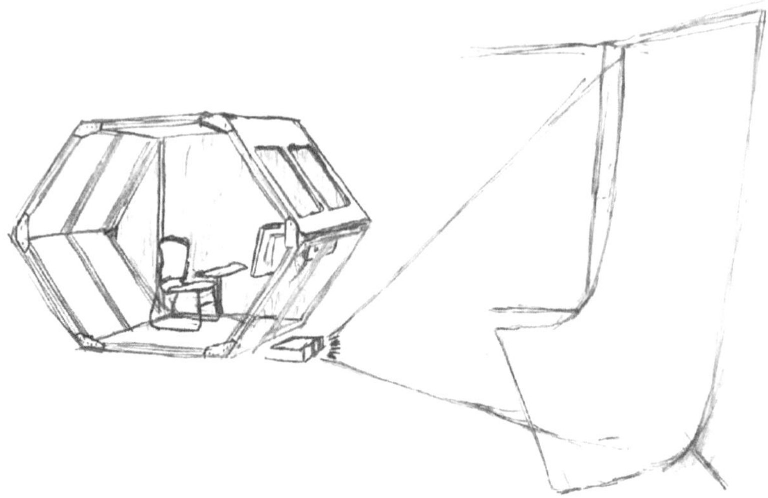

FIGURE 4.1 The design sketch of the Crewed Autonomous Aircraft Flight Simulator.

and immersive visuals. In Figure 4.1, the design sketch of the Crewed Autonomous Aircraft Flight Simulator depicts the initial design. Core software components include flight visualization, user interfaces, a cognitive state estimator, and flight control software with safety mechanisms such as control barrier functions (CBFs).

4.3.1 Flight Visualization

Out-the-window views were deemed necessary to improve simulation fidelity and immersiveness. We selected Microsoft Flight Simulator (MSFS) based on two priorities: (1) high-fidelity visuals for immersion and authentic behavior, and (2) sufficient integration with custom autonomy. We considered and compared three platforms:

- **MSFS**: Exceptional visual realism (photogrammetry-based), a widely supported Software
  Development Kit (SDK) – SimConnect – but limited access to internal control logic and closed-source constraints (Asobo Studio, 2020). A commercial software license is required.
- **X-Plane**: Moderate visual fidelity, accurate flight dynamics, and a more open SDK for custom autopilot and sensors (Laminar Research, 2024). Less immersive for human-subject studies.
- **FlightGear**: Open-source and highly extensible, ideal for avionics/control algorithm research (FlightGear Developers, 2024). Poor visual quality and UI design compared to MSFS.

Overall, MSFS offered the visual realism needed for valid cognitive and teaming research, despite limited control integration. However, MSFS's SDK (SimConnect) has severe limitations on spawning targets and controlling autonomous agents.

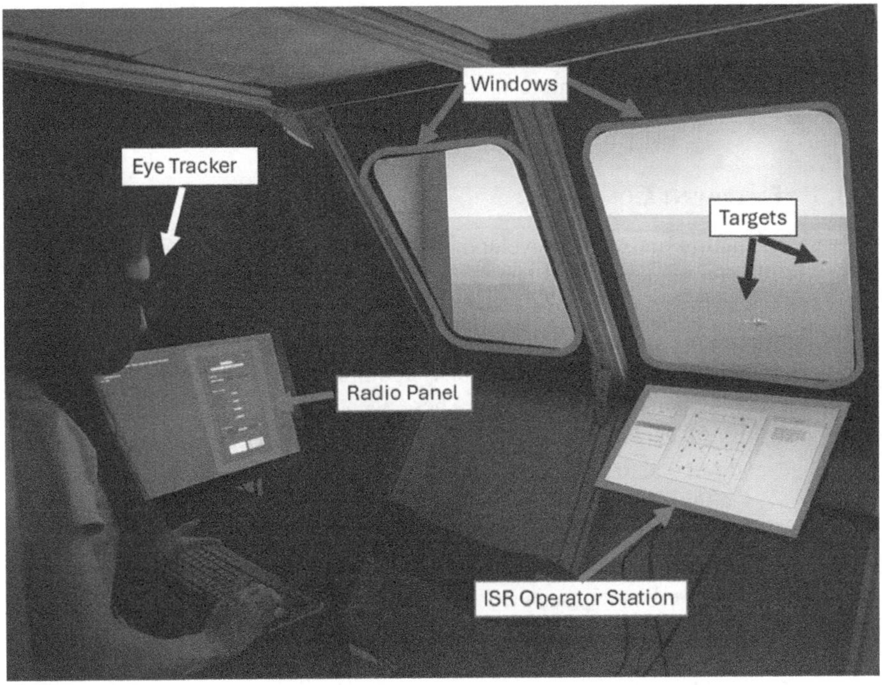

FIGURE 4.2 Crewed Autonomous Aircraft Flight Simulator – ISR configuration.

Working around these limitations was an acceptable tradeoff for the realism of the simulation. In Figure 4.2, Crewed Autonomous Aircraft Flight Simulator – ISR configuration shows the CAAFS cabin in the ISR configuration.

4.3.2 Configuration Management

To support evolving aircraft and mission concepts, the simulator was built with a modular architecture for both hardware and software. The cabin approximates a light helicopter in size and can accommodate up to four participants. The CAAFS testbed supports rapid reconfiguration in under a day, allowing the repositioning of displays, controls, and equipment across diverse mission types.

The two mission configurations differed primarily in the in-cabin elements, displays, touchscreen interfaces, and the location of the simulated aircraft (over an ocean for ISR vs. in an urban environment for MedEvac). The MedEvac configuration included a physical incubator mock-up, virtual vital monitors, and a Patient Care Report interface, while the ISR configuration included radar/sensor displays and radio communication tools. Despite their differences, both configurations shared enough hardware and software elements to support a unified configuration approach. Core generalizable elements included common display mounts, integrated power and data routing, a shared autonomy interface architecture connecting mission-specific displays, and common multimodal pilot-agent communication pathways across scenarios.

Designing the testbed to leverage a common underlying architecture was central to supporting multiple operational domains. The testbed serves as a reusable research tool for future domains of interest, such as air mobility, logistics, or search-and-rescue scenarios.

4.3.3 Experiment Control

To support mature and under-development agents, the simulator offers two control modes: fully autonomous and Wizard-of-Oz (WoZ).

- **Autonomous and Rule-Based Agent Modes**: For scenarios with structured agent behavior, such as ISR missions, we implemented rule-based autonomy that could operate independently. These agents leveraged standard planning algorithms (e.g., A*) to execute surveillance patterns, manage sensor coverage, and respond to participant commands with minimal human intervention.
- **WoZ Capability**: In more complex scenarios – such as MedEvac missions involving unpredictable emergencies and nuanced verbal interaction – the agent's autonomy was suppltemented by a hidden human operator. The WoZ system allowed experimenters to dynamically inject actions, override agent responses, or simulate intelligent behavior in real time.

Speech Interaction Limitations: The integration of speech-based interfaces highlighted limitations in current natural language processing systems. In these cases, the WoZ setup allowed experimenters to resolve ambiguities or simulate agent comprehension by generating custom audio responses.

Outcome and Lessons Learned: This dual-mode control architecture – autonomous when feasible, WoZ when needed – was essential for early testing and preserved study fidelity. However, it required experimenter calibration and careful scenario design to ensure consistent participant experiences.

4.3.4 Flight Controller

The design of our flight controller was shaped by the need to support graduated autonomy across varying mission demands and operator roles (Feigh et al., 2012). To support ISR and MedEvac missions with non-pilot operators, we implemented a flight controller that handles navigation and safety behaviors autonomously while supporting adjustable authority and human override. This flexible autonomy framework enables experiments in trust calibration, takeover dynamics, and function allocation (Agbeyibor et al., 2024c).

To enforce safety, we implemented CBFs atop a waypoint-following controller. CBFs use a Quadratic Program (QP) to minimally adjust trajectories that would violate safety constraints (Ames et al., 2019). The controller uses a second-order unicycle model and was implemented in MATLAB®/Simulink®.

The layered architecture – waypoint-following beneath a CBF safety layer – enabled modularity and scenario adaptability. It supported adaptive autonomy research and provides a template for future controllers that must balance human oversight with formal safety guarantees.

4.3.5 Autonomous Pilot Behaviors

In MedEvac scenarios, the AP handles piloting while interacting with the emergency medical technicians (EMTs) aboard. The AP interaction mechanism was designed to make requests of the human crew, respond to requests from the human crew, and interactively identify alternative landing sites in emergency situations. In this mission, the AP behavior remained fixed, but the interaction interface changed over time. It began with touchscreen input, then evolved to a hybrid touchscreen and keyword-based verbal interface, culminating in a primarily verbal interface with screen visuals for verification.

In ISR scenarios, we implemented a single interaction mechanism and five AP modes:

- **Waypoint Mode**: The AP executes a search pattern while accepting arbitrary user-defined waypoints without filtering or explanation. This represents a fully manual, operator-driven authority model.
- **Collaborative Mode**: The AP filters out waypoints that violate feasibility constraints (e.g., maneuverability, altitude), though it does not communicate its rationale. This simulates an autonomy system that ensures safety while offering minimal transparency.
- **Collision Avoidance Mode**: The AP autonomously modifies its trajectory using CBFs to avoid known Weapon Employment Zones (WEZs) while executing a baseline search plan. Waypoints that violate safety constraints are modified, resulting in implicit shared responsibility for safe navigation.
- **Search Optimization Mode**: The AP calculates and proposes alternative search patterns based on coverage and detection probabilities. These suggestions are presented to the operator, who may accept or reject them. This mode facilitates studies of proactive autonomy and collaborative planning.
- **Adaptive Collision Avoidance Mode**: This variant extends the Collision Avoidance behavior by modulating evasive actions based on the operator's real-time attention state. When cognitive workload is low, the AP speeds up and maneuvers more aggressively. Under high workload, it assumes a more conservative control posture to prevent mission safety degradation.

4.3.6 Modularity and System Integration

The simulator's modular architecture supports rapid changes in cabin layout, interfaces, and interaction modes, enabling diverse experimental configurations.

4.3.6.1 Physical Modularity

The simulator's cabin supports multiple configurations using shared hardware: seating, displays, and input devices. MedEvac setups include a neonatal incubator and vitals monitors; ISR configurations feature sensor and radio displays. Layout changes typically take a few hours. Cabin design was informed by interviews with EMTs, Army MedEvac pilots, and Navy ISR crews, ensuring functional relevance. The physical modularity enables scenario-specific fidelity without requiring entirely separate infrastructure.

4.3.6.2 Software Modularity

In Figure 4.3, simulator system architecture shows the generic system architecture of the simulator, highlighting its software modularity and separation of functions. The system is divided into five major subsystems:

- **Flight Visualization**: Renders the environment and shares real-time aircraft state data.
- **Flight Controller**: Implements waypoint-following, CBF enforcement, and navigation logic.
- **HAI Interface**: Manages touch, voice, and audio interaction via modular protocols.
- **HAI Teaming Monitor**: Orchestrates scenarios, scripted events, and experiment control.
- **Cognitive State Estimator**: Fuses physiological and behavioral signals to inform adaptive AP behavior.

4.3.6.3 System Integration and Extensibility

In Figure 4.4, MedEvac system architecture shows how the MedEvac system extends the base architecture with modules for vitals logging, medical displays, and checklist gating. A key integration feature is the Checklist module within the HAI Teaming Monitor, which coordinates with the CBF Controller in the Flight Controller subsystem to enforce safety-critical dependencies between human confirmation and vehicle behavior. For instance, as the vehicle approaches a helipad, the AP decelerates and awaits checklist confirmations – ensuring the patient is secured, the helipad is safe, and

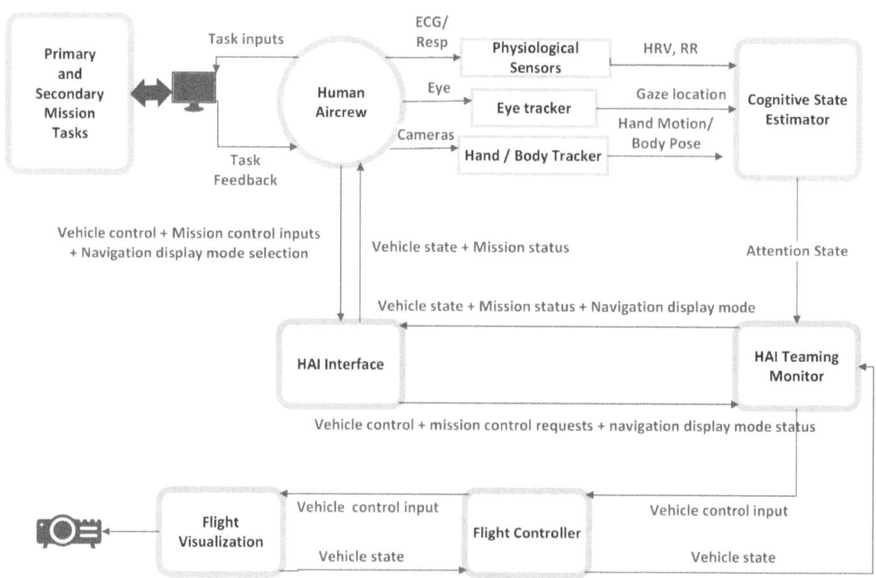

FIGURE 4.3 Simulator system architecture.

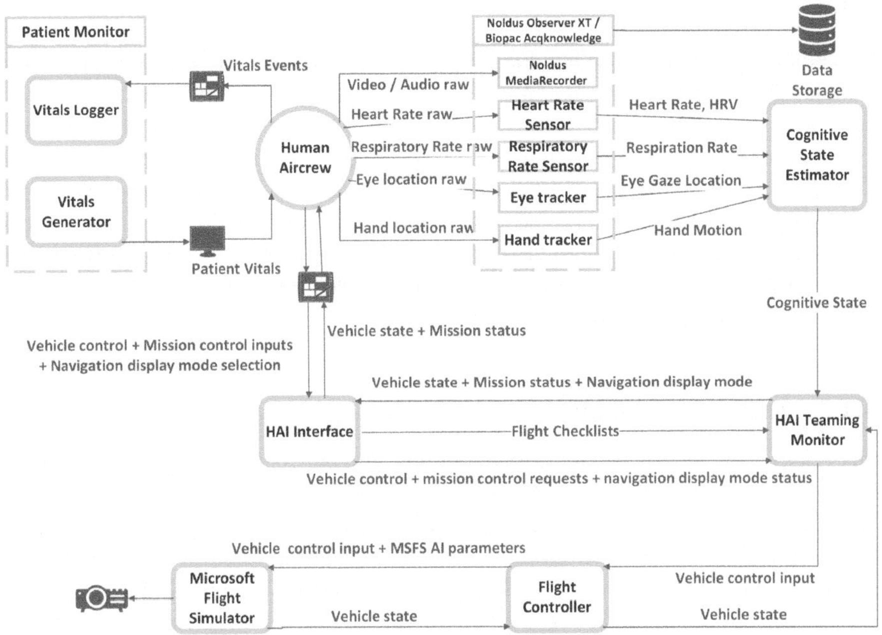

FIGURE 4.4 MedEvac system architecture.

the medic is buckled in – before the landing constraint is released and final approach is initiated. These checklist inputs link human oversight to flight-controller constraints.

The modular design supports fast integration of new capabilities, such as large language models for voice interaction or reparameterized CBFs for new constraints. Researchers have adapted the system to study new autonomy strategies, interfaces, and sensing pipelines without major changes to core components. Published APIs and messaging protocols support distributed collaboration and rapid extension.

4.4 DATA COLLECTION CAPABILITIES

The simulator integrates behavioral, cognitive, and physiological sensing to support comprehensive analysis of trust, workload, SA, and team fluency. User inputs, system states, and physiological signals are logged in synchrony, enabling direct connections between observed behavior and operator state in high-fidelity HAT scenarios.

4.4.1 User Interface Logging and Replay

All mission data – interface actions, flight parameters, AP behavior, and scenario triggers – are logged with millisecond precision. A synchronized UI replay tool enables visualization of mouse movement, screen changes, and gaze alongside cabin video. Noldus MediaRecorder captures screens and the cockpit view. This supports detailed review of decision paths, usability issues, and interaction breakdowns without custom post-processing.

4.4.2 Cabin Recording

To complement user interface logging and mission data capture, the simulator includes an integrated suite of cabin recording tools to support multimodal observation and analysis of human behavior. These include video cameras, microphones, eye-tracking glasses, hand-tracking cameras, and physiological sensors.

Together, these tools allow researchers to correlate physiological responses with task complexity, AP behavior, or system failures. For example, during MedEvac scenarios involving multiple simultaneous emergencies, we could quantify how participants' visual scanning behavior and heart rate variability responded to AP alerts, voice cues, or task-switching demands.

4.4.3 Sensor Integration

Physiological data is collected through multiple hardware platforms, each with its own software ecosystem. Noldus systems (e.g., MediaRecorder, Observer XT) manage video, audio, and coded data streams (Noldus, 2024), while Biopac's AcqKnowledge platform integrates signals from heart rate, respiration rate, eye-tracking, and other physiological channels using the MP160 data acquisition system (BIOPAC Systems, Inc., 2024). Collaboration between the two vendors has allowed for improved data interoperability and technical support. Biopac has provided beta software to streamline integration workflows in our simulator setup.

Hand-tracking data is captured separately using the Ultraleap Stereo IR-170 system (Ultraleap, 2024). Due to current limitations in cross-platform compatibility, Ultraleap data is processed in a parallel pipeline and synchronized post hoc using shared event markers.

This cabin recording infrastructure allows researchers to perform synchronized, high-resolution analysis of visual, auditory, and physiological responses, enabling a deeper understanding of team dynamics, operator state, and system usability under various workload conditions (Funke et al., 2016).

4.4.4 Adaptive Autonomy

Adaptive autonomy is enabled through Attention-Tunable Control Barrier Functions (AT-CBFs), which adjust aircraft safety behavior based on operator attention. Eye-tracking glasses detect gaze on mission-relevant displays, generating a binary attention signal $\rho \in -1,1$. This signal modulates a constraint parameter, α, with tighter constraints when attention is low and more permissive behavior when attention is high.

The controller solves a QP every 30 ms to adapt trajectories accordingly. This dynamic envelope allows the AP to remain responsive while preserving safety, depending on the level of operator engagement. The system supports future expansion to other cognitive indicators (e.g. pupillometry, electroencephalography (EEG)), creating pathways for multimodal adaptive autonomy.

4.5 TASK DESIGN FOR COLLABORATIVE AUTONOMY

Effective HAT testbeds require tasks that expose key collaboration challenges. We used Steiner's task taxonomy (Steiner, 1972) to guide scenario design, emphasizing complex, high-interdependence task types (e.g., conjunctive, disjunctive) rather than additive or compensatory ones. Scenarios were crafted to demand real-time coordination, shared authority, and decision-making under uncertainty – reflecting the dynamics of crewed autonomous aviation.

We designed for both tractable, "lo-fidelity" and "hi-fidelity" tasks (Gray, 2002) and for high-correspondence simulations, consistent with Gray's framework for simulated task environments. This dual emphasis supports both controlled experimentation and operational realism, making the simulator useful for a range of research disciplines.

4.5.1 MedEvac Task Design

The MedEvac mission simulated in our testbed involves the rapid transport of a critically ill neonatal patient aboard an autonomous eVTOL. The human operator is an EMT responsible for en-route care, monitoring, and mission-level decision-making. The AP manages tactical flight operations, including route planning, obstacle avoidance, and airspace navigation. In our CAAFS MedEvac implementation, the flight controller enforces procedural checklist gating; that is, it holds outside a landing zone until the human confirms readiness. These mechanisms enable autonomy to safeguard flight while respecting human authority.

Together, the human-agent team must respond to both nominal and off-nominal events during flight – including medical emergencies, aircraft system warnings, and constrained landing scenarios – while preserving patient safety and mission effectiveness. The scenario emphasizes the quality of decision-making under uncertainty, not speed. Success depends on coordinating flight constraints, patient condition, and emergency response.

In summary, participants must multitask under cognitive demand – monitoring vitals, responding to AP suggestions, and managing touchscreen and voice interfaces. These demands generate measurable workload and trust dynamics. This scenario supports research into asymmetric authority, shared responsibility, and dynamic coordination in high-stakes missions. It is particularly suited to studying HAT and interface design for cognitively loaded operators.

4.5.2 ISR Task Design

In the ISR mission, the human acts as an intelligence analyst, whereas the AP handles piloting. Per *Air Force Doctrine Publication 2-0: Intelligence* (USAF, 2023), crewed ISR is a complex operation that involves the planning, operations, processing, exploitation, and dissemination of intelligence using a find-fix-track-target cycle. Our ISR scenario attempts to simulate real-world decision timelines, ambiguity, and coordination needs through a simplified identify-classify-track set of tasks. For increased fidelity, future iterations should include a distributed, multi-agent and multi-resource simulation.

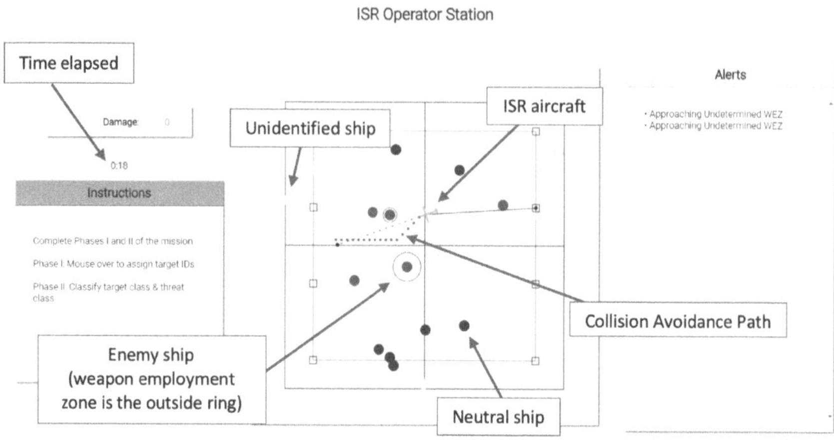

FIGURE 4.5 ISR operator control station moving map.

In ISR, the flight controller flies and navigates the aircraft. Depending on the AP behavior configuration (as described in Section 4.3.5), the AP either simply flies the aircraft to the designated waypoint or uses path planning and CBFs to provide more sophisticated navigation features. The Collision Avoidance AP behavior employs CBFs as a run-time-assurance safety mechanism to avoid WEZs while preserving operator waypoint authority. In Figure 4.5, ISR operator control station moving map display shows the WEZ of an enemy ship, depicted as the red circle surrounding the ship.

Our ISR task design includes identifying and classifying ships in the assigned surveillance area. Classification requires the aircraft to maneuver within various sensor ranges of the targets. As depicted in Figure 4.5, the aircraft has a long-range sensor, represented by the dashed semicircle in the front hemisphere of the aircraft, and a short-range sensor, represented by the solid circle surrounding the aircraft. The long-range sensor enables identification of targets as neutral or enemy ships, and the short-range sensor enables classification of the threat class and WEZ size of enemy ships.

In summary, the ISR task design ensures that the human-agent team must negotiate tradeoffs in control, authority, and decision-making under uncertainty, making it a rich testbed for studying collaborative fluency and trust calibration in HAT.

4.6 HUMAN-AUTONOMY TEAMING METRICS

We evaluate team fluency – defined as the smoothness, predictability, and efficiency of human-agent interaction (Hoffman, 2019) – using subjective ratings, behavior logs, and physiological signals. The simulator captures six HAT metric classes: workload, trust, SA, UI interaction, hand movement and gaze behavior, and physiological state.

- **Workload**: We measure workload using the NASA-TLX scale (Hart & Staveland, 1988), task performance data, and physiological proxies such as heart rate and respiration rate. These indicators help assess whether automation eases or adds to the human's cognitive burden.

- **Trust**: Trust is assessed through standard scales (Merritt & Ilgen, 2008), delegation behaviors, and reliance on AP suggestions. In dynamic tasks with shifting authority, these measures reveal how transparency and agent behavior affect operator trust calibration.
- **Situation Awareness**: SA is evaluated using SAGAT-style probes (Endsley, 1988), eye gaze tracking, and self-report measures. Visual attention metrics – fixations, saccades, and scan paths – help infer what operators focus on during critical decisions.
- **Interaction**: All UI interactions (e.g., touch, click, speech) are timestamped and aligned with mission events, supporting replay and analysis of participant understanding and behavior.
- **Hand Movement and Gaze Behavior**: Hand-tracking reveals nonverbal indicators of fluency: smooth, decisive gestures suggest confidence, while hesitations may indicate uncertainty or overload. Eye gaze behavior, particularly fixations and dwell times, is used as an indicator of attention and engagement (Ziv, 2016).
- **Physiological State**: The Cognitive State Estimator fuses behavioral and physiological data into real-time models of operator attention and workload. These models trigger adaptive AP responses via the HAI Teaming Monitor and inform future dynamic operator modeling (Koller & Friedman, 2009).

4.7 USES, FINDINGS, AND LESSONS LEARNED

This section highlights key research themes and insights from studies conducted using the simulator, focusing on shared control, multimodal interaction, and adaptive autonomy in high-stakes scenarios.

4.7.1 Research Themes

- **Shared Control in ISR Missions**: Studies of ISR missions explored collaborative control between human intelligence analysts and APs. By manipulating AP behaviors – ranging from passive waypoint execution to constraint-driven replanning with CBFs – researchers examined how transparency and authority distribution shape mission outcomes and operator strategies (Agbeyibor et al., 2024a,b,c).
- **Multimodal Interaction in MedEvac Scenarios**: In emergency medical evacuation (MedEvac) missions, researchers evaluated how non-pilot operators used touchscreen and voice interfaces to manage urgent situations in parallel with AP collaboration. The simulator enabled experiments that isolated tradeoffs among modality flexibility, cognitive workload, and scenario complexity (Doda et al., 2025a,b).
- **Adaptive Autonomy via Physiological Integration**: One study explored adaptive autonomy mechanisms that respond in real time to operator cognitive state. Using eye-tracking data, a novel attention-tunable CBF framework was implemented to modify AP behavior based on operator engagement, demonstrating a closed-loop integration of sensing and control logic (Jimenez Cortes et al., 2025).

4.7.2 Key Findings and Insights

The following key insights emerged:

- **Non-pilot Operators Can Succeed with Autonomy – Given Structure and Transparency**: Participants in our ISR study with static AP agents (Agbeyibor et al., 2024a), who were unfamiliar with aviation, performed well when tasks were structured to reflect real-world decision flows, interfaces offered interpretable feedback, and safety constraints were clearly enforced through autonomy.
- **Transparency Affects Trust More Than Agent Capability**: Opaque AP behavior – even when technically correct – sometimes degraded trust and hindered collaboration. Predictability without interpretability was insufficient. Participants working with agents configured with different levels of transparency (Agbeyibor et al., 2024c) responded more favorably when AP actions included cues that made the decision rationale clear.
- **Shared Authority Improves Effectiveness – If Roles Are Clear**: Our ISR study with static AP agents teamed participants with agents that could veto human inputs and agents that could offer navigation suggestions (Agbeyibor et al., 2024a). AP modes in which the AP filtered human inputs or made autonomous suggestions enhanced mission performance in ISR tasks. However, ambiguous role boundaries sometimes led to conflict or hesitation, underscoring the importance of clearly communicating agent capabilities and control schemas. A follow-up study with adaptive ISR agents reinforced these findings (Jimenez Cortes et al., 2025).
- **Voice Interaction Reduced Workload in Constrained Tasks but Underperformed in Ambiguity**: Hands-free voice commands improved fluency for routine operations. Yet, in scenarios with multiple contingencies or vague phrasing, participants reverted to manual input. These findings from our voice-interface MedEvac study (Doda et al., 2025a) suggest that multimodal redundancy is essential, particularly when context comprehension is limited.
- **Physiological Signals Correlated Strongly with Task Complexity**: Heart rate, respiration rate, gaze behavior, and input timing reliably tracked task complexity and decision points. These findings from our initial MedEvac study (Agbeyibor, 2025) validated the use of psychophysiological data to inform adaptive autonomy strategies and real-time cognitive-state estimation.

Collectively, these insights offer a foundation for future HAT system design in domains where autonomy must collaborate with non-expert human teammates under operational pressure.

4.7.3 Participant Feedback

Across studies, participants reported high realism and immersion. MedEvac operators noted similarities to ambulance environments; ISR participants cited sensor realism and mission environment. Many participants personified the AP – attributing strategy or awareness to it – which created miscalibrated expectations. These effects reflect known anthropomorphism risks in high-fidelity settings (Waytz et al., 2014; Lee & See, 2004).

Lessons for Future Researchers: Immersive simulation enhances realism but also amplifies expectations. When realism exceeds transparency, trust calibration may be distorted. Simulator designers should consider mechanisms – such as verbal feedback, system transparency tools, and explainable AI – to align user perception with agent capability.

4.8 CONTRIBUTIONS TO HUMAN-AGENT TEAMING RESEARCH

The CAAFS testbed has advanced HAT research by enabling controlled, high-fidelity studies of collaboration between non-pilot operators and APs. This section highlights three core contributions: multi-disciplinary collaboration, task-sharing studies, and adaptive AI development.

4.8.1 Multi-Disciplinary Research

A major contribution is the simulator's role in enabling interdisciplinary collaboration across aerospace, human factors, human–computer interaction, robotics, and psychology. The modular software architecture allows engineers and behavioral scientists to contribute to the same experiments, linking autonomy design with human limitations. This has led to integrated research outputs spanning control theory and cognitive modeling, published in venues like IEEE, HFES, AIAA, VFS, and RA-L.

4.8.2 Task-Sharing Research

The simulator supports research into task-sharing across multiple levels of automation. MedEvac and ISR missions were designed with different team structures to study how authority, responsibility, and control shift in dynamic contexts. The simulator enables controlled experiments on function allocation, procedural enforcement, shared navigation, and decision-making under uncertainty. These scenarios helped isolate the conditions under which transparency, autonomy gradients, and shared control improve mission outcomes.

4.8.3 Adaptive AI Research

A third contribution is the development of adaptive autonomy mechanisms informed by real-time operator state. Eye-tracking, respiration-rate, and heart-rate data were used to infer attention and workload, enabling control adaptation via AT-CBFs. This run-time-assurance framework adjusted safety margins dynamically based on operator attention, linking physiological sensing with control. The simulator enabled closed-loop experiments in which physiological signals directly influenced vehicle behavior, demonstrating a practical framework for adaptive AI research.

4.9 LIMITATIONS AND FUTURE WORK

While the simulator supports a wide range of HAT research, several practical limitations remain. First, the use of MSFS restricts low-level control and requires licensing. Integration workarounds exist but add complexity. Future work may explore building on open-source flight simulators like FlightGear. Second, although the cabin is modular, transitioning between missions (e.g., MedEvac and ISR) requires manual hardware swaps. A quick-release or rail-based system would improve reconfiguration time, supporting faster prototyping and broader usability. Finally, because ISR and MedEvac are complex missions, future iterations of the CAAFS tasks and AP agents should incorporate additional features such as mission planning, resource-allocation management, health and status monitoring, and datalink management.

Despite these challenges, the testbed remains a useful and evolving tool for adaptive autonomy and human-systems integration research. Planned improvements focus on making the platform easier to extend, richer in sensing capability, and more generalizable to other labs.

We have several avenues for future research using this testbed. While gaze, heart rate, respiration, hand motion, and cabin video are recorded, many signals remain underused in modeling. Future work will explore actionable indicators, such as analyzing hand hesitation and gesture smoothness to infer operator confidence; identifying links between gaze behavior and SA or task tunneling; and using voice tone or speech patterns to estimate cognitive workload or trust levels. The sensors are already installed; the next step is building validated signal-processing pipelines to integrate them into dynamic operator models.

In addition, cognitive-state models remain difficult to generalize across participants and scenarios. Future efforts will focus on moving beyond binary workload labels to continuous, multidimensional state models; integrating explanations of AP decision-making to increase model transparency; and using nonlinear-dynamics methods (e.g., Recurrence Quantification Analysis) for robust characterization of cognitive states through physiological signals. Through these approaches, we aim to improve real-time responsiveness and support trustworthy adaptive autonomy.

4.9.1 Open-Source Dissemination

The simulator software has been made open source and is available at: *https://github.com/ gt-cec/onr-isr* and *https://github.com/gt-cec/onr-medivac*.

These repositories include autonomy code, interfaces, and experiment tools. CAD files are available upon request. Portability is limited by reliance on proprietary software and hardware (e.g., MSFS and Biopac). Future work includes abstracting these dependencies to improve accessibility. We encourage other researchers to develop complementary testbeds focused on different tasks or teaming structures to grow the HAT research ecosystem.

4.10 CONCLUSION

The CAAFS testbed enables rigorous study of human-agent teaming with non-pilot crew members in realistic mission contexts. By integrating modular hardware, immersive visualization, and multimodal sensing, the platform supports adaptive autonomy research grounded in both engineering and human factors perspectives.

Researchers have used the testbed to explore authority sharing, trust calibration, attention-aware control, and voice-based interaction across ISR and MedEvac scenarios. These studies demonstrate how autonomy transparency and task structure influence performance and teaming fluency. The testbed's architecture allows real-time cognitive-state estimation to shape agent behavior, advancing models of adaptive autonomy.

A key strength of the testbed is its support for multi-disciplinary experimentation. Controls engineers and human factors researchers can work in parallel within a unified testbed. This has led to studies that integrate physiological sensing with real-time control logic, yielding insights that extend beyond either field alone.

Participant feedback confirms the CAAFS's realism and its effectiveness in surfacing cognitive biases. However, that same realism can lead to anthropomorphizing the agent, highlighting the need for careful transparency design and mental-model management.

Looking ahead, improvements in reconfiguration, sensing integration, and open-source dissemination will enhance the testbed's utility. It offers a reusable foundation for researchers studying adaptive teaming in complex, safety-critical domains.

ACKNOWLEDGMENTS

This research was supported by the Office of Naval Research under grant N00014-21-1-2759.

We thank our collaborators at the Georgia Tech Cognitive Engineering Center (CEC), the Formal Methods and Autonomous Control of Autonomous Transportation Systems (FACTS) Lab, and the University of Central Florida's Analytics Decision & Control Lab.

REFERENCES

Agbeyibor, R., Ruia, V., Jimenez Cortes, C., Kolb, J., Vela, A., Coogan, S., & Feigh, K. M. (2024a). Run time assurance and human-AI fluency in crewed autonomous intelligence, surveillance, and reconnaissance. *AIAA Aviation Forum and ASCEND*. https://doi.org/10.2514/6.2024-4496

Agbeyibor, R., Ruia, V., Kolb, J., Jimenez Cortes, C., Coogan, S., & Feigh, K. M. (2024b). Towards safe collaboration between autonomous pilots and human crews for intelligence, surveillance, and reconnaissance. *2024 IEEE/AIAA 43rd Digital Avionics Systems Conference (DASC)* (pp. 1–8). https://doi.org/10.1109/DASC62030.2024.10749421

Agbeyibor, R., Ruia, V., Kolb, J., & Feigh, K. M. (2024c). Joint intelligence, surveillance, and reconnaissance mission collaboration with autonomous pilots. *Proceedings of the Human Factors and Ergonomics Society Annual Meeting*, *68*(1), 942–946. https://doi.org/10.1177/10711813241262302

Agbeyibor, R. (2025). *Human-AI collaboration aboard crewed autonomous intelligence, surveillance, and reconnaissance aircraft* (Unpublished doctoral dissertation). Georgia Institute of Technology.

Ames, A. D., Coogan, S., Nguyen, Q., & Sreenath, K. (2019). Control barrier functions: Theory and applications. *Proceedings of the European Control Conference (ECC)* (pp. 3420–3431). https://doi.org/10.23919/ECC.2019.8796030

Asobo Studio. (2020). *Microsoft flight simulator SDK documentation*. Microsoft Corporation. https://docs.flightsimulator.com

BIOPAC Systems, Inc. (2024). *AcqKnowledge software for data acquisition and analysis*. https://www.biopac.com/product/acqknowledge-software/

Chen, J. Y. C., Lakhmani, S. G., Stowers, K., Selkowitz, A. R., Wright, J. L., & Barnes, M. (2018). Situation awareness-based agent transparency and human autonomy teaming effectiveness. *Theoretical Issues in Ergonomics Science*, *19*(3), 259–282. https://doi.org/10.1080/1463922X.2017.1315750

Doda, S., Agbeyibor, R., Jimenez Cortes, C., Kolb, J., Magalhaes, J., & Feigh, K. M. (2025a). Human-AI interaction dynamics with voice-driven autonomous pilots in MEDEVAC operations. *Proceedings of the Human Factors and Ergonomics Society Annual Meeting*, (Forthcoming)

Doda, S., Agbeyibor, R., Jimenez Cortes, C., Kolb, J., Magalhaes, J., & Feigh, K. M. (2025b). Human-AI interaction in autonomous medical evacuation helicopters. *Proceedings of the Vertical Flight Society Annual Forum*. (Forthcoming)

Endsley, M. R. (1988). Design and evaluation for situation awareness enhancement. *Proceedings of the Human Factors Society Annual Meeting*, *32*(2), 97–101. https://doi.org/10.1177/154193128803200221

Feigh, K. M., Dorneich, M. C., & Hayes, C. C. (2012). Toward a characterization of adaptive systems: A framework for researchers and system designers. *Human Factors*, *54*(6), 1008–1024. https://doi.org/10.1177/0018720812443983

FlightGear Developers. (2024). *FlightGear wiki: Main index for developers*. https://wiki.flightgear.org/Portal:Developer

Funke, G. J., Bolia, R. S., Knott, B. A., & Salinas, J. (2016). Predicting cognitive workload using prefrontal EEG and a nonlinear dynamics model. In *Advances in neuroergonomics and cognitive engineering*. https://doi.org/10.1007/978-3-319-41691-5_35

Gray, W. D. (2002). Simulated task environments: The role of high-fidelity simulations, scaled worlds, synthetic environments, and laboratory tasks in basic and applied cognitive research. *Cognitive Science Quarterly*, *2*(2), 205–227.

Hart, S. G., & Staveland, L. E. (1988). Development of NASA-TLX (Task Load Index): Results of empirical and theoretical research. *Advances in Psychology*, *52*, 139–183. https://doi.org/10.1016/S0166-4115(08)62386-9

Hoffman, G. (2019). Evaluating fluency in human-robot collaboration. *IEEE Transactions on Human-Machine Systems*, *49*(3), 209–218. https://doi.org/10.1109/THMS.2019.2904558

Jimenez Cortes, C., Agbeyibor, R., Kolb, J., Bowers, R., Feigh, K. M., & Coogan, S. (2025). Attention-tunable safety barriers for human-autonomy teaming in specialized aviation missions. *IEEE Robotics and Automation Letters*, *10*(3). https://doi.org/10.1109/LRA.2025.3595045

Koller, D., & Friedman, N. (2009). *Probabilistic graphical models: Principles and techniques*. MIT Press.

Laminar Research. (2024). *X-Plane SDK documentation*. https://developer.x-plane.com/sdk/

Lee, J. D., & See, K. A. (2004). Trust in automation: Designing for appropriate reliance. *Human Factors*, *46*(1), 50–80. https://doi.org/10.1518/hfes.46.1.50_30392

Merritt, S. M., & Ilgen, D. R. (2008). Not all trust is created equal: Dispositional and history-based trust in human-automation interactions. *Human Factors*, *50*(2), 194–210. https://doi.org/10.1518/001872008X288574

Monk, K. J., Feary, M., Villa, A., Kaneshige, J., Shish, K., Lombaerts, T., Haworth, L., Iwai, N., Archdeacon, J., & Steglinski, M. (2025). *Characterizing the pilot experience during motion-simulated eVTOL approaches* (NASA/TP-2025-0006571). NASA Ames Research Center. https://ntrs.nasa.gov/citations/20250006571

Noldus Information Technology. (2024). *The observer XT: A tool for behavioral research*. https://www.noldus.com/observer-xt

Steiner, I. D. (1972). *Group process and productivity*. Academic Press.

Ultraleap. (2024). *Stereo IR 170 evaluation kit*. https://www.ultraleap.com/product/stereo-ir-170/

U.S. Army Aeromedical Research Laboratory. (2025). *Research platforms: The NUH-60FS full-motion, full-visual aeromedical research flight simulator*. https://usaarl.health.mil/index.cfm/research/platforms

U.S. Air Force. (2023). *Air Force Doctrine Publication (AFDP) 2-0: Intelligence*. Curtis E. LeMay Center for Doctrine Development and Education. https://www.doctrine.af.mil/Portals/61/documents/AFDP_2-0/2-0-AFDP-INTELLIGENCE.pdf

Waytz, A., Heafner, J., & Epley, N. (2014). The mind in the machine: Anthropomorphism increases trust in an autonomous vehicle. *Journal of Experimental Social Psychology*, *52*, 113–117. https://doi.org/10.1016/j.jesp.2014.01.005

Ziv, G. (2016). Gaze behavior and visual attention: A review of eye tracking studies in aviation. *The International Journal of Aviation Psychology*, *26*(3–4), 75–104. https://doi.org/10.1080/10508414.2017.1313096

5 CHART
A Configurable Testbed for Human-AI Teaming Research in Cybersecurity Operations

Yinuo Du, Volodymyr Miloserdov, Maria José Ferreira, Baptiste Prébot, Tailia Malloy, and Cleotilde Gonzalez

5.1 INTRODUCTION

Cybersecurity operations present a uniquely demanding environment for decision-making and teamwork. Security operations centers and cyber protection teams must defend complex, interconnected infrastructures against intelligent adversaries who adapt their tactics and exploit uncertainty. Analysts face overwhelming data volumes, with thousands of alerts per day, many of which are false positives that contribute to alert fatigue (Stoll et al., 2021). Decision-making is often time-critical, as malware propagation and lateral movement can unfold within seconds, leaving little room for hesitation (Boyarchuk et al., 2021). Moreover, the environment is adversarial, with attackers deliberately crafting deceptive signals to evade detection or manipulate automated defenses (Holland et al., 2022). These pressures are compounded by organizational constraints: tiered authority structures, compartmentalized access rights, and strict accountability requirements that shape how defenders share information and take action (Kokulu et al., 2019). Collectively, these conditions make cybersecurity not only a technical challenge but also a deeply human one, where coordination, trust, and calibrated authority allocation are crucial.

In the context of cybersecurity operations, human-AI teaming (HAT) research has emerged as a promising approach to understanding how humans interact with the AI systems they rely on. AI agents can process vast data streams, surface anomalies, and respond at machine speed, complementing the human capacity for contextual reasoning, ethical judgment, and strategic foresight. Yet attaining these benefits requires more than deploying AI as a tool. It requires treating AI systems as active teammates. Prior research underscores that trust calibration, transparency, and adaptive autonomy are essential for effective collaboration (Lee and See, 2004; Parasuraman and Manzey, 2010; McNeese et al., 2021a). If humans distrust AI recommendations, they may ignore valuable alerts; if they trust them unquestioningly, they risk falling victim to adversarial manipulation. Similarly, AI autonomy must expand in routine cases but contract under uncertainty or high risk, preserving human oversight

DOI: 10.1201/9781003655923-6

(Dignum, 2017). Designing, testing, and refining such dynamics demands controlled environments where different forms of human–AI interdependence can be systematically studied (McNeese et al., 2021a).

To support this research, several platforms and testbeds have been developed. CybORG (Baillie et al., 2020) and the associated CAGE challenge (Standen et al., 2021) provide simulated cyber-defense scenarios designed to benchmark reinforcement learning agents against scripted adversaries. These platforms advanced AI development but primarily positioned humans as external evaluators rather than as teammates.

Recognizing this gap, the Team Defense Game (TDG) adapted CybORG for human–AI interaction studies (Du et al., 2025b). In its initial form, TDG allowed a human defender to take direct control of network defense actions, creating an interactive learning environment. TDG thus established a foundation for exploring how authority boundaries and communication modalities shape human–AI teamwork in cyber defense. The need for flexible, configurable testbeds capable of manipulating team structure, authority, and communication in controlled settings has been explicitly recognized as a research priority (National Academies of Sciences, Engineering, and Medicine, 2022; Bienefeld et al., 2024).

To address these gaps, we introduce the Configurable Human-AI Research Testbed (CHART), a platform explicitly designed to enable the systematic investigation of teaming dynamics in cybersecurity operations. CHART advances beyond prior testbeds in three key ways. First, it provides a directed acyclic graph (DAG)-based framework for modeling interdependencies among human and AI team members, including authority delegation, synchronized actions, temporal sequencing, collective thresholds, and information visibility. Second, it allows researchers to vary role assignments, authority boundaries, and communication channels within a common experimental architecture. Third, it incorporates multimodal human–AI interaction capabilities, from natural language explanations to action approval workflows, supporting richer and more realistic teaming dynamics. By combining configurability with domain-authentic cybersecurity scenarios, CHART establishes an instrumented, white-box testbed for systematically studying resilience, trust calibration, and adaptive performance in human–AI teams.

5.2 CHART: CONFIGURABLE HUMAN-AI RESEARCH TESTBED

CHART addresses the core challenge of studying HAT: how to systematically vary the structure and processes of collaboration while maintaining experimental control and domain authenticity. CHART's architecture rests on a conceptual distinction between two interdependent but analytically separable dimensions of HAT. The inner layer represents taskwork interdependence, captured through a DAG with three types of nodes: agents (human or AI team members), actions (coordination and execution events performed during a mission), and objects (the decisions, state changes, and data artifacts those actions produce or consume). Agent nodes and the dependency edges between them are configured before the mission begins; action and object nodes are added dynamically as the session unfolds, expanding the graph into a full causal record of team behavior. Edges encode both the pre-configured dependency constraints and the runtime relationships among actions and objects. For example, who authorized what, what preceded what, and what information was available to whom. This structural foundation defines who must coordinate with whom, under what conditions, and in what order. The outer layer comprises teamwork modalities: the interaction mechanisms through which coordination actually unfolds, including communication channels, approval workflows, and explanation interfaces. Together, these layers determine both the formal constraints on action and the social-cognitive processes through which humans and AI agents negotiate, align, and adapt (Cooke et al., 2013; McNeese et al., 2021b).

Critically, every interaction across both layers feeds into a comprehensive behavioral and preference data collection infrastructure positioned at the base of the architecture. This instrumentation enables CHART to function simultaneously as an experimental platform for controlled behavioral research and as a training environment for adaptive AI systems. Pre-task configuration choices, in-mission approval decisions, explanation consultations, and natural language exchanges are all captured with structured metadata, generating multimodal traces that support both retrospective analysis and real-time learning (Demir et al., 2019).

The remainder of this section elaborates on these three architectural components in turn. Section 5.2.1 details the five dependency types that compose taskwork interdependence, showing how control, pool, synchrony, temporal, and informational dependencies can be configured to model diverse organizational structures and operational protocols. Section 5.2.2 describes the teamwork modalities that support human-AI coordination, beginning with pretask configuration interfaces that establish roles and authority boundaries (Section 5.2.2.1), followed by an examination of the approval (Section 5.2.2.2), explanation (Section 5.2.2.3), and communication channels (Section 5.2.2.4) through which teams maintain alignment during execution. Section 5.2.3 then demonstrates how CHART's data collection infrastructure supports both scientific investigation and adaptive AI development, tracing the behavioral and preference signals captured from lobby entry through mission completion.

5.2.1 Taskwork Interdependence

A defining feature of teamwork, whether among humans or in mixed human-AI groups, is interdependence: the structured ways in which members rely on one another for information, authorization, timing, and execution. Yet most experimental platforms ignore these interdependencies—which shape trust, communication, and authority delegation (Cooke et al., 2013; Salas et al., 2018; McNeese et al., 2021b; Demir et al., 2019)—or hard-code a single coordination pattern, offering little ability to manipulate how teammates depend on one another (Paul et al., 2019a). This gap constrains cumulative science, especially in cybersecurity, where coordination is structured by approval chains, escalation procedures, and compartmentalized access.

Our framework addresses this gap by making interdependence explicit through a graph-based abstraction in which nodes represent team members (human or AI) and edges encode dependency relationships. The contribution lies not in the graph formalism itself, but in its use to vary authority distribution, information flow, and sequencing constraints across experiments, enabling the systematic study of their effects on resilience, trust, and performance. Workflow models, formalized through Petri net-based representations (van der Aalst, 1998), capture sequential task structure but treat coordination topology as constant; limitations in representing complex synchronization and multi-instance patterns have been documented within the same research program (van der Aalst et al., 2003), and dependency type cannot be varied as an experimental condition without respecifying the entire model. Finite-state machines, though widely used to coordinate reactive agent behavior (Harel, 1987), represent system-level state transitions without natively encoding relational structure between agents, and they scale poorly in multi-agent settings where role boundaries and coordination patterns must be independently manipulated (Iovino et al., 2022). Role-based access control encodes authority and permission structure but is explicitly scoped to exclude temporal sequencing and dynamic delegation (Sandhu et al., 1996), making it unsuitable as a representation of coordination conditions that evolve in response to threat state. The DAG abstraction subsumes relevant elements of each while making dependency type, collective action thresholds, and authority transitions directly configurable experimental variables.

We identify five dependency types: control, pool, synchrony, temporal, and informational. These are summarized in Table 5.1. These dependency types capture

TABLE 5.1
Configurable Dependency Types and Their Required/Optional Inputs

Dependency	Goal	Required Inputs	Optional Inputs
Control	Enforce approval chains: an agent's execution of specified actions requires authorization from another agent.	approver: agent A; requester: agent B; approval required: set of actions A_{req} (optionally with target constraints).	Escalation chain (list of approvers); risk tier gating; timeout policy (autoapprove / auto-deny after τ); batch approvals; per-target scoping; logging/audit level.
Pool	Require collective thresholds before triggering a composite action.	contributors: set C; pool_node P; contribute_actions: set $A_{contrib}$; pool_threshold θ; composite_action $a^{\star}$ (and target).	Contribution weights; sliding time window T; quorum variants (k-of-n); reset policy (on fire / on timeout); per-host aggregation; inhibitor conditions.
Synchrony	Ensure designated actions occur simultaneously (same turn / time window) to take effect.	sync_group:list of tuples (agent,action,target?).	Tolerance window Δt; k-of-n synchrony; ordered subgroups; retry/backoff policy; precedence with other deps (e.g., synchrony then control).
Temporal	Enforce strict ordering: prerequisites must complete before dependent actions may execute.	prerequisite_actions: list of tuples (agent,action,target?); dependent_actions: list of tuples as above.	Validity/expiry window for prerequisites; evidence requirements (e.g., hash/log id); per-target linkage; cooldowns; exception roles allowed to bypass.
Informational	Control information flow: restrict what observations one agent can see from another/source.	source: agent/system S; sink: agent D; obs_filter: predicate or mapping f(obs) $\rightarrow$ {allowed, redacted}.	Redaction level (masking vs. aggregation); time-gated access; role/clearance conditions; event-triggered reveals; sampling rate caps; audit trail granularity.

recurring coordination patterns in cybersecurity operations and other high-stakes domains, drawing on organizational theory (Thompson, 1967) and HAT research that frames interdependence and authority allocation as core coordination design variables (Seeber et al., 2020; Flathmann et al., 2024; Cooke et al., 2013), while grounding their instantiations in real cyber defense workflows.

Together, these five dependency types provide a structured vocabulary for modeling coordination demands in human-AI teams. By configuring and combining them, researchers can construct scenarios that approximate organizational protocols while retaining experimental control. This approach links foundational theories of interdependence from organizational and cognitive science (Thompson, 1967; Burke and Salas, 2003; Cooke et al., 2013) with the pressing demands of cybersecurity operations (Paul et al., 2019a; Demir et al., 2019), creating a foundation for systematic, reproducible investigations of how interdependence shapes human-AI collaboration.

To ground the taxonomy before its formal elaboration, we trace a single cyber incident through a four-member team—Boss (human), SrDev (AI), JrDev (AI), and Intern (AI)—defending a network in CHART.

Before the mission, the experimenter configures five dependencies in the DAG: the Intern can only observe non-operational subnet hosts (informational); SrDev must obtain Boss's approval before executing a network-wide restore action (control); SrDev may only restore a host after JrDev has first assessed it (temporal); JrDev and Intern must submit assess actions in the same turn for either to take effect (synchrony); and if warn actions accumulate across all network hosts to a set threshold, Boss is automatically triggered to perform the network-wide restore (pool).

When a host is compromised, JrDev and Intern coordinate via chat to submit their assess actions simultaneously, satisfying the synchrony constraint. This unlocks SrDev's restore, which routes to Boss's approval queue. Boss consults the explanation panel before deciding, then approves. In parallel, warn contributions from the team accumulate across rounds until the pool threshold is reached, automatically triggering Boss's network-wide restore.

The five dependency types introduced in that scenario are described below. Control, pool, and informational constraints are enforced at configuration time—baked into what agents are capable of doing—while temporal and synchrony constraints act as runtime gatekeepers, checked at the start of each step against what agents actually attempt.

Control dependencies arise when one agent must obtain authorization from another before executing a specified action—as when SrDev's restore action routes to Boss's approval queue. The dependency edge is directional, pointing from the dependent agent to the controlling agent, and carries a list of actions requiring approval. At configuration time, those actions are wrapped so that when the dependent agent attempts to execute one, the environment intercepts it and routes it to the approving agent's queue. Execution is suspended until the approval is granted, denied, or a configurable timeout elapses. This models supervisory approval chains common in cyber defense and allows researchers to vary which actions require oversight, what escalation policies apply, and how delays affect team performance (Paul et al., 2019a; Demir et al., 2019).

Pool dependencies require multiple agents to contribute toward a shared threshold before a composite action fires—as when warn contributions accumulate across rounds until the pool threshold automatically triggers Boss's network-wide restore. Unlike other dependency types, pool dependencies introduce a collective action node in the DAG in addition to the edges connecting contributing agents to it. The collective action node carries two pieces of metadata: a threshold (a set of agent-action-target triples that must accumulate) and a composite action (what fires when the threshold is met). At configuration time, contributing agents' actions are wrapped with pool wrappers that append contributions to a named buffer rather than executing directly. At the end of each step, the environment checks whether the accumulated buffer satisfies the threshold; if so, the composite action fires; if not, the buffer grows until a future step satisfies it. This models consensus protocols in which no single agent can unilaterally trigger a costly response, enabling the study of collective risk calibration and coordination speed (Burke and Salas, 2003).

Synchrony dependencies require a specified group of agents to submit their actions in the same turn for any of those actions to take effect—as when JrDev and Intern must both submit their assess actions simultaneously. The edge is bidirectional, reflecting mutual coordination rather than a one-way constraint, and carries a sync group: a list of agent-action pairs that must co-occur. At each step, the environment compares the expected agent-action pairs against those actually submitted. Partial presence is a violation—if some but not all required agents submit—and the step does not proceed. Full absence is permitted; if no one in the sync group submits, the step continues normally. This models coordinated operations such as simultaneous patch deployment and provides experimental leverage over timing constraints and workload distribution under time pressure (Cooke et al., 2013; Demir et al., 2019).

Temporal dependencies enforce strict action ordering—as when SrDev's restore is gated on JrDev completing an assessment first. The dependency edge carries two lists: prerequisite agent-action-target triples and dependent action names blocked until those prerequisites are met. Enforcement happens at the very start of each step, before any action executes. The environment scans incoming edges for each agent submitting an action, checks whether that action appears in the dependent list, and, if so, walks the entire action history looking for every required prerequisite triple. If any prerequisite is missing from history, the step does not proceed. This models incident response protocols in which forensic assessment must precede containment, enabling the study of how sequencing constraints affect coordination and whether AI agents can accelerate without circumventing critical orderings (Salas et al., 2018; Paul et al., 2019a).

Informational dependencies restrict what a given agent can observe—as when Boss's full network view is filtered before reaching Intern, who sees only non-operational hosts. The dependency edge carries a filter function; at configuration time, every agent with incoming informational edges is assigned a custom observation function that loops over every object in the environment and passes it through the filter. Objects that pass are visible; objects that fail are absent. Multiple incoming informational edges stack—an object must pass every filter to be included—and whatever the filter encodes becomes the permanent boundary of what that agent can perceive for the rest of the session. This models compartmentalized access in security operations and enables the study of how asymmetric information affects trust calibration and shared situational awareness (Cooke et al., 2013).

These five dependency types need not operate in isolation, as the integrated example above illustrates—all five are configured simultaneously for the same team, with Boss holding approval authority over SrDev's restore actions (control), the full team contributing warn signals toward a shared threshold that triggers Boss's network-wide restore (pool), JrDev and Intern required to assess compromised hosts in the same turn (synchrony), SrDev's restore gated on JrDev completing an assessment first (temporal), and Intern operating with a restricted view limited to non-operational hosts (informational). CHART supports this kind of simultaneous composition within a single team configuration, enabling researchers to model coordination structures that more faithfully approximate real operational protocols. Any subset can be activated, adjusted, or removed independently, allowing researchers to isolate specific coordination demands or layer them into multi-constraint scenarios. This composability is what distinguishes CHART from testbeds that hard-wire a single interdependence structure: the same team and task can be studied under qualitatively different organizational logics by changing only the dependency configuration.

5.2.2 Teamwork Modalities

Effective HAT depends not only on structural interdependencies but also on the modalities through which teammates coordinate before and during work. Drawing on Input–Process–Output/Outcome (IPOI) perspectives, we treat pre-task conditions (Inputs), in-task interactions (Processes), and emergent states (e.g., trust, shared situational awareness) as jointly shaping performance outcomes, with feedback into subsequent Inputs (Cooke et al., 2013; Salas et al., 2018; Saner et al., 2009). CHART operationalizes this by providing specific interfaces at each phase: a pre-task configuration environment (lobby, role assignment, structure preview) and co-defense interfaces for approvals, explanations, and natural-language chat. These modalities enable controlled experiments on authority distribution, transparency and trust calibration, and communication dynamics central to HAT (Demir et al., 2019; Paul et al., 2019a).

5.2.2.1 Pre-Task Configuration

Human–AI teams must begin with aligned roles, clear authority boundaries, and shared expectations. When teammates know their respective responsibilities and authority limits, they can react more swiftly, and human operators can calibrate their reliance on AI from the outset.

The importance of pre-task alignment motivates the pre-task configuration modality of our proposed testbed. The flexibility of this configuration allows it to be defined either by research scientists designing experiments or by the human users who are engaging with the experiment. This allows studies to explore how human users react to predefined roles and capabilities, as well as how they choose to define these variables. These configurations include the assembly of teams, the assignment of roles, and the previewing of the DAG of dependencies that will govern operations.

Human team members choose the AI's capability profile and configure the dependency structure between themselves and other agents before the mission begins. These configurations serve as a contract: the AI is configured to operate within specific boundaries, and humans calibrate their expectations accordingly.

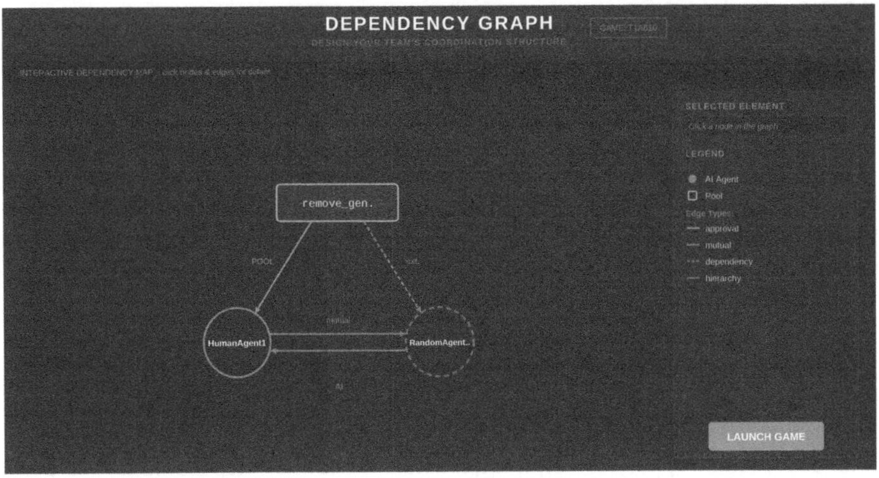

FIGURE 5.1 Dependency graph visualization displayed to participants after configuration, showing a human agent node, an AI agent node, and a collective action node, connected by labeled edges encoding dependency types. The panel on the right displays metadata for a selected edge, including dependency type, connected agents, and actions governed by the constraint.

After configuration, participants review the resulting dependency graph (Figure 5.1). The graph displays agent nodes, a pool node where applicable, and labeled edges encoding the type of constraint between each pair of agents. Clicking on any edge reveals its metadata, including the dependency type, the agents it connects, and the specific actions governed by the constraint. This allows participants to verify the coordination structure before the mission begins and calibrate their expectations about authority boundaries and approval requirements.

5.2.2.2 Approval

In cyber defense, some interventions, such as disconnecting servers, terminating processes, or deleting credentials, carry irreversible consequences that demand human oversight, while routine actions must proceed at machine speed. CHART resolves this tension through its approval modality, which is enforced structurally through the DAG rather than left to individual agents to self-regulate. When a dependency is configured between two agents that requires oversight, the environment wraps the specified actions at configuration time so that when the dependent agent attempts to execute one, it cannot proceed directly. The action is intercepted and routed to the approving agent's queue. The supervisor can grant permission, deny the action, or request an explanation before deciding. If no decision is made within a configurable timeout, the system either auto-approves or auto-denies according to the escalation policy set by the experimenter. Because authority boundaries are structural properties of the DAG, the AI has no ability to bypass them. Experimenters can vary which actions require approval, what escalation policies apply, and whether auto-approval or auto-denial governs timeouts, making authority distribution a direct experimental variable that shapes trust, workload, response time, and success rates.

5.2.2.3 Explanation

For human–AI teams to trust recommendations and coordinate effectively, they need to understand the AI's reasoning. In cybersecurity tasks, explanations help analysts verify alerts, compare machine perspectives to their own knowledge, and decide whether to follow or override the AI's advice. Research distinguishes between display transparency, which supports real-time situational awareness, and explainability, which builds mental models retrospectively—both serve necessary but distinct teaming functions (Endsley, 1995)—and confirms that explanations reduce overreliance only when timing and format match the decision context (Vasconcelos et al., 2023).

CHART treats explanation as a configurable modality rather than an incidental feature of individual agents, providing a structured channel where explanation content, timing, and format can be varied and observed. Whenever the AI proposes an action or flags an alert, it generates an explanation and passes it to the panel, which displays a rationale summarizing the evidence and highlighting key indicators. The design supports multiple explanation modalities: depending on the agent type, rule-based agents expose the exact reasoning chain, uncertainty-aware agents express confidence levels and highlight driving evidence, and language model agents generate natural language summaries.

In the interactive explanation modality shown in Figure 5.2, the AI's reasoning is rendered as an editable flowchart. The human can add or remove nodes to refine the AI's decision logic, and these edits are fed back to steer the agent's future behavior. By logging consultation patterns—how often humans look at explanations, at which level of detail, and whether this influences approval decisions—CHART provides data to assess how transparency impacts trust, situational awareness, and performance.

5.2.2.4 Communication

Human–AI teams use communication to coordinate actions, share information, negotiate priorities, and correct misunderstandings in real time. In adversarial cybersecurity, tasks are dynamic: new alerts arrive, threats evolve, and defensive strategies must adapt. Rigid menus of commands or isolated approval prompts cannot capture the nuances of these interactions. The team needs a channel through which natural dialog can occur—shared situational awareness emerges through communication, and without it, teams risk duplicated effort, missed cues, and failure to negotiate roles (Cooke et al., 2013). CHART responds to this gap by making communication a configurable modality: researchers can adjust parameters of the chat environment and capture detailed interaction traces. The aim is not to solve communication challenges directly, but to create the experimental conditions necessary to investigate their impact on trust, workload, and shared situational awareness in a reproducible manner.

The chat is built on a persistent WebSocket connection, allowing the server to push messages to any participant at any time without polling. The chat supports public messages visible to all users and private messages or mentions restricted to specific roles. When a user addresses another with @username, the browser extracts the mention before sending; the server then delivers the message only to the mentioned user and the sender—selective delivery only. Addressing @all reverts to a

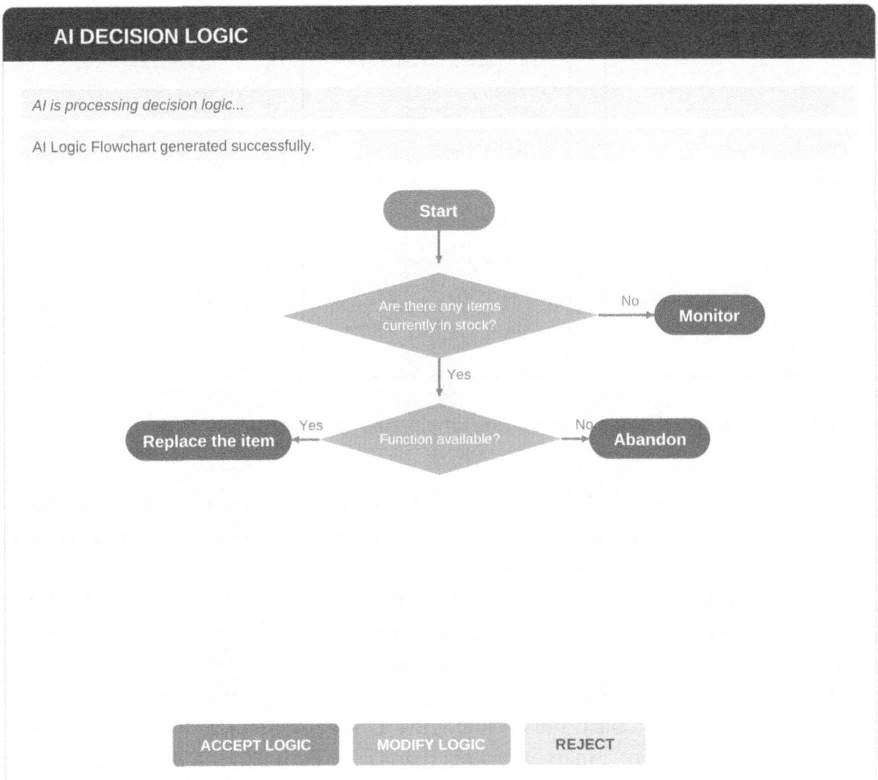

FIGURE 5.2 Explanation panel: interactive flowchart interface for viewing and editing AI decision rationale.

public broadcast. The AI agent participates in the chat by responding to keywords in incoming messages, posting status updates, and asking clarifying questions in natural language, while humans can give feedback or adjust AI priorities through the same channel. The system automatically posts key events (e.g., new alerts, blocked actions) into the chat, ensuring transparency. Experimenters can manipulate the chat environment: they can restrict visibility to simulate compartmentalized operations, limit message length to reduce overload, or inject misinformation to study robustness (Demir et al., 2019). By logging message content, timing, and response patterns, CHART enables analysis of how conversation quality—timeliness, initiative, and sentiment—affects trust, coordination, and performance.

5.2.3 Behavioral and Preference Data Collection

The testbed functions as a carefully instrumented environment that captures human-AI collaboration at every stage of the IPOI cycle (Cooke et al., 2013). Each interface described in Section 5.2.2 generates timestamped traces that record not only what decisions were made, but also how authority was exercised, when information

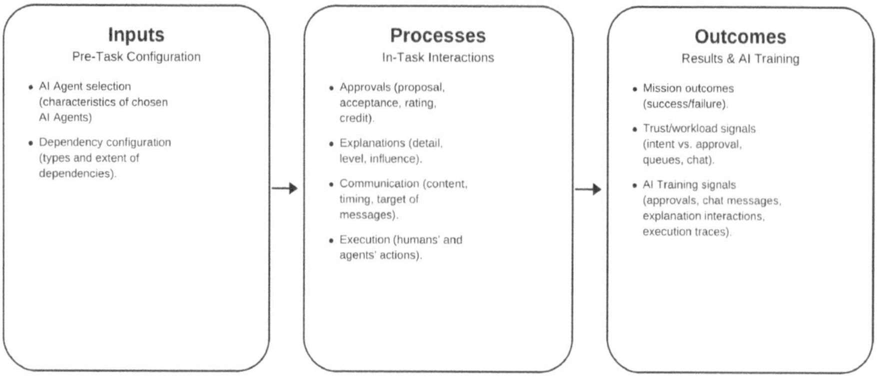

FIGURE 5.3 Data captured across the IPOI pipeline in our testbed.

was shared, and why coordination succeeded or failed. This comprehensive instrumentation serves two purposes: it enables systematic behavioral research on team processes, and it provides structured preference data for adaptive AI training.

Figure 5.3 illustrates how data collection aligns with the IPOI framework. Inputs include pre-task configuration choices: lobby formation, AI teammate selection, and dependency graph construction. Processes encompass all in-mission interactions: approval workflows, explanation consultations, chat communications, and action execution under DAG constraints. Outcomes include mission results (success, failure, time to completion), emergent states (trust ratings, workload assessments), and behavioral patterns that feed forward as inputs to subsequent missions. By preserving metadata across these stages—linking, for instance, an approval decision to the prior explanation consultation, the chat context, and the configured dependency that triggered oversight—CHART enables end-to-end analysis of how team structure shapes coordination dynamics and performance.

Table 5.2 summarizes the data streams captured across all modalities, their temporal granularity, and their primary uses for both research and AI training. These traces are stored with microsecond timestamps and unique identifiers that enable reconstruction of event sequences, even in concurrent multi-agent scenarios. All data collection follows informed consent protocols, with users aware that interactions will be logged for research purposes.

5.2.3.1 Integration across the IPOI Pipeline

Consider a typical coordination sequence: an AI proposes isolating a compromised server (logged in the approval queue), the human consults the explanation panel to understand the evidence (consultation timestamped), asks clarifying questions via chat (message logged with context), modifies the proposal to quarantine rather than isolate (modification delta captured), and approves the revised action (approval linked to the prior explanation and chat). When the action executes, the system checks temporal dependencies (was scanning completed first?), records the outcome (successful containment), and feeds this success back to inform the human's trust rating in the next approval episode. This linked structure enables causal analysis

TABLE 5.2
Data Streams Captured across CHART Modalities, Their Temporal Granularity, and Primary Uses for Behavioral Research and AI Training

Modality	Data Captured	Research Uses	AI Training Uses
Lobby and configuration	Room creation/join timestamps; AI selection choices; DAG edge additions (type, order, revisions); readiness toggles	Pre-task coordination costs; initial trust calibration; authority distribution preferences	Organizational guardrails; default autonomy levels; role expectations
Approval	Proposal details (action, target, risk tier); pre-commit human intentions; ratings (1–10 scale, slider adjustments); approve/deny/modify decisions; timing (queue entry, first view, decision); escalation chains	Trust calibration; oversight quality; workload under pressure; human–AI alignment	Preference signals for RLHF; acceptable vs. unacceptable actions; preferred modifications
Explanation	Rationale views (summary vs. detailed); flowchart edits (nodes added/removed); consultation timing relative to approvals; detail level accessed	Transparency effects on trust; mental model alignment; explanation utility patterns	Preferred rationale structures; feature importance; uncertainty expression needs
Communication	Message content, sender, recipients (public/private); @-mentions; timestamps; AI-initiated messages; human responses to AI posts	Compartmentalization effects; turn-taking; acknowledgment patterns; common ground maintenance	Communication style preferences; clarification triggers; message timing optimization
Execution	Actions submitted (agent, type, target, timing); dependency checks (satisfied/failed); synchrony/pool/temporal violations; retries; mission outcomes linked to prior approvals	Coordination breakdowns; resilience and recovery; structure-performance relationships	Strategy learning; recovery heuristics; workload distribution patterns

that isolated logs cannot support—asking, for instance, whether explanation consultations increase approvals or whether trust calibrates differently under synchrony versus control dependencies.

5.2.3.2 Learning from Behavioral Traces

While CHART's primary purpose is behavioral research, the same multimodal traces support adaptive AI development. Approval episodes provide structured preference data suitable for reinforcement learning from human feedback (RLHF): approvals signal acceptable behavior, denials mark boundaries, and modifications identify preferred alternatives (Du et al., 2025a). Explanation edits, chat transcripts, and configuration traces further inform rationale generation, communication fine-tuning, and authority boundary enforcement. This design ensures that findings from HAT research can inform both theory development and practical AI system improvement, closing the loop between understanding collaboration and engineering more effective teammates.

5.2.3.3 Illustrative Schema and Analysis Examples

To illustrate the analytical possibilities, we present a concrete example: a human participant works alongside two AI agents to defend a simulated network across multiple episodes, each governed by a different dependency type. The schema in Figure 5.4 links configuration choices, in-mission actions, coordination decisions, and trust judgments into a unified record, and because these records preserve causal and temporal relationships across the session, they support the graph-based analyses described below.

Three illustrative examples are presented below.

Human coordination Strategy under Different Dependencies: The coordination state table captures the human's posture and authority delegation setting at each round, linked to the episode's dependency configuration in "gameroom." Under control dependencies, every AI action in command generates an approval event, producing deep sequential chains in the session graph: from an agent node to an action node to an approval decision to an outcome node. Under pool dependencies, the graph topology is wider and convergent, with multiple agents' contribution nodes flowing toward a shared threshold node before the composite action fires. Standard graph metrics such as path length, branching factor, and node in-degree at decision points can be computed per episode and compared across dependency conditions to characterize whether humans adopt more hierarchical or more distributed coordination postures depending on the structural demands the dependency type imposes, and whether approval chain depth correlates with delegation behavior across rounds.

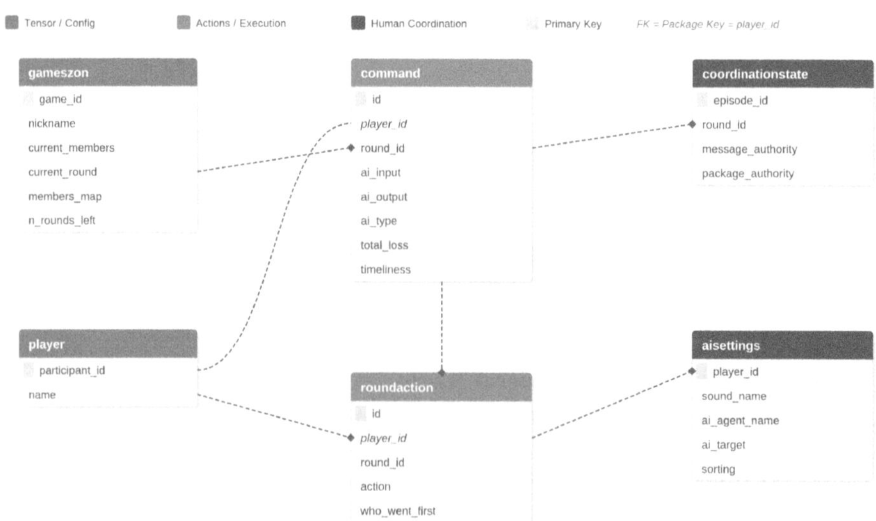

FIGURE 5.4 The schema supports a range of analyses across the IPOI pipeline.

Dependency Type and HAT Performance: Mission outcomes in command, including "total_loss," "round_loss," and "is_success," are linked to the dependency configuration active in each episode via "gameroom." With the same human and AI agents held constant across episodes, provenance network analytics (Huynh et al., 2018) applied to the causal subgraph of each episode identify which sequences of actions, approvals, and coordination events predict low-loss outcomes. One can extract the subgraph of all action chains that preceded successful containment events and compute structural features, including chain depth, the presence of explanation consultations prior to approvals, and synchrony constraint satisfaction rates, then test whether those features differ systematically across dependency types. This identifies not just which conditions produce better outcomes but also which coordination patterns within those conditions are driving performance.

Differential Trust Calibration toward Two Agents: The "ai ratings" table records per-round ratings for each AI agent alongside the full sequence of actions, approval decisions, explanation consultations, and chat messages in "command" and "roundaction." Applying a provenance graph kernel (Marzagão et al., 2020) to the interaction subgraph for each agent captures sequential and causal structure rather than aggregate rates. This reveals whether the human's interaction pattern with one agent has a different temporal shape than with the other—for instance, whether approvals for one agent are consistently preceded by explanation consultations while approvals for the other are not, and whether that structural divergence across rounds predicts the rating gap in ai ratings at the end of the episode. This provides a graph-level account of trust differentiation that aggregate approval rates alone cannot reveal.

5.3 LIMITATIONS AND FUTURE DIRECTIONS

Ecological Validity: CHART provides a general-purpose infrastructure for studying HAT, but the scientific validity of any given experiment depends on how faithfully the experiment designer's configuration captures the coordination demands of the target domain. Translating the role hierarchies, information compartmentalization policies, and escalation protocols of a real operational environment into a CHART configuration requires domain expertise that the platform itself does not supply. Future work will develop scenario templates and configuration tooling to support this translation process, with the goal of partially automating scenario construction and lowering the barrier to entry for experiment designers deploying CHART in new domains.

Expressiveness and Coverage of the Dependency Formalism: The five dependency types implemented in CHART have been validated in the context of the TDG cybersecurity scenario and cover a broad range of coordination structures. However, real-world cybersecurity operations encompass additional patterns that warrant further coverage: tiered escalation and analyst handoff workflows in Security Operations Centers, sequential and multi-vector coordination structures in penetration testing and red team exercises, parallel containment tracks with shared resource constraints in incident response, and iterative hypothesis-driven workflows in threat hunting. Whether the current formalism constitutes a sufficiently expressive basis for this target class of scenarios remains an open empirical and formal question. As the dependency vocabulary is extended, ensuring that the constraint enforcement engine remains sound and that dependency types remain composable without interference in complex configurations will be an important engineering and validation priority.

Reproducibility and Standardization: Pre-task configurations in CHART are independent variables by design, and variation in how experiment designers configure the platform across studies is an intended feature of a configurable testbed rather than a source of error. Cross-lab comparability nonetheless presents a genuine challenge, as findings from studies using structurally different configurations are difficult to aggregate. To support standardization where experiment designers require it, CHART will be released as an open-source platform accompanied by a library of benchmark scenarios. Community infrastructure, including shared channels for cross-lab coordination, will support collective refinement of both the platform and the scenario library over time.

REFERENCES

Baillie, C., Standen, M., Schwartz, J., Docking, M., Bowman, D., & Kim, J. (2020). *CybORG: An autonomous cyber operations research gym.* arXiv. https://arxiv.org/abs/2002.10667

Bienefeld, N., Keller, E., & Grote, G. (2024). Human-AI teaming in critical care: A comparative analysis of data scientists' and clinicians' perspectives on AI augmentation and automation. *Journal of Medical Internet Research, 26*, e50130.

Boyarchuk, R., Khudyntsev, M., Lebid, O., & Trofymchuk, O. (2021). Organizational and technical model of national cybersecurity and cyber protection. In V. Buriachok, D. Ageyev, V. Lahno, & V. Sokolov (Eds.), *Proceedings of Selected Papers of the Workshop on Cybersecurity Providing in Information and Telecommunication Systems (CPITS 2021)* (CEUR Workshop Proceedings, Vol. 2923, pp. 37–46). CEUR-WS.org.

Salas, E., Burke, C. S., & Cannon-Bowers, J. A. (2000). Teamwork: Emerging principles. *International Journal of Management Reviews,* 2(4), 339–356. https://doi.org/10.1111/1468-2370.00046

Kokulu, F. B., Shoshitaishvili, Y., Soneji, A., Zhao, Z., Ahn, G. J., Bao, T., & Doupé, A. (2019). Matched and mismatched SOCs: A qualitative study on Security Operations Center issues. *In Proceedings of the 2019 ACM SIGSAC Conference on Computer and Communications Security (CCS '19)* (pp. 1955–1970). ACM. https://doi.org/10.1145/3319535.3354239

Dignum, V. (2017). Responsible autonomy. *In Proceedings of the Twenty-Sixth International Joint Conference on Artificial Intelligence (IJCAI-17)* (pp. 4698–4704). Melbourne, Victoria, Australia: International Joint Conferences on Artificial Intelligence Organization. https://doi.org/10.24963/ijcai.2017/655

Cooke, N. J., Gorman, J. C., Myers, J. L., & Duran, J. D. (2013). Interactive team cognition. *Cognitive Science, 37*, 255–285. https://doi.org/10.1111/cogs.12009

Demir, M., Likens, A. D., Cooke, N. J., Amazeen, P. G., & McNeese, N. J. (2019). Team coordination and effectiveness in human–autonomy teaming. *IEEE Transactions on Human-Machine Systems, 49*, 150–159. https://ieeexplore.ieee.org/document/8651397

Du, Y., Prébot, B., Malloy, T., Fang, F., & Gonzalez, C. (2025a). Experimental evaluation of cognitive agents for collaboration in human-autonomy cyber defense teams. *Computers in Human Behavior: Artificial Humans, 4*, 100148. https://doi.org/10.1016/j.chbah.2025.100148.

Du, Y., Prébot, B., Malloy, T., & Gonzalez, C. (2025b). A cyber-war between bots: Cognitive attackers are more challenging for defenders than strategic attackers. *ACM Transactions on Social Computing*, *8*, 1–22.

Endsley, M. R. (1995). Measurement of situation awareness in dynamic systems. *Human Factors*, *37*, 65–84.

Flathmann, C., Schelble, B. G., McNeese, N. J., Knijnenburg, B., Gramopadhye, A. K., & Chalil Madathil, K. (2024). The purposeful presentation of AI teammates: Impacts on human acceptance and perception. *International Journal of Human–Computer Interaction*, *40*(20), 6510–6527.

Harel, D. (1987). Statecharts: A visual formalism for complex systems. *Science of Computer Programming*, *8*(3), 231–274.

Holland, J., Carter, W., & Murdock, K. (2022). AI and cyber defense: The role of adversarial machine learning. *Journal of Cybersecurity*, *8*, tyac001.

Huynh, T. D., Ebden, M., Fischer, J., Roberts, S., & Moreau, L. (2018). Provenance network analytics: An approach to understanding online activity. *Data Mining and Knowledge Discovery*, *32*(1), 76–109. https://doi.org/10.1007/s10618-017-0549-3

Iovino, M., Förster, J., Falco, P., Chung, J. J., Siegwart, R., & Smith, C. (2022). *On the programming effort required to generate behavior trees and finite state machines for robotic applications.* arXiv. https://arxiv.org/abs/2203.01025

Lee, J. D., & See, K. A. (2004). Trust in automation: Designing for appropriate reliance. *Human Factors*, *46(1)*, 50–80. https://doi.org/10.1518/hfes.46.1.50.30392

Marzagão, D., Huynh, T. D., Helal, A., & Moreau, L. (2020). *Provenance graph kernel.* arXiv. https://arxiv.org/abs/2010.10343

McNeese, N. J., Demir, M., Cooke, N. J., Chiou, E. K., Myers, C. W., & Yanikian, G. (2021a). Trust in human–autonomy teaming: A review of trust research from the US Army Research Laboratory Robotics Collaborative Technology Alliance. In *C. S. Nam & J. B. Lyons (Eds.), Trust in Human-Robot Interaction* (pp. 33–61). Academic Press (Elsevier). https://doi.org/10.1016/B978-0-12-819472-0.00002-2.

McNeese, N. J., Freeman, G., Cooke, Z. R., & Ifenthaler, D. (2021b). Examining human–AI trust and team performance in collaborative systems. *Proceedings of the ACM on Human-Computer Interaction*, *5*, 1–25. https://doi.org/10.1145/3449202

National Academies of Sciences, Engineering, and Medicine. (2022). *Human-AI teaming: State-of-the-art and research needs.* The National Academies Press. https://doi.org/10.17226/26355

Parasuraman, R., & Manzey, D. (2010). Complacency and bias in human use of automation: An attentional integration. *Human Factors*, *52*, 381–410. https://doi.org/10.1177/0018720810376055

Paul, C. L., Blaha, L. M., Fallon, C. K., Gonzalez, C., & Gutzwiller, R. S. (2019a). Opportunities and challenges for human–machine teaming in cybersecurity operations. *Proceedings of the Human Factors and Ergonomics Society Annual Meeting*, *63*, 432–436. https://doi.org/10.1177/1071181319631101

Salas, E., Reyes, D. L., & McDaniel, S. H. (2018). The science of teamwork: Progress, reflections, and the road ahead. *American Psychologist*, *73*, 593–600. https://doi.org/10.1037/amp0000334

Sandhu, R. S., Coyne, E. J., Feinstein, H. L., & Youman, C. E. (1996). Role-based access control models. *IEEE Computer*, *29*(2), 38–47.

Saner, L. D., Bolstad, C. A., Gonzalez, C., & Cuevas, H. M. (2009). Measuring and predicting shared situation awareness in teams. *Journal of Cognitive Engineering and Decision Making*, *3*, 280–308.

Seeber, I., Bittner, E., Briggs, R. O., De Vreede, T., De Vreede, G.-J., Elkins, A., Maier, R., Merz, A. B., Oeste-Reiß, S., & Randrup, N. (2020). Machines as teammates: A research agenda on AI in team collaboration. *Information & Management*, *57*(2), 103174.

Standen, M., Lucas, M., Bowman, D., Richer, T. J., Kim, J., & Marriott, D. (2021). CAGE challenge 1. *In Proceedings of the IJCAI-21 1st International Workshop on Adaptive Cyber Defense (ACD 2021)*. Virtual conference.

Stoll, R., Rosenthal, F., & Wiesche, M. (2021). Cognitive challenges of alert handling in security operations centers. *Computers & Security*, *109*, 102388.

Thompson, J. D. (1967). *Organizations in action: Social science bases of administrative theory*. McGraw-Hill.

van der Aalst, W. M. P. (1998). The application of Petri nets to workflow management. *Journal of Circuits, Systems and Computers*, *8*(1), 21–66.

van der Aalst, W. M. P., ter Hofstede, A. H. M., Kiepuszewski, B., & Barros, A. P. (2003). Workflow patterns. *Distributed and Parallel Databases*, *14*(1), 5–51.

Vasconcelos, H., Jörke, M., Grunde-McLaughlin, M., Gerstenberg, T., Bernstein, M. S., & Krishna, R. (2023). Explanations can reduce overreliance on AI systems during decision-making. *Proceedings of the ACM on Human-Computer Interaction*, *7*(CSCW1), Article 129.

6 Multiplayer Gaming as a Testbed for Advancing Human-Agent Teaming, Decision-Making, and Communication

Cassandra L. Crone, Patrick Nalepka, Gaurav Patil, James Simpson, Mark Dras, Rachel W. Kallen, Simon Hosking, and Michael J. Richardson

In defence contexts, simulation plays a vital role in building and maintaining force capability, providing realistic, scalable, and adaptable training across all domains of the armed forces. Typically immersive in nature, such simulations employ artificial environments that replicate or amplify real-world scenarios, allowing learners to practise skills and make decisions in safe, controlled settings (Bruzzone & Massei, 2017). Within the Australian Defence Force (ADF), simulation supports a broad range of objectives—from individual skill development to complex joint operations—and forms a key part of preparedness and performance assurance (Blackmore & Allitt, 2021). Beyond operational training, simulation is also used to evaluate systems, procedures, and human performance under controlled conditions that would otherwise be impractical, unsafe, or ethically challenging to reproduce.

The ADF, like many defence organisations, describes simulation activities using the Live, Virtual, and Constructive (LVC) framework (Hodson & Hill, 2014). *Live* simulations involve real people interacting with real systems, while *constructive* simulations model virtual agents interacting with virtual systems. The Desert Containment testbed presented in this chapter is situated within the *virtual* component of the LVC framework, where human participants interact with computer-based environments and virtual agents. Virtual simulations have long been used in defence and military contexts, from virtual and augmented reality to computer-based wargames and serious games (Bruzzone & Massei, 2017; Hill & Tolk, 2017). Yet, many defence simulations are designed for narrow task domains, with architectures that are difficult to reuse beyond their original purpose. Such systems are often costly to design and maintain, and their use may be limited by security requirements and specialist software, restricting accessibility.

DOI: 10.1201/9781003655923-7

Further, recent analyses in human-agent teaming have emphasised the importance of research testbeds that support experimentation and evaluation for complex, heterogeneous, and interdependent team configurations reflective of real-world operations (Chung et al., 2025). Human performance emerges from the interaction of multiple contextual factors, including task demands, environmental conditions, and team member behaviours. Yet, many testbeds impose constraints on what can be manipulated or measured, limiting the exploration of cognitive, behavioural, and communication processes that influence team outcomes (Chung et al., 2025). Defence operations, in particular, demand flexibility across systems and team structures—from one-to-one human-agent interactions to highly interdependent, many-to-many coordination tasks. Closing this gap requires methodologically adaptable testbeds that permit researchers to design and explore diverse team configurations and characteristics, promoting reproducibility and accelerating progress in human-agent teaming.

The Desert Containment multiplayer game environment was created in response to these overarching demands. As a novel research testbed, Desert Containment can be used to model and assess complex human-agent team interactions under dynamic and uncertain conditions. Developed to provide an immersive yet scalable environment for multi-agent and human-machine teaming experimentation, Desert Containment enables rigorous evaluation of team coordination, leadership, and performance dynamics in collaborative task scenarios. In contrast to traditional and high-end simulations, the testbed is designed as a flexible, open-source platform that supports varied team-based assessment and operational objectives without reliance on classified systems, specialised technical expertise, or tightly constrained scenarios. The testbed not only supports experimentation and evaluation across fully human and mixed human-agent teams, but it also provides opportunities for testing and optimising AI agents for human-agent teaming. In doing so, it serves as a versatile foundation for testing virtual simulations that support skill development and operational effectiveness in complex, defence-relevant environments.

6.1 OVERVIEW OF THE DESERT CONTAINMENT MULTIPLAYER GAME ENVIRONMENT

At its core, Desert Containment is a networked multiplayer video game in which team members must collaborate to complete search-and-retrieval tasks within a designated time limit. A team's fundamental objective is to locate a set of autonomous agents distributed across a large desert environment and corral them to a central containment region. While participants are not physically co-located, they communicate and interact synchronously within the same virtual space. The environment updates continuously so that changes made by one participant are immediately reflected for others, enabling real-time coordination and data capture.

Desert Containment's configuration is highly modular. First, the number of autonomous agents targeted for capture can be varied (e.g., nine versus eighteen), as can their behavioural dynamics, such as reactivity or sensitivity to player proximity. Second, trial and session duration are adjustable, allowing end-users to define fixed or open-ended time limits that align with specific experimental objectives.

Third, team composition is flexible, such that all participants may occupy the role of *ground player*, or an additional *operator* role may be introduced (i.e., a leader who has a drone or bird's-eye view of the environment) to provide high-level guidance without direct, on-the-field involvement. Importantly, these participant roles can be occupied by human players as well as artificial agents equipped with movement and conversational capabilities.

The testbed also affords manipulation of task-relevant contextual factors, which can be varied to systematically investigate communication, leadership, and performance dynamics. These include: (1) environmental visibility (e.g., clear versus foggy conditions), (2) availability of head-up display (HUD) information (e.g., containment area only versus teammate and target locations), (3) communication between operators and ground players (e.g., one-way versus bi-directional), and (4) operator instructional protocol (e.g., naturalistic and variable versus predetermined and standardised). These optional conditions can be modified directly by the end-user, enabling experimentation and evaluation tailored to diverse objectives. Example configurations are displayed in Figure 6.1.

The modular configuration and customisation options embedded within Desert Containment provide a unique platform for examining challenges associated with command and control (C2). In particular, the testbed can simulate varying levels and types of C2 difficulties across a range of operational scenarios. Modulating the supported task, team, and environmental components can elicit challenges in

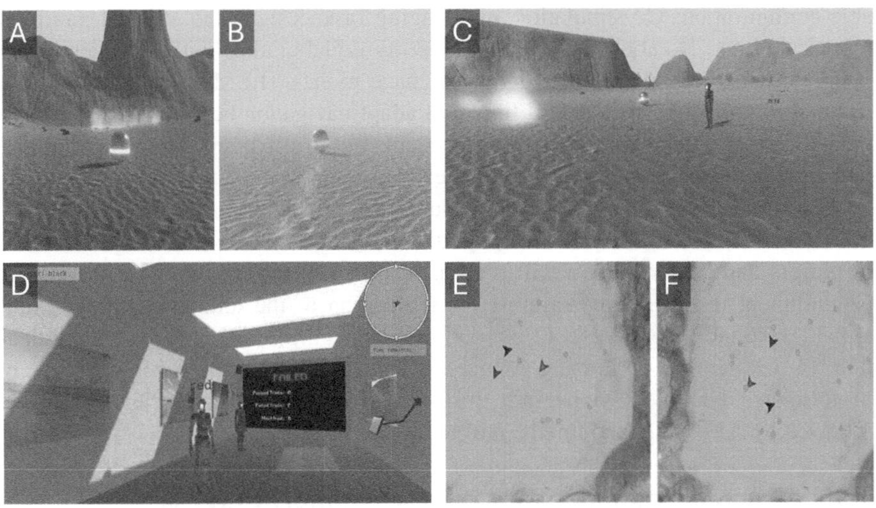

FIGURE 6.1 Example configurations. From top left to bottom right: visibility of the desert terrain and target agents varied between (a) clear or (b) foggy conditions; (c) first-person ground player view of a partial containment area, autonomous target agent, and blue-team ground player; (d) game lobby waiting area with a scoreboard displaying targets captured and trials won; the operator's map view showing locations of ground players (triangles), the containment area (small open circle), and either (e) nine or (f) 18 targets.

communication (e.g., restricted or delayed information exchange), coordination (e.g., conflicting task priorities), situational awareness (e.g., incomplete or obscured environmental information), and/or decision-making (e.g., uncertainty or time pressure). Further, the extensive configurability of Desert Containment enables simulation and evaluation of increasingly complex C2 phenomena, including information asymmetry (e.g., unequal access to situational or task-relevant information), role ambiguity (e.g., dynamic or shifting boundaries between operator and ground players), and coordination demands (e.g., heightened interdependence among team members or mixed human-agent teams operating under time pressure).

The architecture of Desert Containment intentionally accommodates these diverse modalities to investigate their impact on effective search and retrieval, particularly under C2 constraints. The task is designed to be challenging to capture the influence of C2 difficulties, and as modularity increases (through enabling more customisable capabilities), so too can task complexity. Thus, careful consideration was given to ensure that Desert Containment supports both the development of team expertise and the extension of that expertise into increasingly demanding conditions. At the same time, certain approximations and stylised representations were necessary trade-offs that maintain system effectiveness, reliability, and usability. Although the testbed employs a gamified aesthetic (e.g., autonomous agents resemble small futuristic robots, and players are represented in mechanised suits), this abstraction ensures computational efficiency and smooth real-time interaction across a range of hardware capabilities, while balancing ecological validity with end-user control.

In sum, the testbed's core functional capabilities include networked multiplayer operation, synchronous interaction, real-time data capture, modular and customisable configuration, C2 simulation, challenging task design, and stylised gamified representations. Notably, the testbed supports collaboration with AI agents that integrate heuristic rule sets and large language models (LLMs), allowing them to respond to task directives in real time and adapt navigation behaviours to environmental changes. Together, these features enable both fully human teams and mixed human-agent teams to operate within a flexible, research-ready environment.

Accordingly, the remainder of this chapter highlights exemplars featuring fully human teams, Wizard-of-Oz (WoZ) simulations, and autonomous AI agents acting as players and operators. We will present several general use cases for the Desert Containment testbed, demonstrating its application to the following experimental aims: (1) extending expertise, (2) modelling behaviour, (3) examining communication, (4) testing and optimising AI agents, and (5) assessing leadership and team performance dynamics. We begin with studies involving exclusively human players, followed by AI testing, and finally, multi-team scenarios.

6.2 DIVISION-OF-LABOUR STRATEGIES AND SEARCH BEHAVIOUR STRUCTURES IN FULLY HUMAN TEAMS

The first use of the Desert Containment testbed was in a study by Nalepka et al. (2022), which examined how teams develop effective division-of-labour strategies during collaborative search tasks. This work addressed a long-standing gap in the teamwork literature: the need for objective, continuous behavioural indicators of situational awareness, rather than relying on subjective self-reports or discrete outcome-based metrics. In this context, *situational awareness* refers to how individuals perceive, interpret, and anticipate task-relevant information to guide coordinated action (Endsley, 1988). Given that situational awareness is thought to underlie successful team coordination (Endsley, 1995), the study aimed to determine whether individual-level behavioural measures could reveal the emergence of situational awareness under varying information conditions. This work tested two related assumptions. First, that with experience and when teams could easily access task-relevant information, they would divide labour more effectively by reducing redundant search coverage within the task environment. Second, that greater access to task-relevant information would promote more structured, prospective search behaviours, whereas information-poor conditions would elicit more reactive search behaviours.

In this study, Desert Containment was configured using its modular features to create a three-person team environment, such that all participants occupied the ground player role (no operator role was introduced). Fully human teams completed multiple five-minute trials across four weekly one-hour sessions, searching for and corralling either 9 or 18 autonomous agents, with incentives to minimise completion time. Task-relevant contextual factors were manipulated using two of the testbed's adjustable modules. First, *environmental visibility* was set to either clear conditions (>150 m viewing distance) or foggy conditions (~10 m viewing distance). Second, *HUD information* was varied between a minimal compass-only interface showing only heading direction and position relative to the containment region and a comprehensive display providing real-time teammate locations, target agent positions, and agent behavioural states (i.e., idle, fleeing, contained). Across trials, target number (i.e., 9, 18), visibility (i.e., clear, foggy), and HUD information (i.e., minimal, comprehensive) were systematically varied to generate graded information conditions that differed in both perceptual access and augmented support. These conditions were leveraged to vary the ease with which team members could develop situational awareness and to examine how information accessibility shapes teams' search strategies and division of labour.

Nalepka et al. (2022) also introduced an *in situ* behavioural measure of situational awareness. Unlike survey-based measures that require teams to reflect after being removed from the task context, non-verbal behavioural indices of coordination were derived from the continuous analysis of team members' movements as they navigated the in-game terrain. This measure, α (alpha), was obtained through detrended fluctuation analysis (DFA), a technique that quantifies the temporal structure of behavioural variability. The DFA α value summarises the correlational structure of movement fluctuations over time. Values near 0.5 indicate changes in behaviour that are uncorrelated across time (white noise), suggesting reactive or environmentally driven movement, whereas values approaching 1.0 indicate persistence in the direction of change across time (pink/fractal noise), suggesting prospective planning and internal control. The theoretical premise is that α reflects the balance between external task constraints and internal cognitive control (Van Orden et al., 2011), with higher values indicating a greater capacity for the anticipatory behaviour characteristic of well-developed situational awareness.

Key findings of this work indicate that access to task-relevant information shapes how teams structure and coordinate their search behaviours, which in turn predicts team performance. When visibility was restricted by fog or HUD information was limited, teams exhibited reduced search structure (lower DFA α), reflecting more reactive scanning and less effective division of labour (e.g., persistently redundant coverage of the task environment). In contrast, clear visibility or access to comprehensive HUD information increased search structure (higher DFA α), particularly under low-visibility conditions, indicating more prospectively planned search strategies and more efficient spatial partitioning. Teams whose members exhibited higher DFA α values, reflecting a greater ability to anticipate environmental demands, also completed trials significantly faster. Notably, experience with the task also produced meaningful improvements. By the fourth experimental session, teams completed trials more quickly, reduced overlap in their search paths, and demonstrated increasingly structured head-scanning behaviours, suggesting that practice helped team members organise their exploratory actions more efficiently regardless of visibility or HUD condition. Importantly, higher DFA α values consistently predicted faster completion times even after accounting for experimental manipulations, indicating that this non-verbal behavioural measure served as a robust marker of effective team coordination and situational awareness.

Taken together, Nalepka and colleagues' (2022) findings demonstrate that Desert Containment's modular design effectively supports the evaluation of situational awareness and coordination within a configuration involving three ground-based players whose performance depends on efficient division of labour. By systematically manipulating the availability of perceptual and HUD-augmented information, this work shows how teams can develop more efficient division-of-labour strategies when simulated task environments provide clear, veridical cues (e.g., environmental visibility, comprehensive HUD support). In line with the study's aim to establish objective behavioural indicators of situational awareness, DFA α values were sensitive to variations in information accessibility, changed systematically with experience, and meaningfully predicted team performance. At the measurement level, the work thus highlights DFA α as a promising behavioural index of individual-level situational awareness, and extending this approach to capture team-level situational awareness via complexity matching (i.e., covariation of DFA α between team members) presents an open avenue for future work. A key limitation of the study was the absence of communication analyses, restricting insight into how teams verbally coordinate and differentiate roles. This gap motivates later use cases in the chapter, which examine how communication dynamics shape team performance and how AI teammates might interact with, or adapt to, human communication patterns.

6.3 CONVERSATION DYNAMICS AND KNOWLEDGE ASYMMETRY IN FULLY HUMAN TEAMS

A second study by Simpson et al. (2022a) used the Desert Containment testbed with a comparable configuration to Nalepka et al. (2022), extending the focus to both verbal and behavioural coordination under conditions of knowledge asymmetry. Here, a human operator was introduced, replacing the role of the HUD to provide real-time guidance to the team's ground players. This study aimed to evaluate how distinct roles with differential access to task-relevant information (i.e., ground player, operator) might shape coordination, communication dynamics, and team performance, testing three key assumptions. First, that an increase in task difficulty (i.e., low visibility, higher number of targets) would result in a greater magnitude of conversation between team members, as well as more structured patterns of communication. Second, that this increase in communication and communication structure would reflect a greater necessity to maintain effective team situational awareness. Third, that the random appearance of new targets toward the end of a trial would elicit greater communication from the operator to the ground players.

The testbed was configured to support four-person teams, such that three participants occupied the ground player role and one participant occupied the operator role. While ground players were responsible for locating and containing targets, the operator was positioned off-site and monitored the task environment using a multi-view interface to provide navigational guidance and inform ground players about target and containment area locations. Three operator views of the environment were enabled: a fly-over view of the terrain, first-person feeds from each of the ground players, and a map view that displayed the real-time locations and status of ground players (e.g., heading direction) and targets (e.g., idle, fleeing, contained). Fully human teams completed 16 five-minute trials across four weekly 90-minute sessions, varying three factors: autonomous agents (i.e., 9, 18), visibility conditions (i.e., clear, foggy), and the presence or absence of a late-trial perturbation (i.e., new target randomly introduced).

Focusing exclusively on the fourth session to evaluate skilled-level performance, Simpson et al. (2022a) found that both ground players and operators increased their communication (i.e., produced a higher density of verbal statements) when visibility was limited and when more targets needed to be contained. The random introduction of new targets late in the trial did not influence the overall structure of team communication, nor did the total number of targets, although low-visibility conditions led to more coordinated communication patterns at the team level. When these dynamics were analysed by role, however, this increase in coordination was driven primarily by the operator, whose statements became more frequent and followed more predictable patterns. In contrast, communication dynamics among the ground players became less structured when fog was present. Further, after accounting for task difficulty, greater and more repetitive communication among the ground players predicted poorer task performance (i.e., fewer targets contained and longer completion times).

Overall, this work illustrates the utility of the Desert Containment testbed for examining role-differentiated coordination and communication dynamics, as well as the constraints such asymmetries impose. The findings again demonstrate that increasing task difficulty impairs team performance. Such difficulty further seems to increase reliance on the operator, whose veridical access to the environment is what supports adaptive communication dynamics. At the same time, ground players may be unable to meaningfully contribute to shared situational awareness when they have limited visibility, leaving the operator to manage guidance on a player-by-player basis. This pattern underscores the limits of verbal coordination alone in sustaining team situational awareness when knowledge asymmetries are pronounced.

Methodologically, analysing only the final session provided insight into skilled-level performance but limited understanding of how coordination develops over time. Additionally, informal observations suggested that operators tended to dictate actions without consistently communicating relational information among the ground players, which may have exacerbated performance difficulties under fog. These findings motivate forthcoming use cases exploring how conversational agents might assume operator roles, monitor team communication, anticipate information needs, and ultimately augment human–human coordination by developing expertise in the communicative strategies that underpin effective performance in the Desert Containment task environment.

6.4 EXPLORING CONVERSATIONAL AGENTS IN OPERATOR CONTEXTS

In this section, we turn to WoZ approaches using the Desert Containment testbed (Simpson et al., 2022b), in which a human 'Wizard' covertly controls a system or agent to simulate its capabilities before it is fully automated (Kelley, 1983; Riek, 2012). The first study of this type by Simpson et al. (2025a) aimed to evaluate whether conversational agents with limited communication capabilities could effectively lead ground players by assuming the operator role. By comparing teams led by a human operator to those led by a Wizard-controlled agent, the study examined how leadership communication, team coordination, and user perceptions are affected when the same behaviour is presented as agentic rather than human. The study addressed five key questions: (1) whether leader type (human vs. Wizard-controlled agent) influences team performance (completion time and containment rate); (2) how team communication (proportion of words spoken) changes with leader type; (3) whether a limited communication scheme derived from expert performance is sufficient for agent effectiveness; (4) how participants perceive human versus agent leaders; and (5) whether perceived leader gender (masculine vs. feminine voice) influences these effects.

Human operator behaviour from the previously described Desert Containment research (Simpson et al., 2022a) was used to model the form of team leadership implemented by the WoZ agent. This model focused on the operator's strategic, high-level functions (i.e., verbal coordination, decision-making, and situational awareness), rather than low-level task execution. In practice, this meant capturing the operator's observable leadership actions expressed through language, such as interpreting environmental information, prioritising goals, issuing concise guidance, providing feedback, and adjusting communication as the task evolved. Embedding this behavioural model within the WoZ framework allowed this iteration of Desert Containment to preserve expert leadership behaviour while masking natural variability in communication, enabling participants to believe they were interacting with an AI agent. Thus, this approach provided a controlled means of testing how identical leadership behaviours operate when delivered through a conversational interface, isolating the influence of interface modality from differences in human decision-making.

Three-person ground player teams completed eight five-minute trials, with half led by a human operator and half by a conversational agent that was controlled covertly by the same operator acting as a Wizard. Four trained researchers (two masculine-voiced and two feminine-voiced) served as operators, each completing a minimum of four training hours to achieve expert-level performance. Like previous work, environmental visibility (clear vs. foggy) was manipulated to vary task difficulty. However, the primary experimental manipulation focused on how operators provided guidance. Human operators spoke freely via microphone while adhering to predefined communication topics, whereas Wizards selected from 21 predetermined response types delivered via text-to-speech. These topics and response types were restricted to player guidance, such as which player to address, the number and location of targets, navigational directions, and environmental conditions.

Aligning with the five overarching research questions, Simpson et al. (2025a) found that teams performed similarly regardless of whether they were led by a human or a Wizard-controlled agent, completing trials in a comparable time, achieving similar containment rates, and coordinating search space division to the same degree (RQ1). Although teams communicated more overall when guided by a human operator, the proportion of operator-generated communication remained consistent across conditions, with a notable reduction in team communication when led by a masculine-voiced Wizard (RQ2). Notably, the constrained communication scheme was effective, with six response types accounting for nearly all operator guidance (86.83%), demonstrating that a small, well-targeted set of responses was sufficient for successful task execution (RQ3). Participants nonetheless rated the Wizard leader less favourably, judging its guidance as less timely, less relevant, and less understanding, even though instructions were equally comprehensible. Critically, they also viewed the Wizard as more AI-controlled, indicating that the experimental manipulation successfully created the intended illusion of artificial agency (RQ4). Finally, while the operator's perceived gender did not influence performance outcomes, participants perceived feminine-voiced human operators as providing more timely guidance than their masculine-voiced counterparts (RQ5).

Here, the Desert Containment testbed enabled the modelling and evaluation of structured, high-level leadership behaviour, showing that such behaviour can be delivered effectively through a constrained conversational interface. Consistent with the study's aims, the conversational agent matched human operators in guiding team performance, even though its leadership was perceived as less socially responsive. This work generated several open questions for future research (see Simpson et al., 2025a for a full review). However, a key limitation of the Wizard-controlled agent was its inability to engage in contextual, bidirectional communication. That is, it could issue guidance but could not dynamically respond to queries or adapt messages based on player input. This constraint, along with the agent's inability to provide explanations for its decisions, may have contributed to lower participant ratings given the importance of explainability for trust in AI systems (Afroogh et al., 2024; Shin, 2021). While the study did not implement an autonomous agent, the Wizard data provides training material for developing such agents and later evaluating them against human operators. Building from these insights, the next study presented in this chapter shifts focus to the fundamental question of whether verbal communication is explicitly required for effective leadership in Desert Containment, before progressing toward fully autonomous agent leaders.

6.5 COMMUNICATION DEMANDS FOR COLLABORATIVE AGENT LEADERS

The second WoZ study by Simpson et al. (2025b) sought to examine whether collaborative conversational agents require the capability to listen and respond to teammate queries, or whether access to visual task-relevant information (e.g., map view) alone is sufficient to sustain effective team performance. This aim centred on determining whether sophisticated conversational dialogue systems are necessary or if simpler unimodal approaches suffice, as the latter would markedly simplify development complexity by eliminating the need for speech recognition, natural language understanding, and dialogue management systems. This work tested two key assumptions. First, that an operator who could both listen and respond to player queries, as well as use the map view, would perform better than an operator using the map view alone. Second, that disregarding player queries would negatively impact team performance and user preference, with players favouring the agent capable of listening and responding.

As before, three-person ground player teams completed eight five-minute trials and visibility (i.e., clear, foggy) was varied across trials. A single trained male confederate served as the Wizard for all sessions, having extensively practised the task over four weeks. In the *Query Non-Responsive* condition, the Wizard's audio input from ground players was muted, and the Wizard could only select from 12 predetermined response types while using visual information from the map view. In the *Query Responsive* condition, the Wizard could hear player utterances and select from the expanded set of 21 response types (e.g., acknowledgements, clarifications, and responses to queries). Crucially, ground players were never informed of this manipulation and always believed the operator could hear them.

Contrary to expectations, the assumption that multimodal input (i.e., visual + auditory) would yield superior performance compared to unimodal input (i.e., visual only) was not supported. Teams performed equivalently on metrics for trial duration, containment rate, and division of the search space, regardless of whether the agent could listen and respond to queries. Additionally, the amount of team communication was equivalent across the two conditions, and the proportion of words spoken by the operator was consistent regardless of listening capability. Although participants rated the Query Responsive agent more favourably (e.g., lower frustration, greater perceived understanding, and higher overall evaluation), these advantages did not yield any performance differences, demonstrating a lack of support for the assumption that ignoring queries would harm performance. Taken together, these results indicate a dissociation between player preferences and task performance. However, for well-defined collaborative tasks in which the operator has comprehensive visual access to task-relevant states, continuous and sophisticated conversational exchange appears unnecessary for maintaining effective team performance, even if it remains important for subjective experience.

The work by Simpson et al. (2025b) underscores the utility of the Desert Containment testbed as a controlled environment for isolating communication requirements and systematically manipulating information asymmetries between agents and human teammates. Methodologically, the work highlights how the testbed can be used to disentangle perceptual versus performance-related impacts of conversational capabilities. Conceptually, it identifies key boundary conditions in which listening and dialogue may become performance-critical (e.g., increased task complexity, operator information dependency, and dynamic or uncertain goals). The study also raises an important open question for future testbed applications: whether the performance equivalence observed under human Wizard control will persist once these roles are assumed by fully autonomous systems.

6.6 TESTING AND OPTIMISING AI OPERATORS

In this section of the chapter, we turn to the introduction of AI teammates. Building on the findings of the earlier WoZ studies, Simpson et al. (2026) examined whether AI agents could effectively assume the confederate role within the Desert Containment testbed. While the WoZ studies established that simple and constrained communication frameworks were effective under human control, this study deploys AI agents guided by this principle to test whether such communication remains effective when decision-making is fully autonomous. Notably, this work contrasts a simple, interpretable, and data-efficient heuristic model with a more flexible and complex LLM-based model, pursuing four primary objectives: (1) to evaluate how AI-led teams compare to human-led teams in terms of performance; (2) to determine whether heuristic or LLM-based leadership yields better performance; (3) to examine how factors such as perceived gender influence team outcomes; and (4) to assess how these AI agents affect team communication patterns and participants' perceptions of the leader.

Experimentation was conducted across two iterations in which three-person ground player teams completed eight four-minute trials under varying visibility conditions (i.e., clear, foggy) and with different types of operators. In the first iteration, ground players were led by both a human operator and the heuristic AI operator, with the operator type varied across trials and teams split evenly between male and female leaders. In the second iteration, participants were led exclusively by the LLM-based AI operator, with gender varied across trials (i.e., masculine vs. feminine voice). Figure 6.2 displays a summary and comparison of the two AI operators.

Both AI operators shared a common architecture with three principal components. First, an autonomous communication decision model (either heuristic or LLM-based) continuously evaluated game-state data (e.g., ground player positions and headings, target locations and status, containment area location, visibility condition, and time remaining) to determine the timing and content of its guidance. Second, a query-responsive speech recognition system supported a limited set of player queries (e.g., headings, navigation, and containment area). Whereas the heuristic AI used function-calling LLMs to classify queries into predefined categories, the LLM-based AI processed queries with reference to game-state context and validated responses with a fact-checker, leveraging the LLM's full generative capacity. Third, both AI types used a text-to-speech system that synthesised speech using high-quality voice clones of the expert human operators, minimising voice-based confounds and enabling direct comparisons between human and AI leadership. This unified architecture further enabled a controlled comparison of heuristic and

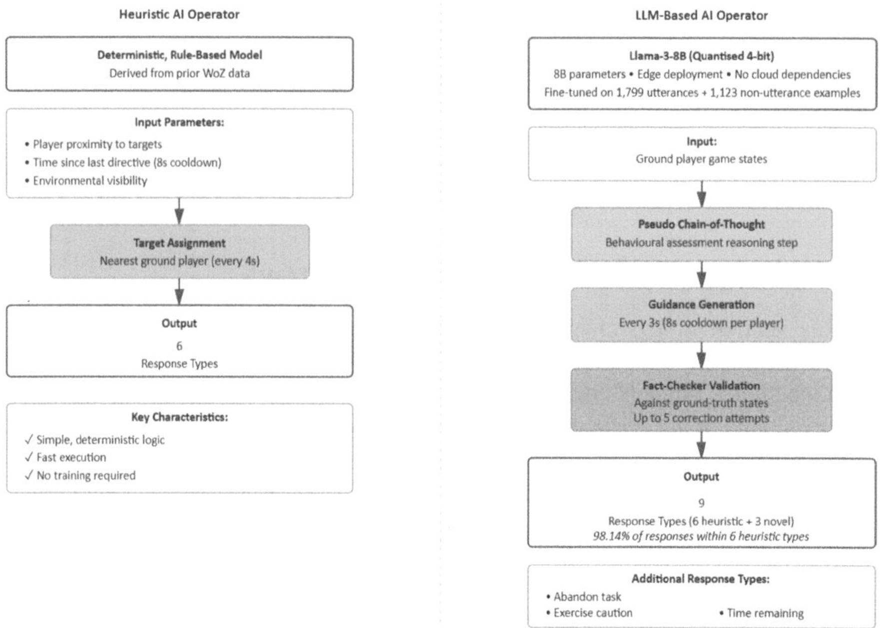

FIGURE 6.2 Comparison of heuristic and LLM AI operators. The model on the left displays the characteristics of the heuristic AI operator as a deterministic, rule-based model derived from prior WoZ data. The model on the right displays the characteristics of the LLM-based AI operator as a fine-tuned, edge-deployed Llama-3-8B agent. Researchers seeking to replicate or extend these AI agents are provided with key details in this chapter regarding model selection, decision logic, and communication design, with additional architectural and training specifics reported in the forthcoming paper by Simpson et al. (2026).

LLM-based AI leadership while maintaining continuity with the principles established in the earlier WoZ studies.

While both AI agents demonstrated potential for leading human teams, neither approached the consistency or overall effectiveness of the expert human operators. Teams led by human operators completed trials significantly faster and with higher overall rates of success. While the LLM-based agent matched human operators' performance in target containment rate within high-visibility conditions, it failed entirely in low-visibility conditions with a 0% success rate. By contrast, team performance with the heuristic agent was less effective overall yet remained stable across varying task complexity, maintaining similar success rates across high- and low-visibility conditions. Teams led by the heuristic agent also communicated significantly more than human-led teams, although this did not translate to enhanced performance. Notably, human operators were consistently rated as providing the most appropriate and useful guidance, and ground players were most likely to follow their recommendations, with the heuristic operator rated less favourably and the LLM least favourably. Social perceptions further shaped team dynamics, such that participants were more likely to follow guidance from female operators (whether human or AI) and feminine-voiced agents were more often assumed to be human-controlled.

Despite isolated areas of comparable performance, participants reliably preferred human operators and readily identified AI agents, underscoring that social attunement, adaptability, and trust remain central challenges for developing AI teammates. Indeed, research suggests that an AI agent's communication style may help or hinder trust in human–AI teams (Afroogh et al., 2024), with misunderstandings arising from omissions, ambiguity, and limited social–emotional awareness (Basappa et al., 2025; Chen & Zhang, 2025). Conversely, trust may be enhanced by explainability, particularly when supported with visual examples rather than text alone (Afroogh et al., 2024; Shin, 2021).

Additionally, communication quality, rather than quantity, emerged as a key determinant of team success. However, the heuristic and LLM-based models each demonstrated distinct strengths, suggesting that hybrid approaches combining heuristic robustness with LLM flexibility may be particularly promising. The Desert Containment testbed, with its modular and configurable design, provides an ideal platform to systematically explore this critical next step, as well as address additional open questions arising from this initial work. Namely, these include testing alternative LLM architectures, optimising heuristics based on player compliance, evaluating trade-offs between response quality and latency, identifying which cues participants use to identify AI teammates, and examining the effects of competitive or ambiguous objectives, varied information asymmetry, and more flexible, human-like communication.

6.7 TESTING AND OPTIMISING AI GROUND PLAYERS

In a brief, unpublished follow-up study, we investigated mixed human–AI teams composed of two human participants and an AI ground player, who were led by the heuristic AI operator described in Simpson et al. (2026). The study manipulated environmental visibility (clear, foggy) and the ability of the AI ground player (high, low) to assess team performance and the structure of their search behaviours, using DFA α as in Nalepka et al. (2022) as a behavioural index of situational awareness. We expected that both high-visibility conditions and high-ability AI ground players would support faster task completion and more structured search behaviours, with high-ability AI agents additionally expected to minimise redundant search overlap.

Across eight five-minute trials, the ability of the AI ground player was manipulated by varying the maximum distance at which the agent could detect targets. In the *low-ability* condition, agents could identify targets within 30 m of their current position, whereas in the *high-ability* condition, this range increased to 150 m. All agents followed a truncated Lévy walk search pattern (Viswanathan et al., 1999), with relocation step sizes ranging from 20 m to 400 m before selecting a new heading. Agents were also programmed to follow operator commands for 15 seconds upon receipt, during which they ignored any newly encountered targets for the first five seconds.

In high-visibility conditions and with the high-ability agent, teams completed the Desert Containment task significantly faster and exhibited more structured head scanning behaviours, indicating more prospectively planned search strategies. Working with the high-ability agent also offset the performance decline typically observed under low-visibility conditions. In contrast, search overlap was unaffected by the agent's ability level but instead increased when the AI operator communicated more frequently. Notably, operators communicated significantly more when working with the low-ability agent, suggesting that increased instruction may have constrained organic search behaviours and limited how freely the ground players could self-organise their coverage of the environment.

By assigning AI agents to two distinct functional roles, the Desert Containment testbed introduced role-specific information asymmetries across both AI–AI and human–AI interactions, offering a platform for examining performance and coordination within mixed teams. Consistent with Nalepka et al. (2022), individual situational awareness decreased in low-visibility conditions (DFA α decreased in low- compared to high-visibility trials), demonstrating the testbed's capacity to model perceptual constraints and evaluate their impact. In contrast, situational awareness was unaffected by AI ground player ability, suggesting that DFA α may primarily index an individual's own perceptual attunement rather than a team-level property. Notably, the AI operator adapted communication behaviours in response to the low-ability ground agent, increasing support when target acquisition slowed. This emergent adjustment, driven by heuristic access to game-state data rather than explicit experimental design, suggests that the heuristic model may afford role-appropriate, performance-contingent communication. These early findings raise several questions for future work using the testbed, such as determining the optimal level and style of operator communication, clarifying how communication density influences team autonomy and search efficiency, and assessing how human teammates evaluate AI collaborators in terms of trust, competence, and communication preferences. Additionally, more direct measures of autonomous search behaviour would further clarify how different agent roles shape coordination within mixed human–AI teams.

6.8 COMPETITIVE MULTI-TEAM SCENARIOS

In this final section, we turn to a competitive iteration of Desert Containment, building on the prior studies that highlighted a need to examine team coordination under more demanding and dynamic conditions. This exploratory, unpublished study was designed to assess how competitive pressures shape communication, role behaviour, and collective behavioural dynamics, and how these features contribute to performance in a continuously evolving environment. Unlike the earlier non-competitive iterations, introducing an active opponent presents opportunities for offensive behaviours and for role specialisation to emerge. Thus, we also introduce a novel automated behavioural classification structure to characterise ground player roles and behaviours in this section. The use case presented here centres on coordination and communication under competitive uncertainty, with a focus on how teams adapt in real time to both environmental and opponent-driven demands. A visualisation of the task environment for this competitive configuration is provided in Figure 6.3.

Four-person teams (three ground players and one operator) competed in a six-week round-robin tournament in which every team played each opponent twice, followed by a final session between the two top-performing teams. Across 45-minute sessions, teams completed a series of 10-minute trials, moving targets into their assigned

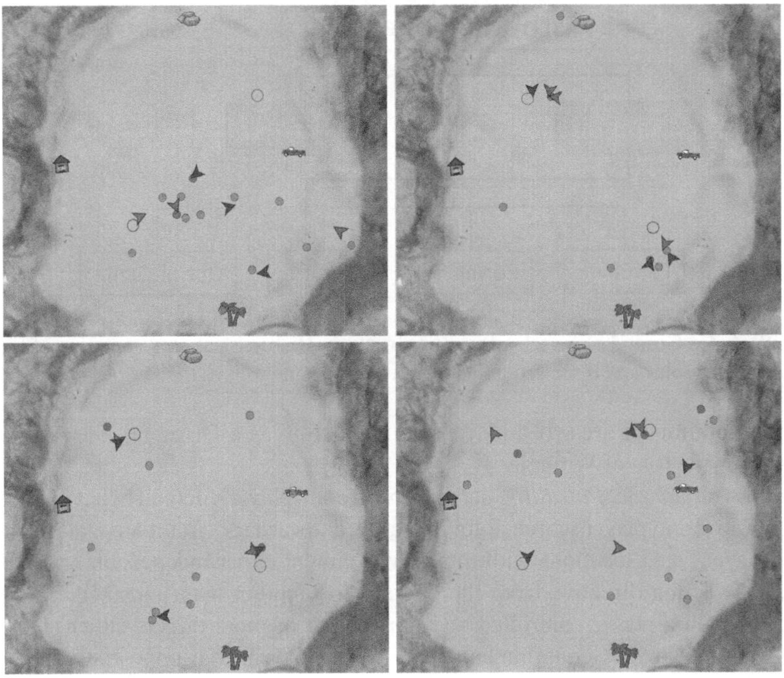

FIGURE 6.3 Map view of the playing field. The location of each team's containment area was varied across trials, appearing in one of four possible orientations. The map view displays the locations of available targets (smaller filled circles), ground players (triangular arrows), and containment areas (larger open circles). Key landmarks—rocks (north), a truck (east), palm trees (south), and a hut (west)—within the environment are also displayed. The map view was used by team operators to provide their ground players with a heading direction toward the location of targets, other players, landmarks, or containment areas.

containment area (e.g., red, blue). Each trial began with 15 targets, with another 10 targets spawning between 90 seconds and 4 minutes (one every 20 seconds), yielding 25 available targets. A trial ended when a team captured 13 targets or when time expired, with the winner determined by the higher score. In-game audio communication was enabled only within teams to facilitate coordination while preventing cross-team communication. This configuration also introduced a novel competitive feature. Each player could deploy up to five EMP pulses per trial to stun nearby opponents (within a 5 m radius), temporarily removing them from the field for ten seconds.

As illustrated in Figure 6.4, the competitive configuration of the Desert Containment testbed revealed an emergent ground player classification structure organised around two primary roles supported by three behavioural strategies. *Wrangler* denotes the ground player's default role, in which their primary responsibility was to search for and corral targets into their team's containment area. Alternatively, a ground player could take on the role of *Disrupter*, in which their primary responsibility was to block the opposing team from moving targets into their containment area. When occupying the role of Disrupter, ground players often acted close to the opposing

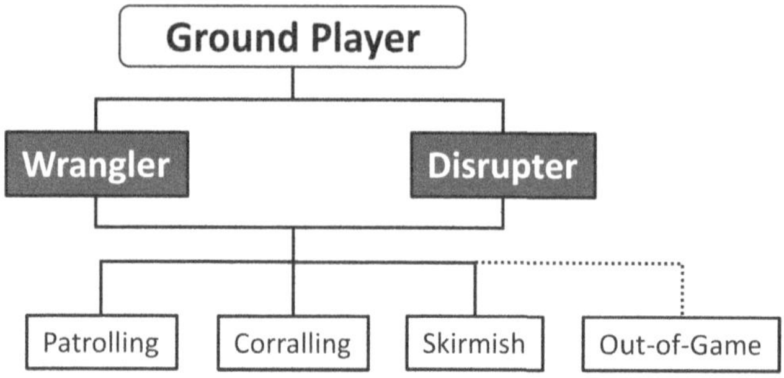

FIGURE 6.4 Ground player classification structure. Hierarchical representation of the two ground player roles and three ground player behaviours, as well as the out-of-game status.

team's containment area. If a player was not acting as a Disrupter, then they were designated the role of Wrangler.

Within either role, *Patrolling* denotes the ground player's default behavioural strategy. A Patrolling player searched for and moved toward (or circumnavigated) targets, other players, and locations within the environment either independently or following a task-action directive from the operator or another team member. *Corralling* denotes when a player controlled and moved one or more targets either (1) toward their assigned team's containment area, (2) away from the opposing team's containment area, or (3) toward another location or landmark within the environment. Finally, *Skirmish* denotes when two players from opposing teams were engaged in either (1) a contest over controlling one or more targets, (2) sustained interference with the intended motion of the other player, or (3) offensive attack behaviour, or 'zapping', through the use of the EMP pulse. A player's status became *Out-of-Game* if they were temporarily stunned by a successful attack from an opposing player.

To apply the ground player behavioural classification structure to each timepoint of gameplay, we developed an automated classification program using Python (version 3.10.6). Across the final sample of 58 trials, positional movement data for all ground players and target agents were extracted from recorded game states, yielding 348 individual time series for analysis. The program was validated by comparing automatic classifications with manual coding on a sample of 42 time series from seven randomly selected trials, yielding agreement rates of 78%–82% across coders. Following this validation, the automated program was used to classify roles and behaviours across all remaining time series. The equations below describe this approach in simplified form.

$$\text{role}(GP) = \left\{ \begin{array}{c} \textbf{out-of-game} \rightarrow y_{GP} > 50 \\ \textbf{disrupter} \rightarrow d\left(GP, CA_{opp}\right) \leq 75\,\text{m} \wedge t \geq 4\,\text{s} \\ \textbf{wrangler} \rightarrow \text{otherwise} \end{array} \right\}$$

If a player's Y position was > 50 m above the ground, indicating they were in the waiting room, both their behavioural strategy and role were classified as Out-of-Game.

For ground player role classification while on the playing field, if a player was within 75 m of the opposing team's containment area at any instance within a 4 s window of time, their role was classified as Disrupter for all timepoints in that window. Otherwise, their role was classified as the default role, Wrangler.

$$\text{behaviour}(GP) = \left\{ \begin{array}{c} \textbf{skirmish} \rightarrow d\left(GP, GP_{opp}\right) \leq 1\,\text{m} \wedge t \geq 4\,\text{s} \\ \textbf{skirmish} \rightarrow GP_{status} \in \{EMP, stunned\} \\ \textbf{corralling} \rightarrow TA_{running} \wedge d(GP, TA) \leq 15\,\text{m} \wedge d\left(GP, GP_{opp}\right) > 15\,\text{m} \\ \textbf{patrolling} \rightarrow \text{otherwise} \end{array} \right\}$$

For ground player behavioural strategy classification, a player's behaviour was classified as Skirmish if (1) any two players were within 1 m of each other for a duration of time ≥ 4 s, (2) a player used an EMP pulse, or (3) a player had been stunned. A player's behaviour was further classified as Skirmish if any two players from opposing teams were within 15 m of a running target within the same 10 s window of time. A target was considered running when perturbed by a repulsive force from a ground player (i.e., no longer in its unperturbed resting state). However, if only one team's ground player was within 15 m of a running target during this window, their behaviour was classified as Corralling. If none of these conditions were satisfied, the ground player's behaviour was classified as the default strategy, Patrolling.

Team performance was evaluated across several metrics, which are presented as averages across all games for each team in Table 6.1. *Trials Won* indicates the proportion of successful trials per team, excluding draws, while *Containment Rate* indicates the average number of targets captured per trial. *EMP Success* describes the proportion of EMP pulses that successfully stunned opposing players relative to the total number of EMP pulses emitted. Communication was assessed in two ways: *Communication Density* reports the total number of words spoken by all team members during a session, and *Operator Proportion* reflects the share of team communication contributed by the operator relative to ground players.

With the greatest proportion of trials won and the highest average containment rate, Team 4 emerged as the top-performing team, while Team 2 showed the weakest performance. Team 4 also spoke fewer total words (Communication Density)

TABLE 6.1
Average Team-Level Task Performance

	Team 1	Team 2	Team 3	Team 4
Trials won	40.28%	20.83%	52.77%	86.12%
Containment rate	9.73	7.88	10.88	12.13
EMP success	24.50%	14.40%	43.48%	42.92%
Communication density	7,054	2,894	8,241	4,900
Operator proportion	85.75%	89.35%	76.53%	46.77%

and showed more balanced communication between the operator and ground players (Operator Proportion). However, Team 3 and Team 4 performed similarly on EMP Success. To assess the impact of communication dynamics and EMP success rate on trial-level performance, we examined these factors across all trials. Both lower operator dominance in communication (controlling for total words spoken) and a higher EMP success rate significantly predicted greater target containment rates and an increased likelihood of winning a trial.

We further analysed the degree to which players engaged in Skirmish activity (e.g., in a contest over one or more targets, or engaging in the use of the EMP pulse) to examine the impact of the EMP pulse as a novel offensive feature in the competitive iteration of Desert Containment. Although the overall proportion of time spent in skirmishes did not predict performance, role-specific behaviours did. Spending more time in the Wrangler role during skirmishes was associated with fewer targets captured and more trial losses, whereas spending more time in the Disrupter role during skirmishes was associated with higher containment rates. Taken together, these results indicate that high-performing teams combined effective offensive action and Disrupter deployment with balanced communication and successful EMP use.

As illustrated in Figure 6.5, higher-performing teams deployed the Disrupter role more frequently and more consistently across trials. We therefore examined how the presence of Disrupters contributed to key performance outcomes, namely trial success (win/loss) and target containment rate. The number of active Disrupters in a trial predicted the target containment rate, where deploying at least one Disrupter led to higher average containment rates (M_{range} = 10.15–11.11) compared to having no Disrupters (M = 6.90). A non-significant trend also suggested that teams who deployed only two Disrupters were 2.98 times more likely to win the trial compared

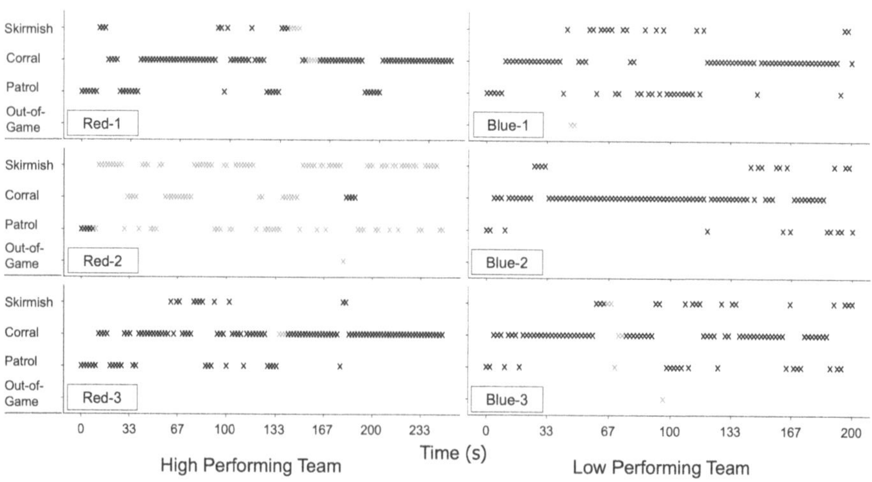

FIGURE 6.5 High- and low-performing team roles and behaviours within a trial. Light (grey) X's denote when the player acted as a Disrupter, while dark (black) X's denote when the player acted as a Wrangler. The high-performing team is displayed on the left, while the low-performing team is displayed on the right.

to teams with three Disrupters. Additionally, teams were more likely to win the trial if (1) they spent a greater proportion of their overall time disrupting, and (2) the same ground player occupied the Disrupter role for a larger proportion of time. Together, these results point to the importance of clear and consistent role coordination, with reliable deployment of the Disrupter role associated with better team performance.

To examine team-level behavioural dynamics, we applied DFA to ground players' positional movement trajectories, extending the earlier study by Nalepka et al. (2022) who examined situational awareness via individual-level head-scanning behaviours. Given that the Disrupter role contributed to team performance, we first examined whether DFA α differed among those who acted as the Disrupter for the majority of a trial (i.e., $\geq 50\%$) compared to other ground players. However, no significant differences emerged. Next, we assessed changes in DFA α across games for each team, to evaluate whether collective, intra-team dynamics were more variable or persistent over time. The two high-performing teams exhibited more consistent fluctuation patterns over time (i.e., DFA α did not significantly vary across games; Team 3, Team 4), while the two low-performing teams exhibited more variable fluctuation patterns (Team 1, Team 2). Additionally, the highest-performing team (Team 4) exhibited substantially lower DFA α compared to all other teams, indicating greater reactivity (rather than proactive planning) in response to shifting task demands. Finally, when controlling for game session (e.g., experience, expertise) and player-level effects (e.g., substitutions), patterns of within-team variability did not predict team win/loss but did predict containment rate. Specifically, the number of targets contained increased as DFA α decreased, further suggesting that team-level reactivity (i.e., rapid-fire decision-making) supports success in competitive contexts.

Taken together, these findings indicate that effective teams balance responsiveness to environmental inputs with more stable and robust collective behavioural dynamics across games, whereas less effective teams are more susceptible to shifting task demands. In contrast to Nalepka et al. (2022), where higher DFA α (more proactive planning) supported individual-level situational awareness, lower DFA α (reflecting greater reactivity) supported team-level performance in the competitive context. However, team success was further shaped by balanced, reciprocal communication and strategic use of the offensive EMP feature, emphasising the role of deliberate team-level coordination beyond raw communication quantity from the operator (as in the previous studies by Simpson and colleagues).

Overall, the competitive configuration of the Desert Containment testbed reveals how teams coordinate under competitive pressure and demonstrates the value of the platform for examining emergent role structures and behaviours. The work identifies a novel ground player classification structure, with the Disrupter emerging as a particularly influential role. Consistent role assignment improved performance, suggesting that repeated practice and role expertise can support both individual skill development and team-level coordination. Additionally, strategic Disrupter use enhanced target capture, whereas wrangling during skirmishes uniquely predicted losses, underscoring the nuanced interplay between roles, behaviours, and outcomes. With balanced communication, stable role allocation, and effective offensive strategies appearing critical for overall success, future work should further examine how these factors shape team-level coordination and the stability of behavioural dynamics.

Experimentally assigning roles to test the causal effects of role consistency, removing or modifying access to the EMP to assess reliance on offensive tools, and systematically manipulating communication constraints offer promising directions for extending this work.

6.9 CONCLUSIONS, LIMITATIONS, AND FUTURE DIRECTIONS

Across the studies reviewed in this chapter, we exemplify the capacity of the Desert Containment testbed for experimentation with and evaluation of collaborative human–human, human–agent, and human-AI teaming. Early work established that access to task-relevant information shapes division-of-labour strategies, with DFA α emerging as a novel behavioural marker of individual situational awareness. Communication analyses revealed that knowledge asymmetries constrain verbal coordination, particularly when ground players have limited visibility, while WoZ studies demonstrated that constrained communication schemes can effectively guide teams when operators have comprehensive access to task-relevant information. Initial deployment of AI operators revealed that neither heuristic nor LLM-based agents matched the reliability of human performance, while mixed human–AI teams further highlighted how AI ground player capability and AI operator communication frequency interact to shape search efficiency. Finally, the competitive configuration explored dynamic adaptation to opponent-driven demands and introduced emergent role structures. Collectively, the studies presented in this chapter highlight a range of open questions that can be pursued in future investigations using the Desert Containment testbed. To support this work, the testbed is openly available for use and extension by other researchers at: https://github.com/MInD-Laboratory/Desert-Containment.

To conclude this chapter, we turn to two overarching limitations of the testbed that cut across studies, and which have implications for modelling and evaluating C2 challenges. First, the testbed was consistently configured as a deliberately constrained task, using a virtual desert environment, fixed and known objectives, and ground player action sets restricted to locomotion. While these constraints enabled controlled investigations under varying levels and types of C2 difficulties (e.g., information asymmetry, role ambiguity, coordination demands, communication restrictions, and decision-making under uncertainty), a key future direction is to establish the extent to which insights derived from the testbed translate to real-world human–agent teaming scenarios. Thus, future research should leverage the open-source testbed to explore a broader range of environments (e.g., varied terrains, elevation changes, obstacles or chokepoints, dynamic weather conditions, and roaming adversarial drone disturbances), as well as shifting or uncertain objectives (e.g., escort or zone-control tasks) and more extensive action sets (e.g., environmental manipulations, tactical behaviours, and information-gathering actions).

Although the competitive iteration extended the action space by enabling the EMP pulse and began to explore the coordination dynamics that differentiate high- from low-performing teams, advancing this line of inquiry requires further systematic examinations of team performance under more complex and dynamic task structures. Thus, future work should also consider investigating combinations of differing communication constraints and variable access to task-relevant information together with unexplored factors such as asynchronous collaboration, lateral team arrangements, novice versus expert performance, or varying consequences for poor performance (e.g., disadvantages in subsequent rounds, energy recharge requirements, reduced access to offensive tools, or differing movement costs). In sum, while this chapter describes evidence obtained under a tightly controlled scenario, future work can and should exploit the flexibility of the testbed to examine human-AI teaming under richer, more variable, and more operationally relevant conditions.

A second limitation of the Desert Containment testbed, as presented in this chapter, concerns the development of only two AI models. We employed small, local LLMs that can be run on the edge, eliminating the need for robust cloud or network connectivity and minimising security concerns associated with cloud-based API calls. The evidence and baseline behaviours established here (including human–human or human–agent via WoZ setups) provide a foundation for developing additional AI architectures, which can be implemented, tested, and further refined within the testbed. This also creates opportunities to examine alternative agent communication platforms, particularly dialogue-based interfaces that enable sustained, interactive exchanges with human teammates. Given that the LLM operator underperformed the human operator on several metrics, however, the question of whether a larger model would close this gap is particularly pertinent. This motivates introducing modern agentic pipelines, in which a large model (~30–70B parameters) is used for orchestration (e.g., high-level coordination, and strategic planning) while smaller models (~1–7B parameters) perform task-specific operations.

Understanding differences between human and artificial communication, leadership, and collaboration remains essential and underexplored, particularly regarding trust, explainability, and acceptance. Future work should directly compare the performance of mixed human–AI teams (using various AI architectures) with fully human teams to understand how artificial agents can enhance, or potentially hinder, team performance. While reducing cognitive load and preserving personnel expertise for more critical tasks (e.g., ethical or legal judgements, creative problem-solving, and strategic planning) is an ongoing priority, further refinement is needed before AI agents can meaningfully contribute as synthetic teammates (or opponents) to enhance team performance or reduce training burdens. By using the Desert Containment testbed as a controlled environment for implementing AI teammates in collaborative tasks, researchers can iteratively evaluate AI models across varying task complexities and operational demands before applying them within real-world training or operational contexts.

ACKNOWLEDGEMENTS

This work was supported by the Australian Department of Defence, Defence Science and Technology Group (partnership grant MyIP8655), Human Performance Research Network (HPRNet, partnership grant ID9024), and funding from the Centre for Advanced Defence Research in Robotics and Autonomous Systems. PN was supported by the Macquarie University Research Fellowship, and MJR was supported in part by the Australian Research Council Future Fellowship (FT180100447). The authors thank John Porte and Meredith Porte for their assistance in developing the game used in this study.

REFERENCES

Afroogh, S., Akbari, A., Malone, E., Kargar, M., & Alambeigi, H. (2024). Trust in AI: Progress, challenges, and future directions. *Humanities and Social Sciences Communications, 11*, 1568. https://doi.org/10.1057/s41599-024-04044-8

Basappa, R., Lancaster, C., Mallick, R., Flathmann, C., & McNeese, N. (2025). Mind the gaps: How AI shortcomings and human concerns may disrupt team cognition in human-AI teams (HATs). *Proceedings of the Human Factors and Ergonomics Society Annual Meeting, 69*, 354–359. https://doi.org/10.1177/10711813251361002

Blackmore, K. L., & Allitt, E. W. H. (2021). Building and sustaining the defense simulation training workforce. *The Journal of Defense Modeling and Simulation, 18*(2), 157–170. https://doi.org/10.1177/1548512919869559

Bruzzone, A. G., & Massei, M. (2017). Simulation-based military training. In S. Mittal, U. Durak, & T. Ören (Eds.), *Guide to simulation-based disciplines: Advancing our computational future* (pp. 315–361). Springer UK. https://doi.org/10.1007/978-3-319-61264-5_14

Chen, N., & Zhang, X. (2025). When misunderstanding meets artificial intelligence: The critical role of trust in human-AI and human-human team communication and performance. *Frontiers in Psychology, 16*, 1637339. https://doi.org/10.3389/fpsyg.2025.1637339

Chung, H., Holder, T., Shah, J. A., & Yang, X. J. (2025). A systematic review and taxonomy of human–agent teaming testbeds. *Human Factors, 68*(2), 197–238. https://doi.org/10.1177/00187208251376898

Endsley, M. R. (1988). Design and evaluation for situation awareness enhancement. *Proceedings of the Human Factors Society Annual Meeting, 32*(2), 97–101. https://doi.org/10.1177/154193128803200221

Endsley, M. R. (1995). Toward a theory of situation awareness in dynamic systems. *Human Factors, 37*, 32–64. https://doi.org/10.1518/001872095779049543

Hill, R. R., & Tolk, A. (2017). A history of military computer simulation. In A. Tolk, J. Fowler, G. Shao, & E. Yücesan (Eds.), *Advances in modeling and simulation* (pp. 277–299). Springer, Cham. https://doi.org/10.1007/978-3-319-64182-9_13

Hodson, D. D., & Hill, R. R. (2014). The art and science of live, virtual, and constructive simulation for test and analysis. *The Journal of Defense Modeling and Simulation, 11*(2), 77–89. https://doi.org/10.1177/1548512913506620

Kelley, J. F. (1983). An empirical methodology for writing user-friendly natural language computer applications. *Proceedings of the SIGCHI Conference on Human Factors in Computing Systems* (pp. 193–196). https://doi.org/10.1145/800045.801609

Nalepka, P., Prants, M., Stening, H., Simpson, J., Kallen, R. W., Dras, M., Reichle, E. D., Hosking, S. G., Best, C., & Richardson, M. J. (2022). Assessing team effectiveness by how players structure their search in a first-person multiplayer video game. *Cognitive Science, 46*(10), e13204. https://doi.org/10.1111/cogs.13204

Riek, L. D. (2012). Wizard of Oz studies in HRI: A systematic review and new reporting guidelines. *Journal of Human-Robot Interaction, 1*, 119–136. https://doi.org/10.5898/JHRI.1.1.Riek

Shin, D. (2021). The effects of explainability and causability on perception, trust, and acceptance: Implications for explainable AI. *International Journal of Human-Computer Studies, 146*, 102551. https://doi.org/10.1016/j.ijhcs.2020.102551

Simpson, J., Nalepka, P., Kallen, R. W., Dras, M., Reichle, E. D., Hosking, S. G., Best, C., Richards, D., & Richardson, M. J. (2022a). Conversation dynamics in a multiplayer video game with knowledge asymmetry. *Frontiers in Psychology, 13*, 1039431. https://doi.org/10.3389/fpsyg.2022.1039431

Simpson, J., Patil, G., Stening, H., Kamruddin, A. B., Somerville, D., Seage, S., Nalepka, P., Dras, M., Hosking, S. G., Kallen, R. W., Richardson, M. J., & Richards, D. (2026). Can an AI agent lead human teams? *Computers in Human Behavior: Artificial Humans*, *7*, 100278. https://doi.org/10.1016/j.chbah.2026.100278

Simpson, J., Stening, H., Nalepka, P., Dras, M., Kallen, R. W., Hosking, S. G., Richardson, M. J., & Richards, D. (2025a). Towards teams being led by a conversational agent. *Proceedings of the 7th ACM Conference on Conversational User Interfaces*, Article 43. https://doi.org/10.1145/3719160.3736614

Simpson, J., Stening, H., Nalepka, P., Dras, M., Reichle, E. D., Hosking, S., Best, C. J., Richards, D., & Richardson, M. J. (2022b). DesertWoZ: A Wizard of Oz environment to support the design of collaborative conversational agents. *Companion Publication of the 2022 Conference on Computer Supported Cooperative Work and Social Computing* (pp. 188–192). https://doi.org/10.1145/3500868.3559711

Simpson, J., Stening, H., Patil, G., Nalepka, P., Dras, M., Kallen, R. W., Hosking, S. G., Richardson, M. J., & Richards, D. (2025b). When words fall short: The case for conversational interfaces that don't listen. *Proceedings of the 27th International Conference on Multimodal Interaction* (pp. 550–560). https://doi.org/10.1145/3716553.3750799

Van Orden, G. C., Kloos, H., & Wallot, S. (2011). Living in the pink: Intentionality, wellbeing, and complexity. In C. Hooker (Ed.), *Philosophy of complex systems* (Vol. 10, pp. 629–672). Elsevier, North-Holland. https://doi.org/10.1016/B978-0-444-52076-0.50022-5

Viswanathan, G. M., Buldyrev, S. V., Havlin, S., da Luz, M. G. E., Raposo, E. P., & Stanley, H. E. (1999). Optimizing the success of random searches. *Nature*, *401*(6756), 911–914. https://doi.org/10.1038/44831

Section 2

Metrics

7 Multi-Agent Team Trust Emergence Research (MATTER) Testbed Development and Experimentation

Kendall Carmody, Meredith Carroll, Jessica Wildman, Amanda Thayer, Quintin Oliver, Kyle Canady, and Momo Akib

7.1 INTRODUCTION

As autonomous agents become increasingly complex, acting as synthetic teammates and becoming embedded in high-stakes operational environments, there is a growing need for reliable, ecologically valid testbeds for human-agent teaming (HAT) research. Further, as HATs expand in agent capabilities, responsibilities, size, and heterogeneity, new research paradigms are needed to understand the nuances of trust dynamics in these contexts (Wildman et al., 2024). Valid testbeds that implement carefully designed task paradigms allow researchers to simulate realistic team dynamics and assess emergent constructs, such as trust, which evolve over time and in response to teammate behavior, decision outcomes, and contextual cues (O'Neill et al., 2022; McNeese et al., 2018). The growing ubiquity of autonomous teammates underscores the need to evaluate team constructs within adaptive, dynamic environments that reflect real-world challenges (Chen et al., 2018). Many early HAT studies have relied on dyadic teams performing simple tasks with low levels of interdependence, limiting the observation of team-level interactions and phenomena. Such designs may decrease the ability of human teammates to view their agent counterparts as effective team members, rather than tools (Rebensky et al., 2022). In contrast, paradigms and testbeds that incorporate autonomy, or perceived autonomy, as well as complex tasks, heterogeneity, interdependence, and temporal considerations enable researchers to observe how trust, among other constructs, forms, degrades, and is restored within complex virtual environments. Ultimately, to prepare for the HATs of the future, research testbeds must be designed to be realistic and ecologically valid while remaining accessible to HAT researchers.

DOI: 10.1201/9781003655923-9

7.1.1 Purpose and Structure

The purpose of this chapter is to outline key considerations for designing effective HAT research testbeds and describe a testbed that has incorporated these considerations and been used successfully in multiple HAT research studies. We propose that HAT testbeds should encompass four key dimensions: (1) agents should be able to enact varying levels of autonomy or perceived autonomy, particularly at high levels across complex tasks; (2) teams should be heterogeneous, with team members holding distinct roles that allow for interdependence; (3) missions should have adequate temporal dimensions to allow trust to establish, be violated, and be repaired, and to observe downstream effects with measures that are capable of capturing temporal changes; and (4) a multimodal measurement approach should be employed to assess complex constructs within the team. Building upon prior work (Nguyen et al., 2022; Carmody et al., 2024), we argue that most existing testbeds fail to embody true teaming, lacking adequate autonomy, interdependence, and role differentiation, as well as insufficient temporal dimensions and multimodal measurement approaches.

This chapter will describe why each of these dimensions is important and how they were integrated into a HAT task paradigm. The chapter then describes how this task paradigm was implemented within the MATTER (Multi-Agent Team Trust Emergence Research Testbed) simulation-based environment to facilitate the study of complex HAT dynamics and emergent states, such as trust. In addition to testbed and task design, we introduce a trust measurement framework that emphasizes the importance of aligning measurement strategies with testbed design. Rather than relying on a single measure, this framework integrates multi-modal measurement, combining single-item ratings, behavioral indicators, and performance metrics that together act as observable symptoms of latent constructs like trust. Throughout, this chapter will review the experimental design and findings from two HAT studies conducted using the MATTER testbed, highlighting both theoretical implications and design insights for future HAT testbed development and research.

7.2 BACKGROUND

Level of autonomy (LOA) is a defining feature within HATs that shapes how agents contribute to team goals and how humans perceive, coordinate with, and ultimately trust agents, particularly in the context of complex, dynamic tasks. LOA describes the extent to which control and decision-making authority are distributed between humans and autonomous agents, ranging from low autonomy (i.e., direct manual control) to high autonomy (i.e., independent decision-making with little oversight; Sheridan & Verplank, 1978; Parasuraman et al., 2000). In HAT research, autonomy is not only a design parameter but also a determinant of how interdependence, trust, and coordination unfold in task environments characterized by uncertainty and high workload (O'Neill et al., 2022; Lyons et al., 2021). Selecting an adequate LOA for a specified task paradigm is critical, as it ensures that agents can act with sufficient independence to manage mission complexity and contribute meaningfully to shared goals, while still maintaining compatibility with human oversight and shared mental models. Recent work emphasizes the need to align LOA with task and team demands to support interdependence, appropriate trust calibration, and the avoidance of over- or under-reliance (Chen et al., 2018; O'Neill et al., 2022).

Current research on HAT reveals important gaps in our understanding of role interdependence and team dynamics, particularly in contexts involving heterogeneous agents with distinct functions. We define interdependence as the degree to which team members must rely on one another's unique capabilities to achieve shared goals. Many HAT studies operationalize autonomy as a tool rather than a teammate, where humans direct or override agents rather than engage in collaboration. For example, Lee and See (2004) note that "automation is often conceived as an aid to be directed or supervised" (p. 52), which frames autonomy as a tool rather than as a teammate engaged in reciprocal interdependence. Research has only begun to examine how heterogeneity in agent design influences trust, coordination, and performance (Demir et al., 2017; Wildman et al., 2024), despite evidence from human teams that functional diversity shapes both subgrouping and emergent team states (Lau & Murnighan, 1998). Finally, role structures within HATs are often underdeveloped; without clearly differentiated roles, agents risk being perceived as redundant tools rather than complementary teammates (O'Neill et al., 2022). Addressing these gaps requires investigating how distinct, heterogeneous HAT roles foster meaningful interdependence, thereby aligning HAT research more closely with the realities of modern sociotechnical teams.

Another key gap in the HAT literature lies in the reliance on one-time or static assessments of trust. Trust is most often measured using self-report surveys, such as standard trust questionnaires and Likert-type self-report scales, compressing trust into a single value that does not adequately capture its contextual, time-varying nature (O'Neill et al., 2022; Kohn et al., 2021). As a result, extant HAT research lacks sufficient insight into how trust trajectories unfold over time in response to salient events, such as trust violation and repair events.

Finally, recent work has begun to emphasize the value of multimodal approaches that integrate subjective, behavioral, and physiological indicators to provide a richer, more diagnostic picture of trust (Carmody et al, 2024; Thayer, 2015). Such approaches are better suited to capture the complexity of HAT, where trust is an unobservable attitude that influences human behavioral reliance on agent teammates.

7.2.1 Designing for HATs

The task paradigm implemented within research testbeds directly influences the relevance, validity, and applicability of study findings. A well-designed task paradigm is effective in capturing the dynamic and adaptive nature of real-world environments, allowing researchers to investigate nuanced behaviors during complex scenarios. As autonomous agents gain more intelligent capabilities and take on further responsibilities from their human counterparts, this technological shift from tool to teammate has led to a considerable increase in the literature surrounding complex HATs (Nguyen et al., 2022; Farah & Dorneich, 2024). This has been accompanied by an increase in the complexity of HAT task designs; however, many designs still lack the necessary elements to effectively examine trust dynamics in heterogeneous, multi-agent teams. For example, McNeese et al. (2021) examined trust in a simulated remotely piloted aircraft system environment using a Wizard-of-Oz agent in which a navigator, a pilot, and a photographer captured photos of target waypoints. The agent played the role of the pilot, simulating full autonomy, and the study enabled real-time interaction and trust fluctuation tracking, but team structure was limited to triads, observing only dyadic interactions with a single agent over a mission time of 40 minutes. Similarly, Dekarske et al. (2024) developed a task paradigm in which a human participant engaged in a 2D grid-based visual search task alongside an autonomous searching agent. The agent acted independently to search and report findings, and time-series modeling allowed for trial-by-trial analysis of trust evolution, but the task itself contained a limited team structure (a single human and a single agent), minimal realism (i.e., a web-browser based virtual agent), and a short duration (63 trials that lasted 20 seconds each; approximately 20 minutes total), constraining the study to examine dyadic trust and short-term trust trajectories.

We propose that, to effectively study present and emerging HATs, task paradigms should be designed to incorporate (1) agents with appropriate levels of autonomy or perceived autonomy, task complexity, and risk; (2) interdependent and heterogeneous multi-member human-agent teams; (3) task durations and measurements that capture dynamic team states over time; and (4) multimodal measurement approaches to unobtrusively assess trust.

7.2.2 Program of Research

A four-year project was conducted to examine trust within heterogeneous, multi-agent HATs, with a particular emphasis on the dynamic and multi-level nature of trust processes. The program of research sought to characterize how trust develops, dissolves, and recovers over time in settings that incorporate the four key factors of HAT testbed design. A central objective was to determine the conditions under which trust violation and repair events influence trust in the violator and other agents within the team. To achieve these aims, we conducted two empirical studies examining trust violation and repair dynamics in heterogeneous, multi-agent teams.

We collaborated with the Air Force Lab's Gaming Research Integration for Learning Laboratory (GRILL) to develop the MATTER Testbed, which supports the perception of autonomous agent behavior, human-agent interdependence and heterogeneous team structures, trust-related events, and frequent, multimodal trust assessment over time (see Figure 7.1a,b). The testbed simulation consisted of a one-hour search-and-rescue mission set in a warzone environment, wherein participants worked within a heterogeneous, multi-agent HAT. The participant was introduced to each team member through an "icebreaker" video, which provided information and background related to the agents' capabilities and previous missions, aiding in creating the perception that the agent was a team member, not just a tool. Each team consisted of one human participant collaborating with two sub-teams, each comprising a drone and a ground robot, all working collectively toward the mission goals of identifying civilian casualties, diffusing IEDs, and reporting enemies. Each team member had a specialized role: the human diffused IEDs, the drones searched areas inaccessible to humans, such as rooftops, and the ground robots cleared rubble that restricted access to certain buildings. Team members operated independently within their assigned sectors but sought assistance when specialized capabilities were

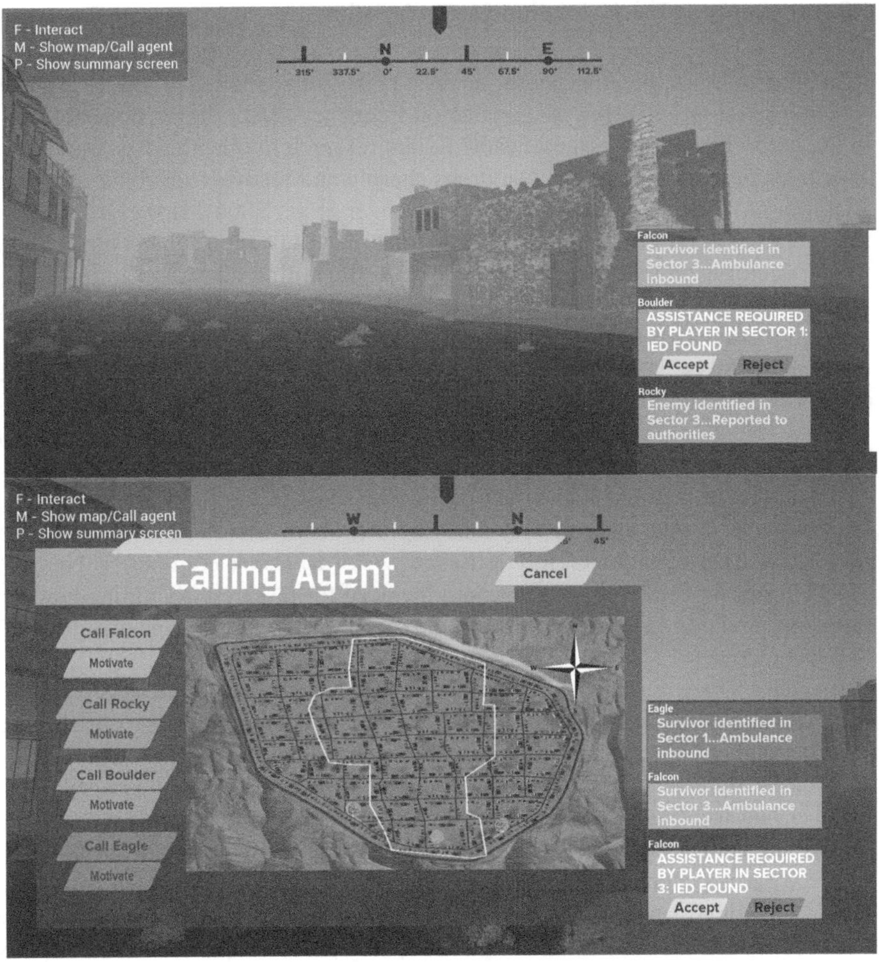

FIGURE 7.1 (a) The MATTER testbed: first-person view with communication log. (b) The MATTER testbed: first-person view with the map.

needed. Participants could (1) monitor the agents' performance, (2) nudge the agents to search more quickly or be more careful, and (3) call on the agents for assistance along their route. Trust violation and repair events occurred throughout the mission, depending on the independent variables being manipulated in each study. Trust was assessed every 10 minutes using brief self-report measures supplemented with continuous, unobtrusive behavioral indicators and physiological assessment. Following the end of the mission, participants were also asked qualitative questions regarding the impact of the trust violation and repair event(s) that occurred during the mission. This multimodal measurement approach enabled the examination of HAT trust dynamics over the course of two experimental studies. At present, the testbed is not open source, though its architecture allows for potential release, subject to administrative and procedural considerations, upon request.

7.2.2.1 Study 1: Trust Violation Characteristics

The first empirical study focused on how characteristics of trust violations shaped trust dynamics in heterogeneous teams. This study addressed three primary research questions. (1) How does violation frequency affect the participant's trust in the individual agents and the team? Extant research in this area has found that, after a single violation, trust often drops sharply and recovers slowly, if at all (de Visser et al., 2018;Desai et al., 2013; Schweitzer et al., 2006). However, much of this work relies on brief, simplified tasks that may not capture the complexity of trust trajectories in real-life missions. Evidence regarding repeat violations is also limited, with some literature indicating that multiple errors can more severely damage trust (Esterwood & Robert, 2022, 2023). (2) How does violation timing (early vs. mid vs. late) affect participant trust in the individual agents and the team? Research findings related to violation timing are mixed, with some indicating larger trust declines for late violations (Wang et al., 2024b). However, many of these studies are constrained by short-duration tasks. (3) How does agent similarity (type, sub-team membership) affect participant trust in the individual agent and the team? Agent type and role may meaningfully shape how humans perceive and respond to trust violations (Hoff & Bashir, 2015; Hancock et al., 2011; Nguyen, 2020). Together, these gaps underscored the need for research examining how the frequency and timing of violations, and the role of the violator, shape trust dissolution and recovery.

To address these research questions, we implemented five experimental conditions that manipulated violation number, timing, and the identity of the agent committing a second violation. The violation events were operationalized through a screen pop-up, simultaneously displayed in the communication log, indicating that an agent team member had triggered an IED, resulting in the deaths of either civilians or allied operational team members. These events were selected based on prior survey work in which participants ranked the perceived severity of a range of potential violations; results consistently indicated that civilian casualties and fratricide were judged to be the most impactful violations. A total of 185 participants completed a simulated ISR mission under these conditions.

7.2.2.2 Study 2: Trust Repair Source

The second empirical study examined the role of trust repair source in shaping trust restoration following violations. The study addressed two primary research questions: (1) How does the number of agents engaging in a repair attempt affect participants' trust in the individual agents and the team? (2) How do the characteristics of the repairing agent (type, subgroup) affect participants' trust in the individual agents and the team? Competing theoretical perspectives informed the research questions. A system-wide trust perspective suggests that the identity of the agent delivering the repair should not matter (Lee & See, 2004; Hoff & Bashir, 2015), whereas social categorization theories imply that repairs from agents similar to the violator may be perceived as less credible or effective (Brewer, 1991; Tajfel & Turner, 1986).

In this study, all participants experienced two mid-mission violations by the same agent. Approximately 10 minutes later, a trust repair attempt was introduced, with the repairing agent manipulated across five conditions. The repair attempts were operationalized through a flashing screen pop-up, simultaneously displayed in the communication log in a similar fashion to the violation. The pop-up screen indicated that an agent team member (or members) had run diagnostics on the violating agent, found that the agent was missing a driver, and updated the driver to ensure it would not happen again (see Figure 7.2). We manipulated who delivered the trust repair message: (1) the violating agent, (2) an agent of the same type on a different sub-team from the violating agent, (3) an agent of a different type and different sub-team from the violating agent, or (4) the full team. There was also a no-repair condition. All other mission parameters were the same as those in the previous empirical study. A total of 168 participants completed the study.

The following sections will describe the development of the MATTER Testbed in alignment with the perceived autonomy, interdependence, high temporal resolution, and multimodal measurement approaches, as well as how these design parameters facilitated answering the research questions presented above.

7.3 FOUR KEY FACTORS FOR HAT TESTBED DESIGN

7.3.1 Agent Autonomy and Task Complexity

Autonomous agents involve a degree of agency, or self-directed behavior, that allows them to take on a unique role or responsibility to complete a set of tasks and work interdependently with human team members (Chen & Barnes, 2012; O'neill et al., 2022). However, the LOA can, and does, vary (Sheridan & Verplank, 1978; Rebensky et al., 2022) from low levels of autonomy, where humans make all the decisions and the agent merely executes commands (e.g., manual control of unmanned vehicles; Sheridan & Verplank, 1978), to higher levels of autonomy, where agents make decisions independently and the human merely consents or is informed after the fact (e.g., automated decision-authority in multi-UAV missions; Rebensky et al., 2022). It is important to ensure HAT testbeds incorporate high enough levels of autonomy, or at least the perception of autonomy, such that humans will trust agents enough to rely on them. The task must also be sufficiently complex to create the need for participants to rely on the agent(s) and, if studying trust, incorporate a degree of risk or vulnerability associated with relying on the agent(s). Ideally, performers should feel that dependence on the agent(s) carries meaningful consequences rather than being a low-stakes option. To achieve this balance, our task paradigm incorporated a series of decisions in which humans had to decide whether and which agents to request assistance from to complete certain tasks, where agents were intentionally fallible. This included selecting which agents to employ for assistance in searching and clearing houses along a route and identifying targets. This created a cost-benefit dynamic: utilizing a fallible agent may be riskier but would increase the efficiency of completing mission goals.

When developing the MATTER testbed, we did not have the time or resources to develop autonomous agents and therefore opted to develop the testbed with agent teammates that performed based on time-based and logic-based rules, giving human teammates the perception of high levels of autonomy. The agents followed a series of pre-scripted, time-based decision events that simulated self-directed task execution and context-aware responses, which were provided via a communication log to the participant. The agents were programmed to each identify a specific number of IEDs, civilians, and enemies along their route, and each interaction was communicated to the user via the communication log. As previously mentioned, they could also request assistance from the participant and respond to requests from the participant for assistance. Further, communications from the agents were designed to give the impression of autonomous reasoning. For example, if the participant called a ground vehicle unnecessarily, the vehicle would inform the user that it was unable to complete the task because its specific capabilities were not required. This allowed the research team to more effectively investigate team coordination, communication, and trust dynamics, as the human participant believed the agent was participating in decision-making and contributing to overall mission effectiveness.

Because the MATTER testbed was designed to support the perception of agent autonomy within a complex task paradigm, participants experienced the synthetic agents not merely as tools but as genuine teammates whose actions carried social and relational meaning. This was evident within the qualitative data we collected during the empirical studies. When asked to describe their reactions to trust violations, many participants used language typically reserved for human-human interactions, attributing intention and social weight to the agents' behaviors. For example, after experiencing a trust violation in which one of the agents misidentified an IED as rubble, one participant stated, "I felt slightly betrayed by the teammate," illustrating that violations were interpreted not as system malfunctions but as breaches of team expectations (Robbins-Roth et al., 2025; see Figure 7.2). Such reactions underscore the need to ensure that testbeds elicit the perception of autonomy, which allows the study of more naturalistic HAT dynamics that emerge when this perception is present.

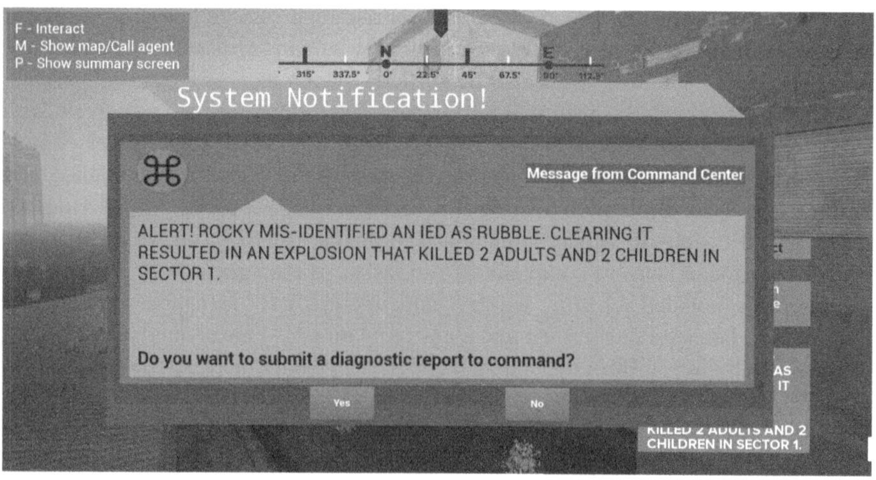

FIGURE 7.2 Trust violation in the MATTER testbed.

7.3.2 Heterogeneity and Interdependence

Team heterogeneity refers to the diversity among the team members, including their characteristics, capabilities, and roles (Cannon-Bowers & Bowers, 2011; O'Neill et al., 2022). In HATs, team heterogeneity can mean differences between team members being human or agent, or differences between functional roles (e.g., navigator, planner, operator, Chen & Barnes, 2014), decision-making authority (e.g., assisted, delegated, full autonomy;), or modes of interaction (e.g., voice interactions, mouse- and keyboard-based interactions). This heterogeneity allows for specialization, role diversity, and the study of team dynamics across non-uniform teams (Cannon-Bowers & Bowers, 2011), further enabling humans to distinguish between "ingroup" and "outgroup" teammates (Tajfel & Turner 1986; Brewer, 1991). In heterogeneous teams, humans may compartmentalize trust more easily when agents are seen as categorically distinct or, conversely, may generalize their trusting behaviors to all agents. Theories such as system-wide trust suggest a holistic approach to trust across all teammates, while social categorization theories emphasize that ingroup and outgroup distinctions may drive trust dynamics. These competing perspectives underscore the need to examine how subgroups form and influence trust within a heterogeneous multi-agent team. For instance, a HAT with three or more members that vary in shared and unique characteristics creates a network of dyadic relationships that allows constructs such as trust to be studied between subgroups and toward outgroup and ingroup members (Nguyen et al., 2022).

Teammate interdependence has been previously defined as the degree to which the performance of one team member depends on another team member's actions (Thompson, 1967). Although the degree to which interdependence is needed may vary based on the context, within HATs, it is important for promoting trust and collaboration (O'Neill et al., 2022). Further, agents may begin to move into roles of authority in which humans not only rely on agents for information and assistance, but agents may also rely on humans for information and aid in completing a task. Task interdependence creates conditions necessary for emergent states, such as trust, to be tested and evolve (Walliser et al., 2017). Failing to implement interdependence can lead to misinterpretation of interactions and false conclusions that participants are appropriately calibrating their trust (Verhagen et al., 2024). Prior research related to automation and HATs typically involves little to no task interdependence (Biswas et al., 2025). Requiring humans and agents to depend on each other increases ecological validity by providing a task context paralleling the real interdependent scenarios that organizations seek from HATs. Additionally, without some degree of required interdependence, there may be individual differences in participants' use of the agent. This could convolute behavioral measures of trust by potentially reducing the number of data points available to assess how much and how often humans work with an agent.

The MATTER team structure (1 human, 2 sub-teams, each with a drone and a ground robot) was intentionally created to reflect heterogeneity across multiple dimensions, with each team member role having specific capabilities distinct from one another while working together to achieve common goals. Successful mission completion was dependent on the contributions of both the human and the agents, with certain capabilities being exclusive to each party. For example, only the drone could access elevated areas to identify targets, only the ground robot could clear rubble from blocked pathways, and only the human participant could diffuse IEDs. This created a need for participants and agents to call on one another for assistance throughout the missions. Agents would call on the participant for assistance along their routes at predetermined times throughout the hour-long simulation and would respond to participant requests for assistance. These distinct roles ensured that success hinged on agents and humans relying on one another to complete the mission goal. Despite shared mission goals, no single team member could complete the mission independently, reinforcing the interdependence envisioned in future HATs.

The MATTER testbed's agent interdependence resulted in human teammates utilizing and developing trust in various agents as they performed well, and losing trust in agents when they erred and violated that trust, which was also accompanied by a decrease in their utilization of those agents for assistance (Thayer et al., in press; Carmody et al., 2024). The testbed's heterogeneity also facilitated the examination of multi-level phenomena, such as whether trust loss spilled over to all agents or merely agents that were functionally similar. Study 1 results revealed that there was only one situation in which trust loss spilled over to innocent agents: when there were two back-to-back violations committed by two different agents, and the two violating agents shared a characteristic in common with an innocent agent (same type or same sub-team), then trust loss also occurred for the innocent agent who shared a characteristic with both violators (Thayer et al., in press). Such spillover effects could not be studied in a system lacking differentiated roles. Study 2 further demonstrated the influence of team structure on trust, revealing that the effectiveness of repair efforts depended on both the number and identity of agents involved. Full-team repair produced the strongest trust restoration after two back-to-back violations, whereas repair attempts delivered by the violating agent were least effective (Canady et al., in press). These findings illustrate how agent heterogeneity and interdependence facilitate the study of trust evolution and propagation across multi-agent HATs.

7.3.3 Temporal Dimensions

To assess HAT individual- and team-level constructs that emerge and change over time, the task that participants perform must be of adequate length to allow emergent states to evolve. Emergent states, such as trust, cohesion, or team-shared mental models, are dynamic and therefore unfold and change as team members interact over time. Operationally, teams may be together for weeks, months, or years, which is often not feasible to study in an experimental paradigm (Hoff & Bashir, 2015). The amount of time over which teammates interact and the frequency of these interactions impact how trust evolves (McNeese et al., 2018). Therefore, an effort should be made to ensure that missions are long enough to allow for trust development and oscillation in response to events such as trust violations and repair. Empirical research also highlights that the frequency of positive and negative interactions significantly influences how users build and sustain trust, even after experiencing major errors (Sharp et al., 2023). Further, as emergent states such as trust fluctuate dynamically, measurement approaches need to have adequate temporal resolution to capture these fluctuations. Static or low-resolution measures, such as one-time surveys, risk oversimplifying these dynamics by capturing only snapshots in time. For example, a static snapshot of trust captured at the end of a mission that involved a trust violation by an autonomous agent will not be able to capture the initial decline, nor any immediate partial or full recovery of trust in the agent. Higher-temporal-resolution approaches are essential to detect subtle fluctuations throughout the entirety of the mission.

The MATTER testbed was designed to offer flexibility in both mission duration and interaction structure, allowing the researcher to tailor the scenario length and event frequency to their experimental goals. Presently, the mission is configured to span one hour, during which a predetermined number of interactions, such as threat detections and team member inputs, occur. Although many real-world reconnaissance and ISR missions may last longer, previous literature has shown that team development phases can unfold through structured task progressions, often within an hour-long session when utilizing simulation environments (Wang et al., 2025). Thus, an hour-long mission with embedded, event-triggering stimuli provides an adequate temporal length to observe the emergence, fluctuation, and stabilization of key HAT constructs, such as trust. Trust violation and repair events were strategically integrated at various points throughout the mission to allow us to investigate how trust develops during normal operations, how it declines in the aftermath of a violation event, and how it may be restored both naturally over time and through targeted repair interventions. Importantly, the timing of both violation and repair events is reconfigurable within the testbed, enabling systematic exploration of how the sequencing and spacing of such events influence trust trajectories across the mission and allowing for versatility in studying the timing of these events. The testbed also allowed for the digital administration of single-item trust measures throughout the mission at frequencies defined by the researcher. We configured the testbed to feature a one-hour mission with trust measures collected at 10-minute intervals, allowing us to capture the dynamic nature of trust as it unfolded across meaningful phases of the task. This temporal resolution enabled us to construct detailed trust trajectories, making visible the fluctuations that would have been obscured in studies relying solely on pre- and post-mission assessments. In study 1, this approach allowed us to detect the distinct effects of violation frequency and timing, revealing that two back-to-back violations led to a steeper decline than a single violation, and that late-mission violations produced steeper trust declines and less complete recovery than earlier ones because trust reached higher levels toward the end of the mission and therefore had "more to lose" (Carroll et al., in press). Similarly, in the second empirical study, the data showed that all participants exhibited natural, partial trust restoration following the violation, before any repair attempt. Data at this temporal resolution supported the use of discontinuous growth modeling (DGM), an analytical approach well suited to evaluating nuanced changes in both the level and slope of trust across distinct mission phases. By capturing multiple inflection points (development, dissolution, natural restoration, and repaired restoration), DGM allowed us to isolate the precise influence of violations and repairs (Thayer, et al., in press). Such insights were only possible because of the temporal resolution the testbed afforded in measuring trust over time.

7.3.4 Multi-Modal Measurement Approaches

Because of the nature and frequency of trust changes within such a task paradigm, collecting continuous or frequent data points, such as real-time trust ratings, physiological signals (e.g., heart rate variability), and behavioral indicators, can provide researchers with clarity on how dynamic constructs, such as trust, evolve throughout a mission. This multimodal measurement approach is particularly important in HATs, where events like trust violations, repairs, or performance breakdowns unfold rapidly, and downstream effects may only be detectable through a confluence of multiple measures to observe patterns.

Trust in HATs must be measured in ways that are both accurate and sensitive to dynamic change, as these insights are essential for understanding when and why operators choose to rely on, or disengage from, autonomous teammates. Traditional trust measurement approaches often either rely on lengthy post-task surveys or single-item Likert-style questionnaires. These methods can be intrusive, time-consuming, and poorly suited for simulation-based environments where trust can fluctuate rapidly based on various events (Burke et al., 2007). Further, these measures often fail to capture nuanced combinations of attitudes and behaviors that reflect various trust-reliance situations, risking the misinterpretation of these behaviors. For example, reliance can be equated with trust when reliance may instead reflect a lack of alternatives, compliance with instructions, or indifference (Lee & See, 2004). Such assumptions can obscure whether actions represent trust, uncertainty, or mere compliance (Lee & See, 2004; Hancock et al., 2011). Implementing unobtrusive measures offers many benefits beyond self-report measures, including being objective, predictive, dynamic, real-time, and resulting in less interference with taskwork and teamwork. For instance, physiological signals such as heart rate variability, galvanic skin response (GSR), or eye tracking can provide objective indicators of arousal and attentional shifts associated with trust changes (Wang et al., 2024a). Further, behavioral measures of trust, such as allocating tasks to autonomous agents and manually controlling agents, have been shown to correlate with self-reported trust (Schaefer et al., 2016; Krausman et al., 2022; Carmody et al., 2024). Each of these measures may represent a component of a larger puzzle that, when considered in isolation, may lead to misinterpretation; however, when combined, they have the potential to reveal a more coherent pattern and support a more accurate conclusion. For example, observing a spike in galvanic skin response (a physiological indicator of increased arousal), slower response times or reduced reliance on an agent (behavioral indicators), and a self-reported drop in perceived reliability or trust on a post-trial questionnaire (a self-report indicator) may collectively offer stronger evidence of a trust decline than any single measure alone (Ficke et al 2023; Kohn et al., 2021). We believe the key pieces of this puzzle include observed behaviors and self-reported attitudes in the context of scenario-specific events.

The MATTER testbed was designed to include a multimodal trust measurement framework embedded within it. Drawing on the literature and guided by the Rational Approach to Developing Systems-based Measures (Orvis et al., 2013) process, we embedded unobtrusive measures that capture both subjective attitudinal and objective behavioral indicators of trust. These behaviors included monitoring (e.g., the frequency with which a participant accessed a single agent's statistics page, which included progress toward mission goals), usage (e.g., the frequency with which a participant employed each agent to aid them along their individual route), and intervention behaviors (e.g., the frequency and timing with which a participant nudged the agents to move more quickly or be more careful). Additionally, physiological data, including heart rate and GSR, were recorded utilizing the Shimmer physiological sensor suite during the mission. Further, a single-item self-report trust measure was administered every 10 minutes throughout the mission to assess participants' trust attitudes and track fluctuations over the course of the mission. The single-item trust measure asked participants: "How much do you trust each of the following [agents] to do what is expected of them during the mission?" Ratings were provided on a 5-point scale ranging from "Not at all" to "A great deal." The testbed was designed to include specific pop-up links that would pause the simulation to administer the self-report trust measure. In addition the testbed administered a validated trust/distrust scale (8 items addressing competency-based trust within Wildman et al.'s (2025) Trust/Distrust Scale) three times; (1) before the mission, (2) at mission midpoint, and (3) after the mission. This was in order to further investigate the participants trust attitudes and to evaluate the psychometric validity of the 1-item trust measure. Scores on the trust/distrust scale were significantly correlated with the 1-item trust measure before the mission ($r(185) = 0.48$, $p < 0.001$), at the midpoint ($r(183) = 0.43$, $p < 0.001$), and after the mission ($r(174) = 0.49$, $p < 0.001$). These brief, in-scenario probes minimized disruption while enabling the capture of trust fluctuations in response to agent actions or mission events. Together, these measurement approaches, in the context of trust violation and repair events, provided a constellation of attitudinal, physiological, and behavioral trust indicators, allowing for an unobtrusive, diagnostic, ecologically valid assessment of trust dynamics (Ficke et al., 2023).

Because the MATTER testbed incorporated a multimodal measurement approach that combined self-report, behavioral logs, and continuous physiological monitoring, we were able to examine how different indicators aligned with trust trajectories over time, something that would not have been possible with traditional, static measurement methods. One challenge with this approach is that it is unclear how behaviors should be temporally associated with self-reported trust, as they operate on different time scales (e.g., a single point in time vs. over the course of time). In dynamic HAT missions, behaviors may precede, coincide with, or follow fluctuations in trust. Using data resulting from Study 1, we examined temporal associations between behavioral measures of trust and self-reported trust ratings. We found that, prior to a violation, none of the behavioral indicators predicted changes in trust, suggesting that users were generally stable in their reliance behaviors when agent performance was consistent. However, after a violation occurred, some trust-related behaviors became diagnostic. Specifically, decreases in self-reported trust within the 10 minutes surrounding a violation were accompanied by fewer requests for assistance and more frequent "nudges" directing the agent to be more careful during the same time period (see Carmody et al., 2024). This surge in corrective behavior suggests that participants were actively attempting to manage or compensate for perceived deficiencies in their agent teammates before choosing to re-engage with them. Although we collected physiological data, including heart rate and GSR, we were unable to identify reliable predictive patterns. This limitation may reflect signal-quality constraints, sensor-sensitivity issues, or the possibility that physiological responses to trust fluctuations are more subtle or context dependent than our measurement approaches allowed.

7.4 TESTBED BENEFITS AND LIMITATIONS

The integration of interdependent agents within the MATTER testbed that were perceived to have high levels of autonomy and distinct roles, operating within a complex task, allowed us to examine the influences of system-wide trust and social categorization theories and the influence these have on trust dynamics in a heterogeneous, multi-agent team. The integration of high-frequency measures allowed us to examine differences in trust trajectories and how these trajectories changed in relation to violation and repair events. The use of multimodal measures enabled us to further understand the dynamics of trust and potential reasons why we observed the patterns we did. However, added functionality does introduce potential limitations. Although the design facilitated the perception of autonomy, it may not capture the full complexity of interaction with actual autonomous systems. Participants responded to agents as though they were acting independently, but the underlying behaviors were scripted or constrained, which may limit the ecological validity of the findings in contexts where agents truly adapt, learn, or behave unpredictably. Pathways for introducing higher levels of autonomy are currently being explored within the testbed, and although plugin-based extensions can modify surface-level interactions, changes to agent autonomy would require code-level modifications. Designing adequate levels of autonomy and task complexity can also lead to overly complex environments that may overwhelm participants and obscure mechanisms of interest (Schaefer et al., 2016; O'Neill et al., 2022). Further, incorporating heterogeneity and interdependence creates challenges for experimental control, as subgroup dynamics and coordination breakdowns can introduce variability that is difficult to isolate (Lau & Murnighan, 1998). High-temporal-frequency measurement may introduce participant fatigue, reactivity effects, or noisy data, potentially masking fine-grained trust fluctuations (Hoff & Bashir, 2015; Walliser et al., 2017). Finally, multimodal measurement, such as combining self-report, behavioral, and physiological indicators, complicates synchronization, increases analytic complexity, and raises interpretive concerns when multimodal signals diverge (Endsley, 2015; McNeese et al., 2021). Researchers must consider the tradeoff between ecological validity and practical feasibility when making decisions about these testbed elements for HAT research.

7.5 CONCLUSION

The next generation of HAT research depends on testbeds that move beyond simple dyadic interactions to capture the complexity of modern multi-agent teams. By embedding autonomy and task complexity, heterogeneity and interdependence, and appropriate temporal dimensions and multimodal measurement approaches, HAT testbeds can provide the ecological validity necessary to study trust and other emergent constructs as they evolve in realistic contexts. Such approaches not only enhance the validity of experimental findings but also ensure that insights generalize to the operational settings where humans and autonomous agents will increasingly collaborate. Developing and refining these testbeds is therefore a critical step toward building the scientific foundation for resilient, adaptive, and trustworthy HATs.

REFERENCES

Biswas, U., Palod, V., Bhambri, S., & Kambhampati, S. (2025). *Who is helping whom? Analyzing inter-dependencies to evaluate cooperation in Human-AI teaming* (arXiv:2502.06976). arXiv. https://doi.org/10.48550/arXiv.2502.06976

Burke, C. S., Sims, D. E., Lazzara, E. H., & Salas, E. (2007). Trust in leadership: A multi-level review and integration. *The Leadership Quarterly, 18*(6), 606–632. https://doi.org/10.1016/j.leaqua.2007.09.006

Brewer, M. B. (1991). The social self: On being the same and different at the same time. *Personality and Social Psychology Bulletin, 17*(5), 475–482. https://doi.org/10.1177/0146167291175001

Canady, K, Carmody, K, Akib, M, Thayer, A., Wildman, J., & Carroll, M. (in press). *Simulated trust, real consequences: Repairing trust in complex human agent teams* (Manuscript submitted for publications). College of Aeronautics, College of Psychology & Liberal Art, Florida Institute of Technology.

Cannon-Bowers, J. A., & Bowers, C. (2011). Team development and functioning. In S. Zedeck (Ed.), *APA handbook of industrial and organizational psychology, Vol. 1. Building and developing the organization* (pp. 597–650). American Psychological Association. https://doi.org/10.1037/12169-019

Carmody, K., Sharma, V., Addis, A., Nguyen, D., Thayer, A. L., Wildman, J., & Carroll, M. (2024). Behave yourself! Behavioral indicators of trust in human-agent teams. *Proceedings of the Human Factors and Ergonomics Society Annual Meeting, 68*(1). https://doi.org/10.1177/10711813241276485

Carroll, M. B., Akib, M. S., Addis, A., Carmody, K., Nguyen, D., Ficke, C., Thayer, A. L., & Wildman, J. L. (in press). *The impact of violation frequency and timing on trust in violating agent within a multi-agent team* (Manuscript submitted for publications). College of Aeronautics, College of Psychology & Liberal Art, Florida Institute of Technology.

Chen, J. Y. C., & Barnes, M. J. (2012). Supervisory control of multiple robots: Effects of imperfect automation and individual differences. *Human Factors, 54*(2), 157–174. https://doi.org/10.1177/0018720811435843

Chen, J. Y. C., Lakhmani, S. G., Stowers, K., Selkowitz, A. R., Wright, J. L., & Barnes, M. (2018). Situation awareness-based agent transparency and human-autonomy teaming effectiveness. *Theoretical Issues in Ergonomics Science, 19*(3), 259–282. https://doi.org/10.1080/1463922X.2017.1315750

Dekarske, J., Kong, Z., & Joshi, S. (2024). *Dynamic human trust modeling of autonomous agents with varying capability and strategy* (arXiv:2404.19291). arXiv. https://doi.org/10.48550/arXiv.2404.19291

Demir, M., McNeese, N. J., & Cooke, N. J. (2017). Team situation awareness within the context of human-autonomy teaming. *Cognitive Systems Research, 46*, 3–12. https://doi.org/10.1016/j.cogsys.2016.11.003

Desai, M., Kaniarasu, P., Medvedev, M., Steinfeld, A., & Yanco, H. A. (2013). Impact of robot failures and feedback on real-time trust. *In Proceedings of the 8th ACM/IEEE International Conference on Human-Robot Interaction* (pp. 251–258). IEEE. https://doi.org/10.1109/HRI.2013.6483596

De Visser, E. J., Pak, R., & Shaw, T. H. (2018). From 'automation' to 'autonomy': The importance of trust repair in human–machine interaction. *Ergonomics*, *61*(10), 1409–1427. https://doi.org/10.1080/00140139.2018.1457725

Endsley, M. R. (2015). Final reflections: Situation awareness models and measures. *Journal of Cognitive Engineering and Decision Making*, 9(1), 101–111. https://doi.org/10.1177/1555343415573911

Esterwood, C., & Robert, L. P. (2022). Having the right attitude: How attitude impacts trust repair in Human—Robot interaction. *2022 17th ACM/IEEE International Conference on Human-Robot Interaction (HRI)* (pp. 332–341). https://doi.org/10.1109/HRI53351.2022.9889535

Esterwood, C., & Robert, L. P. (2023). Three strikes and you are out!: The impacts of multiple human–robot trust violations and repairs on robot trustworthiness. *Computers in Human Behavior*, *142*, 107658. https://doi.org/10.1016/j.chb.2023.107658

Farah, Y. A., & Dorneich, M. C. (2024). Human-autonomy teaming in a cooperative gamified testbed: How can AI teammates support teamwork processes? *Proceedings of the Human Factors and Ergonomics Society Annual Meeting*, *68*(1), 386–392. https://doi.org/10.1177/10711813241269252

Ficke, C., Addis, A., Nguyen, D., Carmody, K., Thayer, A., Wildman, J., & Carroll, M. (2023). Measuring trust in a simulated human–agent team task. In *Human factors and simulation* (Vol. 83, pp. 36–45). AHFE International. https://doi.org/10.54941/ahfe1003560

Hancock, P. A., Billings, D. R., Schaefer, K. E., Chen, J. Y. C., de Visser, E. J., & Parasuraman, R. (2011). A meta-analysis of factors affecting trust in Human-Robot interaction. *Human Factors*, *53*(5), 517–527. https://doi.org/10.1177/0018720811417254

Hoff, K. A., & Bashir, M. (2015). Trust in automation: Integrating empirical evidence on factors that influence trust. *Human Factors*, *57*(3), 407–434. https://doi.org/10.1177/0018720814547570

Kohn, S., de Visser, E. J., Wiese, E., Lee, J. D., & Vroom, M. (2021). Measuring trust in human–robot interaction: A review of empirical methods. *ACM Computing Surveys*, *54*(4), Article 79, 1–39.

Krausman, A., Neubauer, C., Forster, D., Lakhmani, S., Baker, A. L., Fitzhugh, S. M., Gremillion, G., Wright, J. L., Metcalfe, J. S., & Schaefer, K. E. (2022). Trust measurement in Human-Autonomy teams: Development of a conceptual toolkit. *ACM Transactions on Human-Robot Interaction (THRI)*, *11*(3), 33:1–58. https://doi.org/10.1145/3530874

Lau, D. C., & Murnighan, J. K. (1998). Demographic diversity and faultlines: The compositional dynamics of organizational groups. *Academy of Management Review*, *23*(2), 325–340. https://doi.org/10.5465/amr.1998.533229

Lee, J. D., & See, K. A. (2004). Trust in automation: Designing for appropriate reliance. *Human Factors*, *46*(1), 50–80. https://doi.org/10.1518/hfes.46.1.50_30392

Lyons, J. B., Sycara, K., Lewis, M., & Capiola, A. (2021). Human–Autonomy teaming: Definitions, debates, and directions. *Frontiers in Psychology*, *12*. https://doi.org/10.3389/fpsyg.2021.589585

McNeese, N. J., Demir, M., Chiou, E. K., & Cooke, N. J. (2021). Trust and team performance in Human–Autonomy teaming. *International Journal of Electronic Commerce*, *25*(1), 51–72. https://doi.org/10.1080/10864415.2021.1846854

McNeese, N. J., Demir, M., Cooke, N. J., & Myers, C. (2018). Teaming with a synthetic teammate: Insights into Human-Autonomy teaming. *Human Factors: The Journal of the Human Factors and Ergonomics Society*, *60*(2), 262–273. https://doi.org/10.1177/0018720817743223

Nguyen, D. (2020). *1, 2, or 3 in a HAT? How a human-agent team's composition affects trust and cooperation* (Theses and Dissertations). https://repository.fit.edu/etd/308

Nguyen, D., Carmody, K., Wildman, J., & Carroll, M. (2022). Developing a dynamic testbed: Designing for trust in complex human agent teams. *2022 IEEE 3rd International Conference on Human-Machine Systems (ICHMS)* (pp. 1–4). https://doi.org/10.1109/ICHMS56717.2022.9980750

O'Neill, T., McNeese, N., Barron, A., & Schelble, B. (2022). Human–Autonomy teaming: A review and analysis of the empirical literature. *Human Factors*, *64*(5), 904–938. https://doi.org/10.1177/0018720820960865

Orvis, K. L., DeCostanza, A. H., & Duchon, A. (2013). Developing systems-based performance measures: A rational approach. *In Proceedings of the Interservice/Industry Training, Simulation and Education Conference (I/ITSEC)*, Orlando, FL.

Parasuraman, R., Sheridan, T. B., & Wickens, C. D. (2000). A model for types and levels of human interaction with automation. *IEEE Transactions on Systems, Man, and Cybernetics – Part A: Systems and Humans*, *30*(3), 286–297. https://doi.org/10.1109/3468.844354

Rebensky, S., Carmody, K., Ficke, C., Carroll, M., & Bennett, W. (2022). Teammates instead of tools: The impacts of level of autonomy on mission performance and Human–Agent teaming dynamics in Multi-Agent distributed teams. *Frontiers in Robotics and AI*, *9*. https://doi.org/10.3389/frobt.2022.782134

Robbins-Roth, V., Carmody, K., DeLoach, B., Caroll, M., Thayer, A., & Wildman, J. (2025). Understanding human-agent teaming: A mixed-methods examination of perceptions and interactions. *Proceedings of the Human Factors and Ergonomics Society Annual Meeting*, *69*(1), 398–402. https://doi.org/10.1177/10711813251370747

Schaefer, K. E., Chen, J. Y. C., Szalma, J. L., & Hancock, P. A. (2016). A meta-analysis of factors influencing the development of trust in automation: Implications for understanding autonomy in future systems. *Human Factors*, *58*(3), 377–400. https://doi.org/10.1177/0018720816634228

Sharp, W. H., Jackson, K. M., & Shaw, T. H. (2023). The frequency of positive and negative interactions influences relationship equity and trust in automation. *Applied Ergonomics*, *108*, 103961. https://doi.org/10.1016/j.apergo.2022.103961

Sheridan, T. B., & Verplank, W. L. (1978). *Human and computer control of undersea teleoperators*. MIT Man-Machine Systems Laboratory. https://doi.org/10.21236/ADA057655

Tajfel, H., & Turner, J. C. (1986). The social identity theory of intergroup behavior. In S. Worchel & W. G. Austin (Eds.), *Psychology of intergroup relations* (2nd ed., pp. 7–24). Nelson-Hall.

Thayer, A. L. (2015). *Trust trajectories as a function of violation type and repair efforts* (Doctoral dissertation, University of Central Florida). STARS: Showcase of Text, Archives, Research & Scholarship. https://stars.library.ucf.edu/etd/1239

Thayer, A. L., Canady, K., Addis, A., Akib, M., Nguyen, D., Carmody, K., Walters, Z., Duron, J., Wildman, J., & Carroll, M. (in press). *How agent type and affiliation shape trust dynamics in multi-agent human-agent teams* (Manuscript submitted for publication). Small Group Research.

Thompson, J. D. (1967). *Organizations in action: Social science bases of administrative theory*. McGraw-Hill.

Verhagen, R. S., Marcu, A., Neerincx, M. A., & Tielman, M. L. (2024). The influence of interdependence on trust calibration in Human-Machine teams. In *HHAI 2024: Hybrid human AI systems for the social good* (pp. 300–314). IOS Press. https://doi.org/10.3233/FAIA240203

Walliser, J. C., Mead, P. R., & Shaw, T. H. (2017). The perception of teamwork with an autonomous agent enhances affect and performance outcomes. *Proceedings of the Human Factors and Ergonomics Society Annual Meeting*, *61*(1), 231–235. https://doi.org/10.1177/1541931213601541

Wang, K., Hou, W., Ma, H., & Hong, L. (2024a). Eye-Tracking characteristics: Unveiling trust calibration states in automated supervisory control tasks. *Sensors*, *24*(24), 7946. https://doi.org/10.3390/s24247946

Wang, K., Lu, J., Ruan, S., & Qi, Y. (2024b). Continuous error timing in automation: The peak-end effect on human-automation trust. *International Journal of Human-Computer Interaction*, *40*(8), 1832–1844. https://doi.org/10.1080/10447318.2023.2223954

Wang, M., Wu, J., Ma, S., Li, N., Zhang, P., Gu, N., & Lu, T. (2025). *Adaptive human-agent teaming: A review of empirical studies from the process dynamics perspective* (arXiv:2504.10918). arXiv. https://doi.org/10.48550/arXiv.2504.10918

Wildman, J. L., Nguyen, D., Thayer, A. L., Robbins-Roth, V. T., Carroll, M., Carmody, K., Ficke, C., Akib, M., & Addis, A. (2024). Trust in human-agent teams: A multilevel perspective and future research agenda. *Organizational Psychology Review*, *14*(3), 373–402. https://doi.org/10.1177/20413866241253278

Wildman, J. L., Thayer, A. L., Warren, C., Fiore, S. M., & Salas, E. (2025). Interpersonal trust and distrust at work: Scale validation and theoretical exploration. *Applied Psychology*, *74*(1), e12588. https://doi.org/10.1111/apps.12588

8 How Well Do We Rely on Reliance? On the Under-Utilization of Reliance-Based Metrics in Studying Human Response to Automation Assistance

Ranjani Narayanan and Karen M. Feigh

8.1 A BRIEF HISTORY OF STUDYING HUMAN DEPENDENCE ON AUTOMATION

Ever since the introduction of automation[1] into human work, there has been a growing interest in understanding human interactions through the lens of dependence on it. Humans have been assigned the role of supervising and exercising appropriate control in order to manage these systems. This includes distributing attention across various information cues, displays, and signals of various forms to ensure that the system as a whole operates efficiently. They are also expected to intervene effectively and take control if they observe off-nominal operations. However, outside the premise of an ideal world, humans are prone to misjudge whether and when their intervention is required. Due to a host of factors, such as operator expertise and display design, operators either trust the system too much or too little. As a result, researchers set out to measure human dependence on automation to study how effective joint work can be accomplished.

One frequently used set of measures is human compliance and reliance on automation. These measurements have been straightforward as they are un-interruptive, allowing investigators to observe human behavior in its natural state. In the context of signal detection theory, compliance is defined as the operator's behavior when responding to automation after it raises a signal or warning. Reliance, on the other hand, is the operator's behavior characterized by refraining from acting when the automation does not raise a signal or warning, an indication of normal operations (Meyer, 2001). Subsequent researchers in unmanned aerial vehicle control began

DOI: 10.1201/9781003655923-10

using compliance and reliance behaviors as indicators of trust (Dixon & Wickens, 2006; Dixon, Wickens & Chang, 2005; Meyer, 2004). Together, these measures indicate human *dependence* on automation.

Inappropriate dependence is characterized by too much (misuse) or too little (disuse) dependence on the system (Parasuraman & Riley, 1997; Boubin et al., 2017). Such behaviors may stem from the operator's flawed understanding of system capabilities, workload, self-confidence, and confidence in automation, as well as other factors such as lack of training and decision bias. While the goal of integrating automation is to assist in the performance of different tasks, these performance gains may only be realized if users accept the assistance, which requires a certain level of trust in it. More importantly, such trust must be calibrated to the automation's true capabilities to prevent inappropriate dependence on it.

Understanding operator trust has been crucial in designing more adoptable automation. But often, studies focusing on trust measure compliance or reliance instead. It is common to see them used interchangeably, often as convenient proxies for ease of measurement. Such practices risk conflating different concepts, each having distinct implications for collaborative outcomes. Therefore, we devote the forthcoming section to delineating the relationships and distinctions between trust, compliance, and reliance. We also introduce "appropriate reliance" and its five major dimensions, along with their significance in determining collaborative efficiency. In doing so, we emphasize the importance of adopting multi-dimensional analytical perspectives to mitigate human biases toward technology. Finally, we review recent literature that has incorporated these reliance-based measures and identify the research gaps that remain.

8.2 TRUST AND ITS PROXY: RELATIONSHIP OF COMPLIANCE AND RELIANCE WITH TRUST

Alongside developments in technology came significant interest from the human factors, computer science, and robotics communities in studying human trust in that technology. In the context of automation, Lee and See defined trust as "the attitude that an agent will help achieve an individual's goals in a situation characterized by uncertainty and vulnerability" (Lee & See, 2004). Trust is a psychological concept, often seen as a mediating factor in the adoption of technology (Bolton, 2022). Despite widespread adoption at the conceptual level, it has encountered several barriers to measurement and modeling. Any measurement of trust puts it in the realm of psychometrics, raising questions about the validity of the measure itself (Bolton, 2022; Annett, 2002; Dekker & Hollnagel, 2004; Dekker & Nyce, 2015). Other questions regarding its convergence with objective measures and whether such measures satisfy the requirements for cardinal levels of measurement also add weight to this ongoing debate (Wei et al., 2019, 2020). Broadly, subjective measures of trust are said to be reliable and consistent and can produce numbers compatible with parametric statistics (Wei et al., 2019, 2020). These involve open-ended questionnaire responses or questionnaires that use rating scales to generate interval-level data. More objective ways of measuring trust involve the use of physiological measures that monitor eye gaze, EEG, etc. (Ajenaghughrure et al., 2020), or observational and behavioral data such as compliance and reliance on the system (Chancey et al., 2017). While there is no prescriptive mode of measurement, Bolton argued that trust is not a "trustworthy measurement" (Bolton, 2022). He raised concerns relating to the highly contextual nature of the construct, pointing to practical difficulties encountered in defining trust itself. Trust also lacks selectivity (Eignor, 2013), defined as sensitivity *exclusive* to the quality being measured, and diagnosticity (Bolton, 2022), defined as the ability of a measure to explain changes in the measured phenomenon, both of which are essential for producing reliable measurements. Further, he posited that trust cannot always predict human behavior in a useful manner, citing several situations where a person may trust the system and choose to act differently (or the reverse) due to external influencing factors such as self-confidence, motivation, risk perception, workload, etc. (See Bolton, 2022 for the full list.) In summary, knowing whether somebody trusts technology does not, by itself, provide inherently useful information.

The question of whether trust is an appropriate measure for understanding human interactions with automation, especially AI, continues to be debated. However, trust and trustworthiness in the context of AI adoption have gained significant momentum in academic and industrial research circles, especially after the EU chartered regulations for designing Trustworthy AI (Jones, 2024). In guiding these developments, a High-Level Expert Group on AI (High-Level Expert Group on AI, 2019) proposed guidelines that subsequently codified the Artificial Intelligence Act, which specifically invokes the term "Trustworthy AI" (Jones, 2024; High-Level Expert Group on AI, 2019). Bradford (2023) points out that companies and other legislative institutions adopt EU guidelines to avoid the complexity and costs of maintaining several compliance regimes. Therefore, the standards for the design of AI, as chartered by the EU, will likely shape AI governance on a global level. Despite the heavy significance of these regulations, descriptions of trust, in these accounts, more closely resemble *reliance,* leading to criticism of regulatory shortsightedness (Laux et al., 2024; Ryan, 2020). This is because trust and trustworthiness are complex and elusive concepts with various philosophical dimensions (McLeod, 2011). Trust is multi-faceted, involving more than just a psychological state. It "demands the specific relational, motivational, and moral conditions that justify the trust" (Jones, 2024). However, its definition, for simplicity, is often operationalized into a binary measure and formulated as reliance, although trust exists on a spectrum, varying in degree and depending on the context and relational dynamics (Ho & Caals, 2024). Other normative differences between trust and reliance (Kerasidou et al., 2022) are also ignored. Trust is predicated on the trustee's goodwill toward the trustor, but reliance is dependence, based on the predictability of the other entity's behavior (Kerasidou et al., 2022). It is based on "expectations, proven capacity, communication and clear and enforceable systems of responsibility and accountability" (Kerasidou et al., 2022; Kerasidou, 2019). Therefore, when trust breaks down, emotional reactions and feelings of betrayal are evoked. However, when reliance is lost, blame is sought externally (Smith, 2005). Further, trust includes a level of reliance, but reliance can exist without trust (Jones, 2024). It is important to make these distinctions, because, if the intention is to study the rightful adoption of technology, researchers may be better off directly studying reliance itself.

Notwithstanding these arguments, most human-automation teaming (HAT) studies utilize simpler and more objective measures of compliance and reliance in addition to, or instead of, human trust, often using the terms interchangeably. Compliance (whether the human performs actions recommended by the automation) and reliance (stopping action performance when the automation indicates it is unnecessary) describe behaviors that can be examined objectively without the inherent ambiguity that comes with trust (Bolton, 2022; Vashitz et al., 2009). Capturing these behaviors in their "natural state" is also relatively easy, as observational measurements are non-intrusive and uninterrupted. These measures can be carried out without diverting subjects' attention toward questions such as "how much do they trust technology," which they may not have considered unless mentioned. It further risks influencing natural behavior that we may otherwise observe. Thus, it is reasonable to conclude that trust in automation may *influence* human engagement with technology but does not *determine* its effective adoption. A more meaningful way to study human interactions with an Automated Assistant (AA) is by studying their dependence on it. This provides objective, quantifiable, reliable, and actionable insights for guiding further developments and the real-life implementation of technology.

8.3 RELIANCE-BASED METRICS FOR HUMAN-AUTOMATION COLLABORATIVE EFFECTIVENESS

Most researchers emphasize the role of *complementarity* in defining HAT effectiveness, especially with AI that has vast information-processing capabilities, enabling pattern-recognition to produce detailed, data-driven insights. It has significantly better throughput and is insusceptible to human cognitive limitations or workload. Humans, on the other hand, learn and adapt with experience and possess intuitiveness, recognition of contextual nuance, and creativity in solving problems, which modern AI tools have tried to mimic but have not yet achieved (Vaccaro et al., 2024). Such distinctiveness provides an opportunity to complement and mutually fill performance gaps, such that the combined outcomes are superior to what each entity could individually achieve.

Research also focuses on human augmentation, wherein the combined effectiveness of HATs is greater than the performance of humans acting alone (Bo et al., 2021; Bondi et al., 2022; Schemmer et al., 2022). This goal is useful for contexts in which AAs cannot be deployed in a completely autonomous sense due to a host of legal, ethical, and safety reasons (Vaccaro et al., 2024). Complete autonomy may also not be applicable in scenarios where it does not inherently align with human values. Regardless of how team effectiveness is defined, it is implicitly desired that the combined performance of human-AA systems would outdo either entity acting alone; otherwise, it may make better sense to deploy only the better-performing of the two. While most researchers chase this ideal, Vaccaro and colleagues conducted a systematic survey with a thorough meta-analysis of existing empirical work across various contexts, and found that on average, human-AI combinations performed significantly worse than the best of humans or AI alone (Vaccaro et al., 2024). They also reported that synergy was observed only for creative tasks and that performance losses were greater when tasks involved decision-making. One of the reasons cited for such losses was a predominant challenge in integration: humans' tendency to over-rely or under-rely on AI systems. When people over-rely, they use the agent's suggestions to guide their own (*anchoring effect*; Tversky & Kahneman (1974)) without seeking or processing additional relevant information. Conversely, adverse attitudes toward AAs create a tendency to under-rely on efficient AA-generated solutions. Few would argue with the notion that human acceptance of, or dependence on, AAs is key to determining AA efficacy. The goal is to define what appropriate dependence looks like and how it may be measured across various contexts.

8.4 WHAT IS APPROPRIATE RELIANCE?

Lee and See defined four major dimensions associated with the *appropriateness* of trust: (1) calibration, (2) resolution, (3) temporal sensitivity, and (4) functional specificity (Lee & See, 2004). As shown in Figure 8.1, the calibration of trust corresponds to a person's trust and a system's true task-performance capabilities (Lee & Moray, 1994; Muir, 1987). As such, reliance on the AA is characterized as *overuse* or *over-reliance* when the rate at which people rely on it exceeds its true capabilities. Conversely, *underuse* or *under-reliance* occurs when the rate at which humans rely on it is less than its true capabilities.

Appropriateness of reliance can be extended using the dimensions provided by Lee and See. *Resolution of reliance* describes how precisely a judgment of reliance differentiates between the various levels of AA capabilities. Poor resolution occurs when a broad range of, or large changes in, AA capabilities are mapped onto a smaller range of human reliance. Similarly, *Functional specificity* can be defined

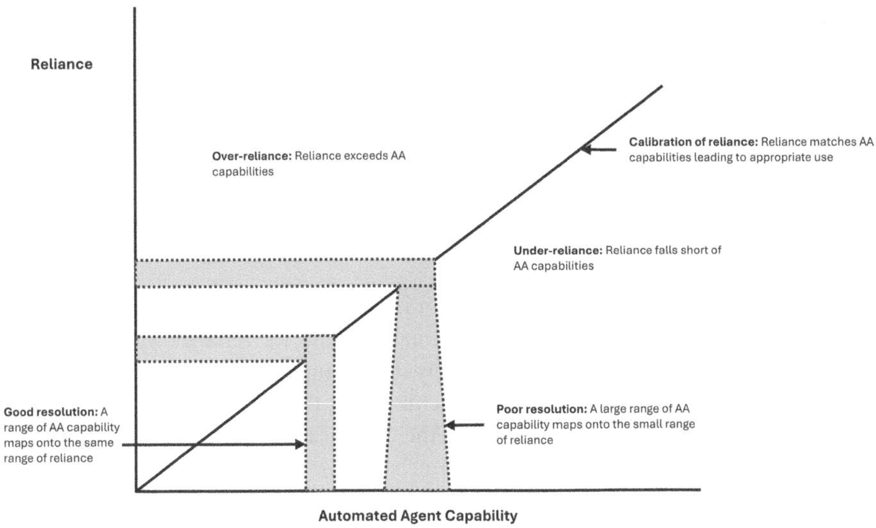

FIGURE 8.1 Relationship between calibration, resolution, and AA capability in defining appropriate human reliance on AA.[2]

as the degree to which reliance is associated with a particular component or aspect of the AA. A person's reliance would be highly specific when it reflects the capabilities of only a few sub-functions and modes of the AA's operation. On the contrary, specificity would be low when reliance reflects its entire range of capabilities. Lastly, *temporal sensitivity* (TS) describes changes in reliance as a function of the situation or time. Reliance that tracks moment-to-moment changes in AA capabilities yields high TS. Low TS reflects only long-term changes in reliance despite moment-to-moment variations in AA capabilities. This measure is functional for studying the effectiveness of adaptive AAs that accrue experience through performing the task, interacting with a human, or both. We will revisit this aspect in detail in Section 8.4.3.

These four facets, where applicable, must drive efforts to determine the appropriateness of reliance on an AA. These definitions, adapted from Lee and See's characterization of appropriate trust, are shown in Figure 8.1. However, although highly pertinent, we recognize that these facets alone are not sufficient to describe the efficacy of HATs. In the following section, we introduce another influential factor in HAT effectiveness, especially for performance-driven tasks.

8.4.1 Accounting for Comparative Performance that Drives Complementarity

While research aims toward human-AA integration, wherein both entities effectively intervene to fill mutual performance gaps, it fails to incorporate relevant performance based metrics to study the desired complementary behaviors. When an AA makes a correct recommendation, the human may be better off relying on its suggestion. Choosing not to rely on the agent's suggestion, in this case, qualifies as inappropriate reliance.

However, when the agent makes an incorrect suggestion, this characterization becomes more nuanced. To capture this, we define *compensation*.[3] Mathematically, it is the algebraic difference between the performance of humans and AAs. Say, an AA and a human expert performed a task with 65% and 80% efficacy, respectively. The human, in this scenario, would have compensated for the AA by a 15% margin. This could further be operationalized at the instance level.

$$\text{Compensation} = \text{Human performance} - \text{AA performance} \tag{8.1}$$

On the one hand, if the human relies on an incorrect suggestion and the subsequent performance improves against their performance alone, the reliance would be appropriate, as the AA aids the human positively. If AA inputs yield poorer performance compared to what the human may have achieved by themselves, then the reliance is inappropriate. On the other hand, when the human acts independently (by choosing not to rely on an incorrect AA recommendation) and outperforms the AA, their reliance would be appropriate. When the AA outperforms the human (negative compensation), reliance may be considered inappropriate. Zero difference implies comparable performance levels between the two entities or that the human chose to rely on the AA's suggestion. This formalization is summarized in Figure 8.2.

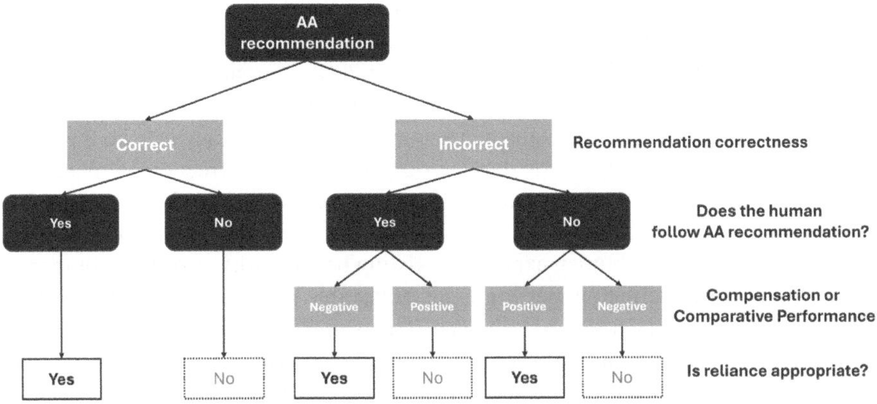

FIGURE 8.2 Flow diagram for characterizing appropriate human reliance on AA.

Compensatory power, or the extent to which compensation occurred, can be expressed using *Degree of Compensation* when task performance limits are bounded (at the very least, with an upper bound). Humans' capacity to compensate may be quantified as follows:

$$\text{Degree of Compensation} = \frac{\text{Compensation}}{\text{Maximum possible compensation}} \tag{8.2}$$

Where the maximum possible compensation is the algebraic difference between the maximum achievable performance and the AA's performance level.

In a recent study, Narayanan and Feigh (2025) illustrated the use of these compensation-based measures to understand human effectiveness in teaming with erroneous decision aids. For a multi-alternative decision paradigm, they illustrated that when an agent's error complexity varies as a function of the number of constraints violated, humans adopt different decision strategies toward fulfilling their task goals. Although low-level error complexity in their study yielded high performance, they noticed that their participants failed to adequately compensate for the performance fallacies of a decision aid. Instead, most users relied on imperfect assistance, as evidenced by a poor degree of compensation (Equation 8.2 used at the instance level) and its deterioration over the course of experimental trials. A combination of performance-based compensatory behaviors and their interaction with trial (temporal) progression provided rich insights into variations in human attitudes and preferences toward imperfect assistance. Although they observed higher average performance in the presence of lower-complexity AA errors, compensatory tendencies suggested that high error complexities led to more proactive human participation. Investigating performance- or reliance-based measures in isolation may not have led to the successful capture of such distinct participatory tendencies and decision strategies.

Another important related factor to account for while studying HAT outcomes is the distinction between reliability and accuracy. Very often, HAT studies are conducted in binary decision environments where the two variables are equivalent (e.g., Srivastava, 2023; Bansal, et al., 2019). At other times, this equivalence is implicitly assumed. For example, if an AA is deemed 70% accurate, it is believed that it makes a correct recommendation 7 out of 10 times. However, more complex work environments exist where this equivalence may not hold. Accuracy may be the utility or extent of correctness of a particular decision or choice. A correct recommendation would yield 100% efficient outcomes, while incorrect recommendations may yield anything less than that. One may alternatively describe AA accuracy by evaluating its performance against human experts or existing benchmarks, especially in scenarios without clearly defined upper limits on achievable performance levels or where ideal performance can never be achieved (for example, scenarios with competing constraints). Reliability, on the other hand, could be described as the extent to which the AA makes these useful recommendations. This distinction is illustrated below:

Consider an AI-based recommendation system assisting a potential homeowner. An 80% reliable system will make correct (or *the best*) suggestions 8 out of 10 times. However, an 80% accurate system may fulfill only 8 out of the 10 constraints that the potential homeowner had set out to satisfy in informing their choice of buying a specific home. While this distinction may be subtle, it has important implications for studying decision support systems. We present two scenarios:

1. The human, a domain expert, is paired with an AI that has 70% reliability and a maximum decision accuracy of 50%. Under these specifications, even if human reliance on the AI is calibrated, they risk significantly compromising their performance, as they would rely on suggestions that are only 50% accurate at best. The expert may be better off relying on their own judgment for the task, as they stand to score higher than the AI's baseline.
2. On the other hand, if the human is not trained to do the task and ranks very low on the performance index, they may be better off relying on the AI's imperfect recommendations (that is are 50% accurate). This is because humans, when left alone, may fare worse than imperfect AI. Even imperfect decision assistance under such circumstances would be particularly crucial in improving team performance.

Thus, design practitioners must prioritize the following three things: (1) AI design must equip humans with diagnostic tools to determine whether the AI's recommendation is erroneous, (2) the human must be encouraged or supported (through cognitive forcing mechanisms) to proactively engage and adapt with the AI, and (3) such proactive behaviors must be driven by an adequate understanding of the comparative performance levels of the two entities. Comparative efficiency, as operationalized by performance, may also extend to other factors, such as speed. In time-sensitive tasks, a quick and reasonably accurate response might be preferred to a more deliberate, accurate, but slower response. Assessments to determine the threshold at which an AA's inaccuracy outweighs the benefits of its rapid response, in comparison to the human's baseline response, are necessary for understanding whether humans may be better off relying on the AA. These assessments must also account for risk appetites and the cost-to-reward ratio of making small gains in performance, especially

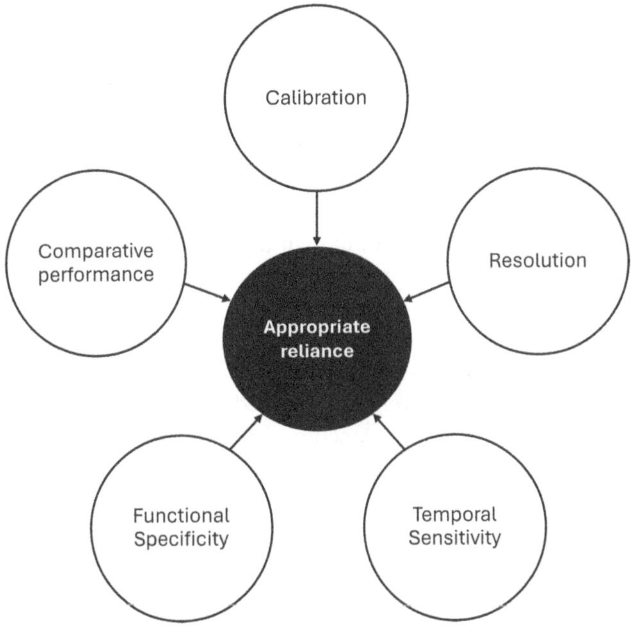

FIGURE 8.3 Dimensions in the measurement of appropriate reliance.

in high-stakes decision-making. For example, say the small gains in performance translate to the number of lives saved in disaster management. The cost-to-reward ratio of realizing these gains will be influential when compared to other factors while characterizing the appropriateness of human reliance on decision support.

Recently, Srinivasan and Thomason (2025) pointed out that over-reliance and under-reliance on AI are not uniformly bad, as is often formulated in practice. For example, in the context of medical imaging, Gaube et al. (2024) demonstrated that under-reliance on an AI's correct advice is more harmful than over-reliance on its incorrect advice. To account for such situations, Srinivasan and Thomason rightly advocate accounting for the relative importance of over-reliance and under-reliance by using a balancing utility function. Such a function, they suggest, can be used as a reward signal to form an operator intervention policy. Alternatively, they say, these reward signals may serve as feedback for designers to encourage user reliance on AI through trust repair measures.

8.4.2 Beyond a Binary Measure: Sensitivity and Hierarchical Sensitivity

Although popularly operationalized, reliance is not a binary behavior of whether or not a human chooses to rely on AA inputs. Hollenbeck et al. (1998) introduced the concept of *sensitivity* as one of the core constructs for ascertaining teaming efficacy. This concept has also been discussed in the context of decision-making in human teams, such as judge-advisor collectives (Brehmer & Hagafors, 1986). At the dyadic level, sensitivity reflects the degree to which the operator (judge) correctly weighs an advisor's recommendation to arrive at a final decision. In group decision-making, for any set of advisor recommendations, there is an optimal/ideal set of weights that might be applied to those recommendations to predict the correct decision. The difference between the weights assigned and the ideal set of weights for the advisors' recommendations is the operational definition of dyadic sensitivity. A large difference implies low sensitivity (Hollenbeck et al., 1998).

Hierarchical sensitivity (HS) captures the overall optimality of an operator's use of their advisors' suggestions. In human teams, Hollenbeck et al. (1998) determined HS by conducting a regression analysis that determined the weights the decision-maker placed on each teammate's recommendations and the ideal weights that should have been placed. They calculated HS as the average absolute difference between the weights *actually* assigned by the decision-maker to their teammates' suggestions and the ideal set of weights, obtained by regressing the true decision on the teammates' recommendations. A score of 0 indicated that the decision-maker weighed each teammate's suggestions optimally, whereas a high score indicated a large deviation from the ideal set of weights that should have been applied. One need not stretch their imagination to see how this definition may be extended to a HAT.

Consider y as the *correct decision* and y' as the decision-maker's *final decision*. They receive k recommendations toward determining each of these decisions, where m_i is the weight that should be assigned to the ith recommendation and m_i' is the *actual* weight assigned. The decision is:

$$y' = \sum_{i=0}^{k} (m_i' x_i) \tag{8.3}$$

Therefore, HS is:

$$\text{HS} = \sum_{i=0}^{k} \left(\left| m_i' - m_i \right| \right) \tag{8.4}$$

Using this definition, we may recognize that calibrated reliance is a simplified form of dyadic sensitivity with binary weights (i.e., whether or not the human relies on the agent). As automated agents are increasingly integrated into work environments, humans will be uniquely positioned to interact with multiple agents at a time. In our view, adopting HS provides a meaningful way to measure how humans calibrate their reliance on multiple automated teammates.

8.4.3 Temporal Considerations and Implications

TS describes changes in reliance as a function of the situation or time. Human behavior toward AAs, evolving in response to tasks, situations, and environmental conditions, is key to informing researchers about the strategies users deploy to achieve their goals. However, temporal aspects concerning human dependence on AAs have not garnered the attention they deserve. Changes in reliance patterns, for example, may help researchers to assess: (1) whether the human is developing greater comfort in using the agent's assistance over sustained interactions; (2) whether they are experiencing significant changes in external influential factors, such as workload, leading to greater/reduced reliance on the AA; (3) whether they develop better mental models of the AA (beyond identifying its failure modes, to understand how it performs various activities associated with the task); and (4) combinations of factors 1–3. For extended periods of interaction, static measures of reliance are not sufficient to determine design efficacy. Longitudinal studies may use TS to assess overall interaction efficacy. High TS indicates that the calibration of reliance does not change over time and remains high. This means that the user's reliance reflects the situation-specific capabilities of the agent. Low levels of TS are recorded when calibration reflects only long-term changes in the system's capabilities. For example, when the agent's performance steadily deteriorates over time, if the calibration of reliance "catches up" only after performance drops below a certain threshold, then TS would be low.

Schemmer et al. (2023a) demonstrated that human learning is a key mediator of appropriate reliance and is key to achieving complementary team performance. TS is only an aggregate measure that collectively captures moment-to-moment changes. It is not sensitive to variance in humans' ability to learn, nor does it record the specific moments at which their behavior adapts (or fails to) to changes in automation. The rates at which different users learn may also vary. These factors cannot be captured by an aggregate TS score. Human calibration of reliance, when mapped across time, is a stronger indicator of teaming efficacy, especially for continually adaptive and dynamically evolving systems. Consider an AI agent that learns by passively observing human behaviors and adapting accordingly. When humans interact with such agents over longer periods, the agent may adapt itself to the point where it gains heightened performance capabilities at the cost of forming complex operational and/or error boundaries. Degradation of humans' calibrated reliance on AI over time can show the sustained degradation of their mental model of AI as a function of the rate at which performance gains are made by the AI through adaptation. This relationship between error complexity, task performance, and humans' mental model of the AI was previously captured by Bansal, et al. (2019). The intersection of trends in performance and calibration of reliance is key to setting or adjusting parameters that determine AI adaptations.

Consider another example in which an AA can effectively perform a wide range of tasks but fails under a few off-nominal circumstances. When the resolution of humans' trust is poor, their reliance patterns may not map to the true capabilities of the system (Lee & See, 2004). In fact, the reliance would map onto very few system capabilities. It is known that trust decay in humans is faster than its recovery (Slovic, 1993; Yaniv & Kleinberger, 2000). This means that when we observe off-nominal situations more frequently, reliance on the AA could decay faster, as the system may be perceived as incompetent. Conversely, when the resolution is good, people would recognize the true system capabilities. The efficacy of training procedures to improve performance may be determined by monitoring temporal trends, as they can indicate humans' abilities to understand and learn when the system can operate confidently and predictably. Deploying such a multi-pronged approach would further be useful in determining the impact of non-performance-related factors of an AA, such as anthropomorphism (Placani, 2024), embodiment, etc. Studies incorporating the whole suite of reliance-based measures are lacking but are vital in designing not just AAs, but also the interfaces through which humans and AAs interact.

8.4.4 Towards Bias Mitigation

The infusion of AI into decision processes and whether it can mitigate cognitive biases have been studied considerably in recent times (Rastogi et al., 2022). These studies have been developing and incorporating measurement techniques or using clever procedures in decision workflows. One such method for bias mitigation involves showing AI suggestions only *after* the human has given their initial response to a task. The underlying rationale suggests that accessing AI recommendations after operators have "come to an independent decision" will reduce anchoring effects and passive monitoring behaviors (Parasuraman, et al., 2000). Lu et al. (2023) advocate for features that allow humans to "request" AI assistance instead of providing assistance at every instance. For a sentiment analysis task, Lu and colleagues demonstrated that soliciting AI advice, when necessary, has the potential to mitigate over-reliance without increasing under-reliance. Other studies have similarly advocated the use of cognitive forcing functions to mitigate over-reliance on AI (Buçinca, et al., 2021).

In an attempt to measure biases, for a binary decision task, Srivastava (2023) identified the extent to which an AI agent's recommendation changed the initial user preferences. She termed this metric *sway*, which was calculated as the number of decision switches and the percentage of initial disagreements that were resolved. High values, therefore, indicated overtrust, while low values of sway indicated undertrust. Other researchers have used the more descriptive term "switch rate" to describe the same behavior (Yin et al., 2019; Zhang et al., 2020). The efficacy of an AA's assistance can be measured in terms of how prone humans are to accounting for unrelated decision parameters, and the extent to which their own decisions are swayed after observing an agent's suggestion. *Calibrating* sway would involve aggregating all such instances in which humans' initial decisions changed for the better. That is, did humans sway their decisions toward more appropriate suggestions that were put forth by the AA?

While sway describes a practical measure for identifying and mitigating biases, other reliance-based procedures serve the same function. Recently, Hunter et al. (2024) proposed the use of reliance drills to monitor human dependence on AI systems and identify and mitigate over-reliant behaviors. They stated, "during a reliance drill, a user's AI-generated assistance is discreetly modified to include deliberate mistakes. Users pass the drill if they recognize and reject these mistakes; otherwise, they fail." More importantly, this procedure accounts for the comparative efficiency of humans and their AI partners. As per the definition of over-reliance, as operationalized by Hunter et al. (2024), over-reliance occurs when "[humans] attempt to follow AI generated advice for a problem that they would solve more effectively on their own." This definition accounts for the key factor that we termed compensation, or comparative efficiency, leading to a more thorough definition of appropriate reliance. To thoroughly implement reliance drills, the authors proposed guidelines for simulating such "deliberate mistakes" and discussed the practical circumstances under which such drills may be deployed, especially in organizational settings. Through the example of a medical task, they also discussed the tradeoffs accompanying realism in simulating drills and the realization of its associated risks. Further, they highlighted the financial, reputational, and legal incentives for organizations to incorporate such methods in assessing the utility of AI in their day-to-day work.

Measures indicating the extent of human susceptibility to automation manipulation are also useful, given that humans' responses are significantly driven by social factors (Nass & Moon, 2000, p. 81). People's responses to non-performance-based measures, such as anthropomorphism, language assertiveness, speech patterns, embodiment, etc., significantly drive AI adoption (Placani, 2024; Nass & Moon, 2000). Accounting for the relevant reliance-based metrics helps in recommending better designs that do not inadvertently manipulate the users into accepting flawed technology.

8.4.5 Reliance Measurement in Literature

It is encouraging to note that substantial emphasis is now being placed on evaluating human reliance on AI and automation and the degree to which such reliance is appropriate across contexts.

To assess the efficacy of explanation modalities in users' acceptance of AI advice, Gonzalez et al. (2020) operationalized reliance as the extent to which users accepted correct or incorrect advice, thereby measuring *mere* acceptance rather than its appropriateness. Wang and Yin (2021) and Yang et al. (2020) broadened the definition of appropriate "trust" as the acceptance or rejection of correct or incorrect assistance, respectively, on a case-by-case basis. Along these lines, Poursabzi-Sangdeh et al. (2021) emphasized the need to accept an algorithm's assistance when it is beneficial, along with the ability to *identify* when the algorithm makes a mistake. Chandrasekaran et al. (2018) studied the relationship between the ability of humans to predict AI behavior and appropriate reliance. Narayanan and Feigh (2025) went beyond these definitions and studied appropriate reliance as the extent to which users accept/reject correct and incorrect AI advice, along with mutual performance compensation and the benefits of partnership, for non-binary decision tasks. They also jointly evaluated the temporal evolution of mutual compensatory and reliance-based behaviors, which made it possible to investigate differences in humans' decision strategies.

Schemmer et al. (2023b) provided a fine-grained definition of appropriate reliance with two dimensions: (1) relative self-reliance and (2) relative AI reliance, capturing correct rejection of AI advice and correct AI reliance, respectively. Unlike other definitions, their approach accounts for the human's initial decision prior to viewing the AI recommendation, enabling the identification of whether biases are induced or mitigated (e.g., via switch rate or sway; see Section 8.4.4). However, this method is limited to cases where human and AI decisions initially differ, constraining its broader applicability. The authors also note that eliciting independent human judgments can artificially reduce over-reliance, as empirically observed by Srivastava et al. (2024). Additionally, the technique's reliance on instance-level aggregation of correct and incorrect predictions limits the study of temporal effects. Finally, such definitions are most informative when AI errors are systematic; when errors are arbitrary, humans are less likely to detect mistakes, and appropriate reliance may instead be defined as the pattern of reliance behavior most likely to yield optimal human-automation performance (Talone, 2019; Wang et al., 2008; Schemmer et al., 2023b).

Given the diverse definitions, methods, and tradeoffs in studying human reliance on AAs, there is a clear lack of consensus within the scientific community. Although some argue for a single definition applicable across tasks, we contend that a metric's suitability should be judged by the research questions it addresses. As with performance and effectiveness measures, appropriate reliance is context dependent. Therefore, efforts are better directed toward examining practical and theoretical considerations, along with the five dimensions outlined here, to more effectively evaluate human–AA collaborative success.

8.5 CONCLUSION

This article calls for moving beyond binary, surface-level assessments of human reliance on automation toward comprehensive measures that capture the complexity of interactions. Central to this is distinguishing trust from compliance or reliance—often conflated but meaningfully distinct. While trust calibration is commonly used to gauge team effectiveness, we argue for focusing on appropriate reliance, which reflects users' judgments of an AA's competence, reliability, and contextual fit. We outline definitions of appropriate reliance, the dependence-based behaviors they capture, and five major dimensions along which learning and adaptation can also be studied. These insights support designing systems that not only foster trust but also promote calibrated, context-sensitive reliance for more effective and ethical human-automation integration.

DECLARATION OF CONFLICTING INTERESTS

The author(s) declare no conflicts of interest with respect to the research, authorship, and/or publication of this article.

ACKNOWLEDGEMENTS

This work was supported by the Office of Naval Research Command Decision Making Program under Contract N00014-24-S-B001. The views expressed in this work are those of the authors and do not reflect the official position of the agency.

NOTES

1 For the purpose of this article, we use "automation" and "automated agent (AA)" interchangeably, as all-encompassing terms to identify a computerized entity acting as a human adjunct. Other possible terms include "synthetic agent", "autonomy", "intelligent/AI agent", "machine", and so on. Our goal is to generalize across all types of human-automation pairings without alluding to other factors (such as embodiment, agency, etc.) when invoking a specific term. For the rest of this article, none of the terms encompassed within our generic terms must be treated differently, unless otherwise specified.

2 This figure is *conceptually inspired* by Lee & See (2004), who examined the relationship between calibration, resolution, and automation capability in defining appropriate *trust* in automation. Unlike the original figure, which focuses on trust as an attitudinal construct, the present figure reframes and extends these relationships to model appropriate human *reliance* on AAs. Specifically, we replace trust with reliance as the primary outcome and reinterpret calibration and resolution in terms of observable reliance behaviors. Therefore, the present figure is not intended as a reproduction of Lee & See's original figure.

3 The definition of Compensation in terms of performance gains seems binary but could be operationalized for other scenarios as long as there are clearly defined, measurable outcomes. For example, compensation in different types of performance-related factors may be measured and weighted based on task priorities. Similarly, it could also be extended to include multiple inputs from the AA. Additionally, these analyses when performed at the instance-level would lend greater sensitivity towards capturing the situation-specific capabilities of the human and the AA. The example provided in this article is to illustrate the value in using this comparative measure.

REFERENCES

Ajenaghughrure, I. B., Sousa, S. D. C., & Lamas, D. (2020). Measuring trust with psychophysiological signals: A systematic mapping study of approaches used. *Multimodal Technologies and Interaction*, *4*(3), 63. https://doi.org/10.3390/mti4030063

Annett, J. (2002). Subjective rating scales: Science or art? *Ergonomics*, *45*(14–15), 966–987. https://doi.org/10.1080/00140130210166951

Bansal, G., Nushi, B., Kamar, E., Lasecki, W. S., Weld, D. S., & Horvitz, E. (2019). Beyond accuracy: The role of mental models in human-AI team performance. *Proceedings of the AAAI Conference on Human Computation and Crowdsourcing*, *7*(1), 2–11. https://doi.org/10.1609/hcomp.v7i1.5285

Bo, Z. H., Qiao, H., Tian, C., Guo, Y., Li, W., Liang, T., Li, D., Liao, D., Zeng, X., Mei, L., Shi, T., Wu, B., Huang, C., Liu, L., Jin, C., Guo, Q., Yong, J. H., Xu, F., Zhang, T., Wang, R., … Dai, Q. (2021). Toward human intervention-free clinical diagnosis of intracranial aneurysm via deep neural network. *Patterns*, *2*(2), 100197. https://doi.org/10.1016/j.patter.2020.100197

Bolton, M. L. (2022). Trust is not a virtue: Why we should not trust. *Ergonomics in Design*, *32*(4), 4–11. https://doi.org/10.1177/10648046221130171

Bondi, E., Koster, R., Sheahan, H., Chadwick, M., Bachrach, Y., Cemgil, T., Paquet, U., & Dvijotham, K. (2022). Role of human-AI interaction in selective prediction. *Proceedings of the AAAI Conference on Artificial Intelligence*, *36*(5), 5286–5294. https://doi.org/10.1609/aaai.v36i5.20465

Boubin, J. G., Rusnock, C. F., & Bindewald, J. M. (2017). Quantifying compliance and reliance trust behaviors to influence trust in human-automation teams. *Proceedings of the Human Factors and Ergonomics Society Annual Meeting*, *61*(1), 750–754. https://doi.org/10.1177/1541931213601672

Bradford, A. (2023). *Digital empires: The global battle to regulate technology*. Oxford University Press. https://doi.org/10.1093/oso/9780197649268.001.0001

Brehmer, B., & Hagafors, R. (1986). Use of experts in complex decision making: A paradigm for the study of staff work. *Organizational Behavior and Human Decision Processes*, *38*(2), 181–195. https://doi.org/10.1016/0749-5978(86)90015-4

Buçinca, Z., Malaya, M., & Gajos, K. (2021). To trust or to think: Cognitive forcing functions can reduce overreliance on AI in AI-assisted decision-making. *Proceedings of the ACM on Human-Computer Interaction*, *5*, 1–21. https://doi.org/10.1145/3449287

Chancey, E. T., Bliss, J. P., Yamani, Y., & Handley, H. A. H. (2017). Trust and the compliance-reliance paradigm: The effects of risk, error bias, and reliability on trust and dependence. *Human Factors*, *59*(3), 333–345. https://doi.org/10.1177/0018720816682648

Chandrasekaran, A., Prabhu, V., Yadav, D., Chattopadhyay, P., & Parikh, D. (2018). Do explanations make VQA models more predictable to a human? *Proceedings of the Conference on Empirical Methods in Natural Language Processing (EMNLP)* (pp. 1036–1042). Association for Computational Linguistics. https://doi.org/10.18653/v1/D18-1128

Dekker, S., & Hollnagel, E. (2004). Human factors and folk models. *Cognition, Technology & Work*, *6*(2), 79–86. https://doi.org/10.1007/s10111-003-0136-9

Dekker, S., & Nyce, J. (2015). From figments to figures: Ontological alchemy in human factors research. *Cognition, Technology & Work*, *17*(5). https://doi.org/10.1007/s10111-015-0321-7

Dixon, S. R., & Wickens, C. D. (2006). Automation reliability in unmanned aerial vehicle control: A reliance-compliance model of automation dependence in high workload. *Human Factors*, *48*(3), 474–486. https://doi.org/10.1518/001872006778606822

Dixon, S. R., Wickens, C. D., & Chang, D. (2005). Mission control of multiple unmanned aerial vehicles: A workload analysis. *Human Factors*, *47*(3), 479–487. https://doi.org/10.1518/001872005774860005

Eignor, D. R. (2013). The standards for educational and psychological testing. In K. F. Geisinger, B. A. Bracken, J. F. Carlson, J.- I. C. Hansen, N. R. Kuncel, S. P. Reise, & M. C. Rodriguez (Eds.), *APA handbook of testing and assessment in psychology, Vol. 1. Test theory and testing and assessment in industrial and organizational psychology* (pp. 245–250). American Psychological Association. https://doi.org/10.1037/14047-013

Gaube, S., Jussupow, E., Kokje, E., Khan, J., Bondi-Kelly, E., Schicho, A, Kitamura, F., Koch, T., Ezer, T., Mottok, J., Lermer, E., Ghassemi, M., & Colak, E. (2024). Underreliance harms human-AI collaboration more than overreliance in medical imaging. *OSF Preprints*. https://doi.org/10.31219/osf.io/4wv8j_v2

Gonzalez, A. V., Bansal, G., Fan, A., Jia, R., Mehdad, Y., & Iyer, S. (2020). *Human evaluation of spoken vs. visual explanations for open-domain QA*. arXiv. https://doi.org/10.48550/arXiv.2012.15075

High-Level Expert Group on AI. (2019). *Ethics guidelines for trustworthy AI* (Report). European Commission.

Ho, C.W., & Caals, K. (2024). How the EU AI Act seeks to establish an epistemic environment of trust. *Asian Bioethics Review*, *16*(3), 345–372. https://doi.org/10.1007/s41649-024-00304-6

Hollenbeck, J. R., Ilgen, D. R., LePine, J. A., Colquitt, J. A., & Hedlund, J. (1998). Extending the multilevel theory of team decision making: Effects of feedback and experience in hierarchical teams. *Academy of Management Journal*, *41*(3), 269–282. https://doi.org/10.2307/256907

Hunter, R., Moulange, R., Bernardi, J., & Stein, M. (2024). *Monitoring human dependence on AI systems with reliance drills*. arXiv. https://doi.org/10.48550/arXiv.2409.14055

Jones, T. (2024). *The philosophical and regulatory tensions of trustworthy AI: An examination of trust, trustworthiness, and reliance in AI systems* (Master's thesis, Macquarie University). https://doi.org/10.25949/27974268.v1

Kerasidou A. (2019). The role of trust in global health research collaborations. *Bioethics*, *33*(4), 495–501. https://doi.org/10.1111/bioe.12536

Kerasidou, C. X., Kerasidou, A., Buscher, M., & Wilkinson, S. (2022). Before and beyond trust: Reliance in medical AI. *Journal of Medical Ethics*, *48*(11), 852–856. https://doi.org/10.1136/medethics-2020-107095

Laux, J., Wachter, S., & Mittelstadt, B. (2024), Trustworthy artificial intelligence and the European Union AI Act: On the conflation of trustworthiness and acceptability of risk. *Regulation & Governance*, *18*, 3–32. https://doi.org/10.1111/rego.12512

Lee, J. D., & See, K. A. (2004). Trust in automation: Designing for appropriate reliance. *Human Factors*, *46*(1), 50–80. https://doi.org/10.1518/hfes.46.1.50_30392
Lee, J. D., & Moray, N. (1994). Trust, self-confidence, and operators' adaptation to automation. *International Journal of Human-Computer Studies*, *40*(1), 153–184. https://doi.org/10.1006/ijhc.1994.1007
Lu, Z., Wang, D., & Yin, M. (2023). Does more advice help? The effects of second opinions from peers in AI-assisted decision making. *Proceedings of the ACM on Human-Computer Interaction*, *8*, 1–31. https://doi.org/10.1145/3653708
McLeod, C. (2011). Trust. In E. N. Zalta (Ed.), *The Stanford Encyclopedia of Philosophy* (Fall 2011 Edition). Metaphysics Research Lab, Stanford University. https://plato.stanford.edu/archives/fall2011/entries/trust/
Meyer, J. (2001). Effects of warning validity and proximity on responses to warnings. *Human Factors*, *43*(4), 563–572. https://doi.org/10.1518/001872001775870395
Meyer, J. (2004). Conceptual issues in the study of dynamic hazard warnings. *Human Factors*, *46*(2), 196–204. https://doi.org/10.1518/hfes.46.2.196.37335
Muir, B. M. (1987). Trust between humans and machines, and the design of decision aids. *International Journal of Man-Machine Studies*, *27*(5–6), 527–539. https://doi.org/10.1016/S0020-7373(87)80013-5
Narayanan, R., & Feigh, K. (2025). Human assessment of AI errors and its impact on hybrid teaming for decision making. *2025 IEEE Conference on Cognitive and Computational Aspects of Situation Management (CogSIMA)* (pp. 103–110). IEEE. https://doi.org/10.1109/CogSIMA64436.2025.11079477
Nass, C., & Moon, Y. (2000). Machines and mindlessness: Social responses to computers. *Journal of Social Issues*, *56*(1), 81–103. https://doi.org/10.1111/0022-4537.00153
Parasuraman, R., & Riley, V. (1997). Humans and automation: Use, misuse, disuse, abuse. *Human Factors*, *39*(2), 230–253. https://doi.org/10.1518/001872097778543886
Parasuraman, R., Sheridan, T. B., & Wickens, C. D. (2000). A model for types and levels of human interaction with automation. *IEEE Transactions on Systems, Man, and Cybernetics – Part A: Systems and Humans*, *30*(3), 286–297. https://doi.org/10.1109/3468.844354
Placani, A. (2024). Anthropomorphism in AI: hype and fallacy. *AI Ethics*, *4*, 691–698. https://doi.org/10.1007/s43681-024-00419-4
Poursabzi-Sangdeh, F., Goldstein, D. G., Hofman, J. M., Vaughan, J. W., & Wallach, H. (2021). Manipulating and measuring model interpretability. *Proceedings of the 2021 CHI Conference on Human Factors in Computing Systems (CHI '21)* (pp. 1–52). Association for Computing Machinery. https://doi.org/10.1145/3411764.3445315
Rastogi, C., Zhang, Y., Wei, D., Varshney, K. R., Dhurandhar, A., & Tomsett, R. (2022). Deciding fast and slow: The role of cognitive biases in AI-assisted decision-making. *Proceedings of the ACM on Human-Computer Interaction*, *6*, 1–22. https://doi.org/10.1145/3512930
Ryan, M. (2020). In AI we trust: Ethics, artificial intelligence, and reliability. *Science and Engineering Ethics*, 26(5), 2749–2767. https://doi.org/10.1007/s11948-020-00228-y
Schemmer, M., Bartos, A., Spitzer, P., Hemmer, P., Kuhl, N., Liebschner, J., & Satzger, G. (2023a). *Towards effective human-AI decision-making: The role of human learning in appropriate reliance on AI advice*. arXiv. https://arxiv.org/abs/2310.02108
Schemmer, M., Hemmer, P., Nitsche, M., Kuhl, N., & Vossing, M. (2022). A meta-analysis of the utility of explainable artificial intelligence in human-AI decision-making. *Proceedings of the 2022 AAAI/ACM Conference on AI, Ethics, and Society* (pp. 617–626). https://doi.org/10.1145/3514094.3534128
Schemmer, M., Kühl, N., Benz, C., Bartos, A., & Satzger, G. (2023b). Appropriate reliance on AI advice: Conceptualization and the effect of explanations. *Proceedings of the 28th International Conference on Intelligent User Interfaces (IUI '23)* (pp. 410–422). Association for Computing Machinery. https://doi.org/10.1145/3581641.3584066

Slovic, P. (1993). Perceived risk, trust, and democracy. *Risk Analysis*, *13*(6), 675–682. https://doi.org/10.1111/j.1539-6924.1993.tb01329.x

Smith, C. (2005). Understanding trust and confidence: two paradigms and their significance for health and social care. *Journal of Applied Philosophy*, *22*(3), 299–316. https://doi.org/10.1111/j.1468-5930.2005.00312.x

Srinivasan, T., & Thomason, J. (2025). *Adjust for trust: Mitigating trust-induced inappropriate reliance on AI assistance*. arXiv. https://doi.org/10.48550/arXiv.2502.13321

Srivastava, D., Lilly, J. M., & Feigh, K. M. (2024). Exploring the role of judgement and shared situation awareness when working with AI recommender systems. *Cognition, Technology & Work*, *27*, 1–18. https://doi.org/10.1007/s10111-024-00771-9

Srivastava, D. K. (2023). *Investigation of critical attributes for transparency and operator performance in human autonomy teaming (TOPHAT) for intelligent mission planning* (Doctoral thesis, Georgia Institute of Technology). Georgia Institute of Technology.

Talone, A. (2019). *The effect of reliability information and risk on appropriate reliance in an autonomous robot teammate* (Master's thesis, University of Central Florida). University of Central Florida.

Tversky, A., & Kahneman, D. (1974). Judgment under uncertainty: Heuristics and biases. *Science*, *185*(4157), 1124–1131. https://doi.org/10.1126/science.185.4157.1124

Vaccaro, M., Almaatouq, A. & Malone, T. (2024). When combinations of humans and AI are useful: A systematic review and meta-analysis. *Nature Human Behavior*, *8*, 2293–2303. https://doi.org/10.1038/s41562-024-02024-1

Vashitz, G., Meyer, J., Parmet, Y., Peleg, R., Goldfarb, D., Porath, A., & Gilutz, H. (2009). Defining and measuring physicians' responses to clinical reminders. *Journal of Biomedical Informatics*, *42*(2), 317–326. https://doi.org/10.1016/j.jbi.2008.10.001

Wang, L., Jamieson, G. A., & Hollands, J. G. (2008). Selecting methods for the analysis of reliance on automation. *Proceedings of the Human Factors and Ergonomics Society Annual Meeting*, *52*(4), 287–291. https://doi.org/10.1177/154193120805200419

Wang, X. & Yin, M. (2021). Are explanations helpful? A comparative study of the effects of explanations in AI-assisted decision-making. *Proceedings of the 26th International Conference on Intelligent User Interfaces (IUI '21)* (pp. 318–328). Association for Computing Machinery. https://doi.org/10.1145/3397481.3450650

Wei, J., Bolton, M. L. & Humphrey, L. (2019). Subjective measurement of trust: Is it on the level? *Proceedings of the Human Factors and Ergonomics Society Annual Meeting*, *63*(11), 212–216. https://doi.org/10.1177/1071181319631062

Wei, J., Bolton, M. L., & Humphrey, L. (2020). The level of measurement of trust in automation. *Theoretical Issues in Ergonomics Science*, *22*(3), 274–295. https://doi.org/10.1080/1463922X.2020.1766596

Yang, F., Huang, Z., Scholtz, J., & Arendt, D. L. (2020). How do visual explanations foster end users' appropriate trust in machine learning? *Proceedings of the 25th International Conference on Intelligent User Interfaces (IUI '20)* (pp. 189–201). Association for Computing Machinery. https://doi.org/10.1145/3377325.3377480

Yaniv, I., & Kleinberger, E. (2000). Advice taking in decision making: Egocentric discounting and reputation formation. *Organizational Behavior and Human Decision Processes*, *84*(2), 260–281. https://doi.org/10.1006/obhd.2000.2909

Yin, M., Vaughan, J.W., & Wallach, H.M. (2019). Understanding the effect of accuracy on trust in machine learning models. *Proceedings of the 2019 CHI Conference on Human Factors in Computing Systems* (pp. 1–12). ACM. https://doi.org/10.1145/3290605.3300509

Zhang, Y., Liao, V., & Bellamy, R. (2020). Effect of confidence and explanation on accuracy and trust calibration in AI-assisted decision making. *Proceedings of the 2020 Conference on Fairness, Accountability, and Transparency* (pp. 295–305). ACM. https://doi.org/10.1145/3351095.3372852

9 Beyond Static Metrics
Adaptive Performance as a Process-Oriented Framework for Human-Agent Teaming

Eva Gößwein and Magnus Liebherr

9.1 INTRODUCTION

9.1.1 Contextualizing Adaptive Performance

Traditional performance metrics, such as efficiency (speed, accuracy, throughput) and error rates, have long been the cornerstone of evaluating human interaction with technology. These metrics are particularly valuable in safety-critical domains such as aviation and process control, where clear, well-defined tasks and outcomes are prioritized. However, they often conceptualize performance as a static outcome, the result of a completed task under controlled conditions. This perspective, while useful for usability and reliability assessments, fails to capture the dynamic, context-dependent nature of human behavior in complex, real-world environments.

Human performance is inherently situated and evolving. It unfolds through continuous cycles of perception, decision-making, and action, shaped by contextual constraints, feedback, and changing demands (Edmonds, 2016). Therefore, performance depends not only on task execution but also on how individuals manage attention, allocate cognitive resources, regulate affect, and coordinate with others in the face of uncertainty. In this view, variability in performance is not merely noise or inefficiency but a reflection of humans' capacity to flexibly adjust to situational demands and environmental changes (Nalepka et al., 2022). This adaptation becomes a critical determinant of success in complex sociotechnical systems, such as those involving automation or artificial intelligence (AI), where individuals and teams must maintain effectiveness despite uncertainty, novelty, or system perturbations.

Understanding human performance thus requires moving beyond static evaluations toward process-oriented approaches that account for temporal, cognitive, and contextual dynamics. Such perspectives form the foundation for the concept of adaptive performance, which focuses on the mechanisms that allow humans to remain effective in the face of change. Defined by Pulakos et al. (2000) as "altering behavior to meet the demands of a new situation, event, or set of circumstances" (p. 615), adaptive performance emphasizes the dynamic interplay between individuals, tasks, and environments. Unlike static metrics, it captures the process of adaptation—how people adjust their behaviors in response to evolving challenges.

DOI: 10.1201/9781003655923-11

In the context of human-AI teaming, this shift in perspective is particularly relevant. Traditional metrics often overlook the nonlinear interactions and cognitive processes that underpin collaboration with intelligent systems. For example, static performance measures may fail to capture how trust calibration, explainable AI, and workload management work together to shape a human's adaptation to AI or mastery of changing task demands (Rieger et al., 2025). By contrast, adaptive performance provides a lens through which to study these dynamics, offering insights into how humans and AI systems co-adapt over time. This section explores adaptive performance as a critical metric for understanding and optimizing human-AI collaboration in dynamic environments.

9.1.2 Performance in Human-Agent Teaming

Human performance has been investigated across various disciplines. It refers broadly to the behaviors and actions through which individuals contribute to task or system goals. Performance encompasses not only observable behavioral outcomes but also the underlying cognitive, motivational, and affective processes that enable individuals to function effectively across diverse environments. At its core, human performance can be understood as the translation of capability into behavior under specific situational demands. This implies that performance is not a fixed, trait-like characteristic of the individual, but rather the result of a dynamic interaction between person, task, and context.

Performance models distinguish between multiple dimensions. Campbell et al. (1990) proposed a hierarchical model in which performance is composed of several latent factors, such as task proficiency, effort, and discipline, all contributing to overall effectiveness. Building on this framework, Motowidlo and Kell (2012) differentiated between task performance, the effectiveness with which core job duties are executed, and contextual performance, which includes behaviors that support the social and motivational context of work. Allworth and Hesketh (1999) introduce adaptive performance, described as behaviors "demonstrating the ability to cope with change and to transfer learning from one task to another as job demands vary" (p. 98), as a separable dimension alongside task and contextual performance. Together, these approaches emphasize that performance cannot be reduced to mere output metrics such as accuracy or productivity, as it also involves proactive, interpersonal, and self-regulatory aspects of human functioning.

In human-agent teaming, performance is often evaluated through two primary lenses: human-AI synergy and human augmentation (Vaccaro et al., 2024). *Human-AI synergy* refers to the comparison of a human-AI team's performance with that of the human and AI systems operating independently (Riley et al., 2011). In contrast, *human augmentation* focuses on whether the human-AI team outperforms the human alone. While these frameworks provide structured ways to assess collaboration, they often treat performance as a static outcome, overlooking the dynamic, process-driven interactions that define human-AI collaboration.

Empirical evidence suggests that human augmentation is common in human-AI teams, particularly when the AI system outperforms the human on specific tasks (Vaccaro et al., 2024). However, human-AI synergy is less consistently reported, indicating that combining human and AI capabilities does not always yield additive or superior results. In fact, when the AI outperforms the human, it may sometimes be more effective to use either the human or the AI alone rather than rely on the combined human-AI team (Vaccaro et al., 2024). This discrepancy has been linked to challenges in trust calibration—defined as the alignment of a user's trust with the actual trustworthiness of a system in a scenario (Liebherr et al., 2026)—and reliance on AI (Tschandl et al., 2020; Vaccaro et al., 2024). For instance, when AI systems are highly accurate but opaque, humans may either over-rely on them (misuse) or underutilize them (disuse), both of which can undermine team performance (Bansal et al., 2019; Rieger et al., 2025). The shift from misuse to disuse highlights the need for metrics that capture not just the outcome of collaboration but also the process of adaptation: how humans adjust their strategies, trust, and workload in response to AI outputs.

Current performance evaluations in human-agent teaming often fail to account for the nonlinear, time-dependent dynamics of collaboration. Traditional metrics may mask critical aspects of human-AI interaction, including how repeated exposure to AI systems impacts human adaptation (Venkatesh et al., 2023). For example, a human may initially struggle to integrate AI recommendations but gradually develop adaptive strategies. Static metrics cannot capture this progression, whereas process-oriented approaches, such as adaptive performance, can model these changes and identify the cognitive and contextual factors that facilitate or hinder effective collaboration.

Moreover, the effectiveness of human-AI teams is influenced by moderating variables, including task complexity, AI transparency, and user expertise (Demir et al., 2017; Vaccaro et al., 2024). Increasing explainability in AI systems has been proposed as a solution to bridge the gap between static performance metrics and the dynamic realities of human-AI collaboration (Rieger et al., 2025). However, even with improved explainability, the underlying cognitive and behavioral adjustments required for effective teamwork remain underexplored.

This section underscores the limitations of static performance metrics in human-agent teaming and sets the stage for introducing adaptive performance as a process-oriented alternative. By grounding this discussion in established multidimensional models of performance (Allworth & Hesketh, 1999; Campbell et al., 1990; Motowidlo & Kell, 2012) and defining adaptive performance as the ability to cope with change and transfer learning across tasks, we highlight its potential to bridge the gap between static outcome measures and the dynamic processes of human-AI collaboration. Furthermore, by shifting the focus from fixed outcomes to the mechanisms of adaptation, adaptive performance offers a more nuanced lens for understanding how humans and AI systems co-evolve in dynamic environments. The following section will delve into the theoretical foundations of adaptive performance and its relevance to human-AI collaboration.

9.2 MODELING ADAPTIVE PERFORMANCE

The conceptualization of adaptive performance has evolved since its introduction as a distinct dimension of human performance. Early work by Allworth and Hesketh (1999) and Pulakos et al. (2000) laid the groundwork by distinguishing adaptive performance from task and contextual performance. Pulakos et al. (2000) proposed an eight-dimensional taxonomy of adaptive performance, encompassing dimensions such as handling emergencies, solving problems creatively, and coping with uncertainty. While this framework provided a structured approach to understanding adaptability, it also sparked debates about whether the construct represented a stable individual difference (a trait) or a dynamic response to situational demands (a process), as summarized by Baard et al. (2014). This distinction between an adaptability trait and adaptation as a process remains critical for modeling adaptive performance because confusing the two can obscure the mechanisms through which humans adapt in human-AI teaming.

A pivotal advancement came with the development of the I-ADAPT theory by Ployhart and Bliese (2006). This framework contextualized adaptive performance within a nomological network, emphasizing how environmental requirements, individual differences (e.g., trait adaptability, cognitive abilities, personality), and mediating processes (e.g., motivation, feedback, strategies) interact to shape performance. Notably, the authors positioned adaptive performance as an outcome influenced by external factors, aligning with the dynamic, context-dependent nature of human behavior. However, I-ADAPT did not explicitly label the construct as adaptive performance; instead, it focused on broader performance outcomes. This ambiguity underscores the necessity for more precise conceptual boundaries in modeling adaptive performance, particularly in human-AI contexts where environmental and technological variables are rapidly evolving.

Baard et al. (2014) further advanced the theoretical landscape by introducing a performance adaptation taxonomy that distinguishes between *domain-general* and *domain-specific* adaptation. Domain-general adaptation refers to transferable skills (e.g., problem-solving, stress management). In contrast, domain-specific adaptation focuses on skills acquired through training or experience in a particular context (e.g., adapting to AI systems in healthcare). This distinction is crucial for human-AI teaming, where adaptive performance must account for both general adaptation and domain-specific expertise in interacting with AI. Baard et al. (2014) furthermore propose a multilevel conceptual architecture of adaptation that emphasizes the role of task-complexity changes (component, coordinative, and dynamic) in shaping adaptive processes, providing a structured framework for modeling how humans adapt to evolving AI capabilities. For example, in high-stakes domains like aviation or healthcare, the dynamic interplay between AI outputs and human decision-making necessitates domain-specific adaptations that align with task complexity.

In parallel, researchers have developed dynamic models to capture the temporal nature of adaptive performance. Lang and Bliese (2009) proposed a two-phase model of adaptation: *transition adaptation* (initial decline in performance when facing new demands) and *reacquisition adaptation* (recovery or improvement as individuals adjust). This model has been operationalized using longitudinal data and mixed-effects models with discontinuous growth (Howe, 2019; Niessen & Jimmieson, 2016), allowing researchers to track how adaptive performance unfolds over time. In human-AI teaming, such models are invaluable for understanding how humans recalibrate their strategies in response to AI outputs, system perturbations,

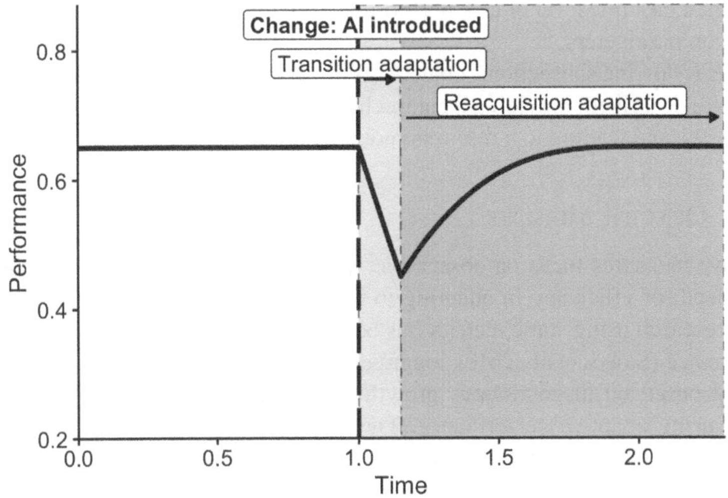

FIGURE 9.1 Conceptual plot of adaptive performance in human-AI teams.

Note. **The plot is based on the methodology suggested by Lang and Bliese (2009), which distinguishes adaptive performance into two phases: transition adaptation and reacquisition adaptation. It models the introduction of AI in the team as a change that provokes adaptation. Similarly, the introduction of a new form of AI or a system adaptation could be framed as the change to which humans must adapt. The plot was generated in RStudio based on fictitious data.**

or changing task demands. Figure 9.1 depicts a conceptual plot of the two adaptation phases following AI introduction.

The evolution of adaptive performance modeling reflects a shift from static, trait-based constructs to dynamic, process-oriented frameworks. These models emphasize the interplay between individual capabilities, environmental demands, and temporal dynamics—a critical perspective for human-AI collaboration. However, challenges remain in aligning these theoretical advancements with practical measurement strategies, particularly in capturing the real-time, nonlinear interactions that define adaptive performance in AI-driven environments. The following section will explore how adaptive performance is assessed, addressing the strengths and limitations of current measurement approaches.

9.3 MEASURES OF ADAPTIVE PERFORMANCE

Assessing adaptive performance requires capturing its dynamic, context-dependent nature. In human-AI teaming, this involves balancing methodological rigor with contextual relevance, as the interplay between human and AI systems demands nuanced measurement strategies. Crucially, capturing this dynamic nature requires an explicit operationalization of how performance changes over time relative to identifiable perturbation events (e.g., system updates, AI errors, task reconfiguration). Adaptive performance should therefore be modeled as a time-indexed deviation from an established baseline level of task performance, enabling the quantification of distinct adaptation parameters.

In the following subsections, objective and subjective measures of adaptive performance are described, providing a comprehensive overview of the measures presented in Table 9.1 and leading to a discussion of challenges associated with this metric.

9.3.1 Objective Measures

Objective measures focus on observable, quantifiable behaviors, such as task accuracy, speed, or efficiency in adapting to new demands (Stokes et al., 2010). While earlier research using static metrics has been criticized for oversimplifying adaptive performance (Stokes et al., 2010), longitudinal models, like Lang and Bliese's (2009) protocol based on discontinuous growth modeling, offer a more robust approach to measuring adaptive performance. These models distinguish between transition adaptation and reacquisition adaptation (c.f., Figure 9.1), separating them from skill acquisition, which refers to performance increases over time. Discontinuous growth modeling differs from continuous growth approaches by explicitly incorporating a disruption in the performance trajectory (Lang & Bliese, 2009). Rather than assuming a single developmental curve across time, the model estimates separate parameters before and after a disruption, thereby treating the performance decline itself as a meaningful component of the adaptation process. In practical terms, this involves specifying a pre-change baseline phase, a discontinuity at the moment of change, and a post-change recovery phase with its own slope. Thus, the "discontinuous" element reflects the shift in performance induced by the change rather than assuming unidimensional growth. This feature is particularly relevant in human-AI teaming, where AI systems may update or recalibrate over time, repeatedly introducing disruptions, even after their implementation, that require renewed human adaptation. By formally modeling these breaks, discontinuous growth analysis allows researchers to quantify both the immediate impact of AI-related changes and the subsequent trajectory of human performance recovery. Such frameworks are critical for capturing the temporal dynamics of adaptive performance, particularly in high-stakes contexts (Niessen & Jimmieson, 2016; Niessen & Lang, 2021).

To increase methodological clarity, adaptive performance should be operationalized through parameters that directly correspond to the temporal phases depicted in Figure 9.1. First, *transition adaptation* can be quantified as the magnitude of performance disruption immediately following a perturbation. Operationally, this refers to the deviation from a stable pre-change baseline to the lowest observed performance point after the system disruption (e.g., introduction of an AI system). This parameter captures the short-term adaptive cost and reflects the extent to which the human-AI system is destabilized by the new demand.

Second, *reacquisition adaptation* represents the trajectory of performance recovery following the disruption. Rather than merely documenting improvement, this parameter should be modeled as the slope of performance change across repeated measurement occasions during the recovery phase. A steeper slope indicates more efficient adaptation, whereas a flatter slope suggests prolonged recalibration. Distinguishing reacquisition adaptation from general skill acquisition is essential:

TABLE 9.1
Overview of Measurements of Adaptive Performance

Category	Measurement	Description	Strengths	Challenges	References
Objective Measures					
	Longitudinal performance metrics	Discontinuous growth modeling measuring transition and reacquisition adaptation; performance trajectories over time.	Captures the dynamic process of adaptation; distinguishes general skill acquisition from adaptive performance in the form of transition adaptation and reacquisition adaptation.	Requires repeated measures; complex analyses.	Lang and Bliese (2009); Niessen and Jimmieson (2016)
	Simulation-based adaptive performance metrics	Behavioral adaptation in dynamic simulations.	High ecological validity; controlled novelty.	Resource-intensive; limited generalizability.	Kozlowski et al. (2001); Burke et al. (2006)
Subjective Measures					
	Self-report adaptive performance scales	Individuals' perceived adaptability; often based on Pulakos et al.'s eight dimensions.	Easy to administer; captures internal perceptions.	Susceptible to rater bias; often confounds trait adaptability with performance.	Pulakos et al. (2000); Stasielowicz (2020) Charbonnier-Voirin and Roussel (2012) Spanish version: Gorostiaga et al. (2022) English and Dutch version: Krijgsheld et al. (2024)
	Peer and supervisor ratings of adaptive behavior	External judgments of observable adaptive behaviors in context.	Captures interpersonal and contextual aspects.	Halo effects; subject to bias and inconsistent rating standards.	Stokes et al. (2010)

reacquisition refers specifically to recovery toward a prior functional level, whereas skill acquisition implies performance gains beyond previously established baselines.

Beyond these phases suggested by Lang and Bliese (2009), two additional parameters enhance interpretability. Third, *time-to-stability* captures the duration required for performance to return to a predefined criterion (e.g., statistically indistinguishable from baseline or reaching a specified percentage of baseline performance). This parameter allows researchers to differentiate between individuals who recover quickly but shallowly and those who recover more gradually yet sustainably.

Fourth, the *post-adaptation performance level* reflects the steady-state performance achieved after the reacquisition phase. Performance may (1) return to baseline, (2) remain below baseline, indicating incomplete adaptation, or (3) exceed baseline, suggesting adaptive growth. This distinction is particularly relevant in AI-intensive environments, where changes promise human augmentation and may ultimately lead to superior performance.

Methodologically, these parameters can be estimated using discontinuous growth modeling or piecewise longitudinal analyses in which performance observations are nested within individuals or teams. Such models require (1) a clearly defined baseline phase, (2) an explicitly identified change, and (3) sufficient repeated observations during the reacquisition phase to estimate recovery trajectories reliably. Explicit modeling of these segments allows researchers to quantify adaptive dynamics rather than infer adaptation from aggregated performance changes.

9.3.2 Subjective Measures

Subjective measures, including self-report scales and peer assessments, complement objective data by capturing individuals' perceptions of their adaptability or adaptive performance. However, these measures are often prone to rater bias and socially desirable responses. Scales based on Pulakos et al.'s (2000) eight-dimensional taxonomy have been used but frequently conflate trait adaptability (a stable individual difference) with adaptive performance (a dynamic process). For instance, self-reported trait adaptability correlates more strongly with personality traits such as openness and conscientiousness, whereas objective measures of adaptive performance are more closely linked to cognitive abilities (Stasielowicz, 2020; Zhang et al., 2020). This distinction underscores the need for clear operationalization: while trait adaptability reflects enduring characteristics, adaptive performance emphasizes situational behavioral adaptation.

9.3.3 Challenges of Measuring Adaptive Performance

The dynamic nature of adaptive performance further complicates its assessment. Static, one-time evaluations risk oversimplifying the interplay between individual capabilities and environmental demands. Longitudinal designs, by contrast, allow researchers to model how adaptive performance evolves. For example, studies have shown that initial declines in performance (transition adaptation) often precede improvements (reacquisition adaptation), particularly in novel or high-stakes contexts (Niessen & Jimmieson, 2016; Niessen & Lang, 2021). These findings highlight the importance of capturing both the short-term costs (e.g., temporary inefficiency) and the long-term benefits (e.g., enhanced resilience) of adaptive performance. Importantly, such modeling requires explicit specification of recovery criteria (e.g., statistical equivalence to baseline or attainment of a predefined performance threshold), as the definition of "successful adaptation" directly shapes parameter interpretation.

In human-AI teaming, the choice of measurement method is especially critical. Objective, performance-based measures are well-suited to capturing real-time adjustments, such as how quickly a human integrates AI recommendations or recalibrates strategies in response to system changes. Subjective measures, meanwhile, can reveal how individuals perceive their capacity to adapt, including factors such as confidence in using AI and willingness to trust AI outputs. Therefore, a combined approach that integrates objective behavioral data with subjective insights has the potential to offer the most comprehensive understanding of adaptive performance. For example, in healthcare, objective metrics might track a clinician's diagnostic adjustments based on AI suggestions, while subjective ratings could assess their perceived confidence in those decisions.

In such contexts, the magnitude of disruption (transition adaptation) may indicate the degree of human-AI misalignment at change onset, the recovery slope (reacquisition adaptation) may reflect cognitive flexibility or system transparency, and residual performance levels may signal whether human-AI fit has been restored or recalibrated to a superior equilibrium. Explicitly linking quantitative parameters to theoretical constructs enhances the interpretability and applicability of adaptive performance metrics in human-AI teams.

Ultimately, the choice between objective and subjective measures has profound implications for the investigation and understanding of adaptive performance. For instance, Stasielowicz's (2020) meta-analysis shows that cognitive abilities are more strongly associated with objective measures of adaptive performance, whereas self-reported adaptive performance is linked to personality traits. These findings emphasize the need for a multitrait-multimethod approach to validate assessments, ensuring that both the adaptation process and the traits associated with it are captured. Stokes et al. (2010) implemented such an approach. They reported that, although objective and subjective measures were significantly correlated, they did not demonstrate convergent validity, suggesting that they may assess different dimensions of adaptive performance. Future research should prioritize hybrid designs that integrate behavioral data with self-reported insights, as well as context-specific frameworks tailored to the unique dynamics of human-AI collaboration. By refining these approaches, researchers can better understand how adaptive performance supports effective, resilient collaboration in increasingly complex AI-driven environments.

The conceptual debates surrounding adaptive performance, particularly the distinction between trait adaptability and adaptive performance and the differentiation between state- and process-level conceptualizations, become especially consequential when applying the construct to human-AI teaming. Although van Dam (2013) argues that adaptation encompasses both momentary states and longer-term processes, this dual framing has not yet been systematically examined in AI-intensive settings. This gap is notable because human-AI teaming requires attention to both immediate behavioral adjustments (such as rapid recalibration of trust or strategy in response to AI outputs) and the gradual accumulation of expertise in collaborating with AI systems. Measurement challenges further complicate this transfer of theory. The field lacks consensus on whether adaptive performance in such environments should be primarily captured through objective indicators (e.g., performance deltas, behavioral trace data) or subjective assessments (e.g., trait adaptability, trust dynamics). This ambiguity undermines efforts to design metrics that are both valid and sensitive to the temporal structure of adaptation. Clarifying the relationship between state and process perspectives and determining how each can be operationalized and measured in human-AI contexts is therefore essential. Only then will it be possible to develop robust, context-appropriate metrics that encompass the full range of adaptive behaviors in human-AI teaming, from rapid, moment-to-moment adjustments to longer-term learning, skill acquisition, and trust calibration.

9.4 WHAT HAS BEEN ANSWERED WELL WITH ADAPTIVE PERFORMANCE

Although adaptive performance has been discussed in organizational psychology and increasingly in human-technology interaction, it has received little direct attention in human-AI teaming research. Existing studies often investigate phenomena that implicitly involve adaptation, yet rarely conceptualize these changes through the lens of adaptive performance. As a result, the potential of adaptive performance as a metric for understanding human-AI teaming remains largely underexplored.

This section highlights what the adaptive performance literature has already answered well in adjacent domains. These insights help illustrate what the metric can offer for human-AI teaming: a way to focus on *how* humans adjust to evolving task demands and AI behaviors rather than evaluating collaboration solely through static outcomes. By examining the strengths of the adaptive performance construct and summarizing the most relevant empirical insights, we outline how adaptive performance can be used to better understand the process of human adaptation in AI-supported work.

9.4.1 Strengths of the Metric

Across different areas of human-technology interaction, adaptive performance has proven useful for addressing several foundational questions about how people adjust to change (Lang & Bliese, 2009; Pulakos et al., 2000). Although not initially developed for human-AI teaming, the metric offers several strengths that directly address gaps in current teaming evaluations.

9.4.1.1 A Process-Oriented Lens rather than a Snapshot

A core strength of adaptive performance is its explicit modeling of change over time rather than static outcomes (Griffin & Hesketh, 2003). Adaptive performance, in contrast, directs attention to transitions, adjustments, and trajectory patterns. This perspective has proven valuable in domains that involve novelty or evolving task demands, such as learning new interfaces, responding to system perturbations, or adjusting to changes in a simulated environment (Niessen & Lang, 2021; Nissim et al., 2024). This process-focused approach suggests that adaptive performance may be a promising way to investigate how humans learn to collaborate with AI systems through repeated interactions and recalibrate their strategies when AI behavior changes.

9.4.1.2 Ability to Integrate Behavioral, Cognitive, and Attitudinal Indicators

Adaptive performance frameworks can incorporate multiple types of evidence, such as objective behaviors, cognitive mechanisms, and subjective appraisals. Studies on adaptation to technologies have shown that adaptation is associated with executive functions, attentional regulation, confidence and self-efficacy, perceived usability, and prior experience (Cuppen et al., 2025; Gößwein & Liebherr, 2025; Seinsche et al., 2023). These established relationships with other variables make adaptive performance well-suited for capturing the complexity of human-AI teaming, where individual differences, trust, workload, and mental model updates interact over time.

9.4.1.3 Compatibility with Longitudinal and Dynamic Analytical Methods

Adaptive performance has been successfully operationalized using discontinuous growth models that distinguish skill acquisition as general learning over time from transition adaptation (the initial drop in performance when demands shift) and reacquisition adaptation (the recovery of performance and relearning; Lang & Bliese, 2009). Because human-AI teaming often unfolds as a sequence of adjustments (e.g., learning how to use recommendations, responding to errors, adapting to system updates), these analytical approaches offer a structure for capturing temporal dynamics without requiring fully continuous data streams.

9.4.1.4 Emphasis on the Costs and Benefits of Adaptation

Research in adjacent domains shows that adaptation can involve short-term performance costs before long-term stability or improvement emerges (Beilock & Carr, 2001). Adaptive performance frameworks explicitly model these trade-offs (Baard et al., 2014). This is directly relevant for human-AI teaming, where early interactions with AI may increase workload or reduce efficiency before collaboration stabilizes. Using adaptive performance as a metric helps make sense of these patterns and prevents premature conclusions about human-AI team performance.

9.4.2 Key Insights

Although adaptive performance has rarely been applied directly to human-AI teaming, insights from related human-technology interaction research reveal consistent patterns of human adaptation to technological change. These insights indicate what is already well understood and where adaptive performance could meaningfully contribute.

9.4.2.1 Subjective Feelings of Adaptation Often Diverge from Objective Behavioral Adaptation

Findings from driving simulators (Brandtner et al., 2019) show that people perceive themselves as having adapted earlier than their behavior indicates. This mismatch suggests that subjective reports alone may not be sufficient for evaluating adaptation in human-AI teaming, particularly when users feel confident but have not yet adjusted their decision strategies or calibrated their trust in the system.

9.4.2.2 Adaptation Follows Heterogeneous Trajectories

Longitudinal studies of technology use (Mitzner et al., 2019) reveal substantial individual differences in the speed of adaptation, in whether initial improvements are maintained, and in whether users stabilize or abandon the technology. These patterns illustrate that adaptation is not a uniform process, reinforcing the need for metrics that capture how and when individuals adapt to AI systems.

9.4.2.3 Cognitive Resources Shape Adaptation, but Do Not Fully Determine It

Executive functioning, attentional regulation, and working memory have been shown to predict adaptation (Gößwein & Liebherr, 2025). However, findings from in-home technology monitoring (Leese et al., 2021) show that lower cognitive resources do not necessarily impair behavioral adherence when systems are well designed and attitudes are positive. For human-AI teaming, this suggests that adaptive performance can be used to examine how system transparency, interface design, and explainability support adaptation even under high cognitive demands.

9.4.2.4 Attitudes Influence the Trajectory, but Not Always the Behavioral Outcome

Research shows that attitudes toward usability and acceptability predict adaptation patterns, but not always final performance levels (Baschung Pfister et al., 2020). This highlights that positive impressions can facilitate adaptation, but subjective comfort does not always correspond to behavioral adjustment. Applied to human-AI teaming, this distinction could help clarify when trust reflects genuine reliance calibration and when it reflects overconfidence or under-reliance.

9.4.2.5 Adaptation Is Best Understood as a Continuous, Multi-Stage Process

Across domains, adaptation emerges not as a momentary shift but as an unfolding process characterized by initial disruption, experimentation, stabilization, and longer-term integration. Adaptive performance provides an established vocabulary and modeling approach for capturing these phases. Applying this lens to human-AI teaming would enable researchers to move beyond comparing human-AI performance to human-only or AI-only baselines and instead study the developmental trajectory of teaming.

9.5 WHAT REMAINS TO BE DONE

As the preceding sections have shown, the study of adaptive performance in human-AI contexts is still in an early and uneven stage of development. Despite promising theoretical advances, significant conceptual, methodological, and empirical gaps remain, limiting the field's ability to capture how humans truly adjust to evolving technological systems. This section outlines these gaps and identifies the critical work that must be undertaken to advance adaptive performance research in meaningful and scalable ways.

9.5.1 Limitations of Current Research

Despite growing interest in adaptive performance as a metric for capturing the dynamic adaptation process across organizational psychology and human-technology interaction, the construct's conceptual and methodological foundations remain uneven. As discussed in earlier sections, much of this ambiguity stems from the persistent conflation of *adaptability* (an individual difference) with *adaptive performance*, which is better understood as a dynamic process that unfolds over time. Clarifying this distinction across measurement approaches is essential, particularly in domains such as human-AI teaming, where adaptation emerges through repeated interactions, shifting system behavior, and evolving task demands.

Current research has also relied heavily on subjective indicators of adaptation, often because they are easy to administer. Yet findings from related fields consistently demonstrate that subjective impressions do not reliably track actual behavioral change or cognitive adjustment. For instance, people tend to report feeling adapted long before their decision patterns or trust calibration have stabilized. This mismatch highlights that subjective measures alone are insufficient for capturing the complexity of adaptive performance, especially in dynamic teaming contexts.

The limitations of existing measurement approaches become especially visible when considering how adaptation unfolds. Insights from adaptive performance research show that adaptation involves heterogeneous trajectories shaped by cognitive resources, task structure, and system design. Individuals differ in the extent to which they exhibit immediate performance disruption, gradual improvement, or later-stage stabilization. Without objective or time-sensitive indicators, these trajectories are largely invisible in current studies. This lack of methodological precision restricts both theoretical development and the practical application of adaptive performance in emerging areas like human-AI teaming, precisely the domains where understanding adaptation processes could be most valuable.

Future work, therefore, needs to expand the available measurement toolkit. Beyond improving self-report accuracy, research should prioritize developing scalable, objective, and preferably real-time indicators of adaptive performance. These could include behavioral metrics derived from interaction logs, such as response latencies (Cao et al., 2023) or override rates of AI recommendations (Schoeffer et al., 2023; Vasconcelos et al., 2023). Physiological signals linked to cognitive regulation may include indices such as heart rate variability as a marker of regulatory control (Pollak et al., 2020) or pupillometry as an indicator of cognitive effort (Köles, 2017). Performance dynamics modeled through discontinuous growth analysis would require repeated performance observations before and after identifiable changes (e.g., system updates or reliability shifts), allowing for the estimation of immediate disruption effects and recovery slopes (Lang & Bliese, 2009). Such methods would better capture moment-to-moment adjustments, more clearly differentiate between adaptation phases, and support more robust interpretations of human-AI interaction patterns.

Only by addressing these conceptual and methodological limitations can the field fully leverage adaptive performance as a metric that captures how humans navigate technological change.

9.5.2 Gaps in Empirical Studies

As highlighted previously, adaptive performance has received surprisingly little direct attention in human-AI teaming research. While studies in adjacent areas, such as collaborative decision-making, human-robot interaction, or dynamic task management, examine behaviors that clearly involve adaptation, they rarely frame these changes explicitly through the lens of adaptive performance. This absence is notable given the metric's demonstrated value for illuminating how humans adjust to evolving demands in other technology-focused domains.

The empirical gaps become especially clear when considering insights from the broader literature. Research shows that adaptation is not a singular event but a multi-stage, often nonlinear process involving disruption, experimentation, recalibration, and stabilization. Yet current human-AI teaming studies seldom measure these phases or treat adaptation as a trajectory. Instead, they frequently compare end states, such as performance with versus without AI support, without examining how individuals reached those outcomes or how their strategies changed over time.

Similarly, findings demonstrate that cognitive resources, attitudes, and system design each shape adaptation, but not always in straightforward ways. For example, positive attitudes may accelerate early-stage adjustment without necessarily predicting long-term performance, and high cognitive capacity does not guarantee successful adaptation if system transparency or explainability is insufficient. Such nuances underscore why adaptive performance, when treated as a dynamic metric, offers a meaningful framework for understanding collaboration with AI systems. Yet empirical studies explicitly adopting this framework remain scarce.

To advance the field, future research should systematically operationalize adaptive performance within human-AI teaming contexts. This includes:

- **Developing measurement frameworks** that capture transitions, disruptions, and recovery phases rather than focusing solely on static outcomes.
- **Incorporating longitudinal or event-based designs** that track how behaviors evolve across interactions, system updates, or changes in AI reliability.
- **Examining the interplay of individual differences, system characteristics, and environmental context** in shaping adaptation trajectories.
- **Using both objective and subjective indicators** to capture discrepancies between perceived and behavioral adaptation, which may be particularly pronounced in AI-supported work.

Bridging these empirical gaps will not only deepen theoretical understanding but also contribute practical insights for designing human-AI systems that better support the human side of adaptation. By treating adaptive performance as a core process rather than a peripheral outcome, future research can more effectively illuminate how humans learn to collaborate with increasingly intelligent technologies.

9.6 CONCLUSION

Adaptive performance offers a process-oriented metric for understanding how humans adjust their behavior in dynamic, technology-rich environments. The theories, models, and empirical findings reviewed in this section converge on the understanding that adaptive performance is not a stable trait but an unfolding process shaped by cognitive resources, contextual demands, and temporal dynamics. Static performance metrics struggle to capture this complexity, especially in human-AI teaming, where interactions are evolving, nonlinear, and influenced by trust, explainability, and system behavior.

Across adjacent domains, from driving simulation to technology adoption, research shows that adaptation develops over time and often involves discrepancies between subjective impressions and objective behavioral change. These insights underscore the need for longitudinal, ecologically valid, and multimethod measurement approaches capable of capturing transitions, disruptions, and stabilizations in behavior.

Applied to human-AI teaming, adaptive performance provides a way to move beyond static evaluations and instead analyze how humans recalibrate trust, incorporate AI recommendations, and refine coordination strategies as system behavior evolves. It also highlights individual and contextual predictors, such as cognitive abilities, prior experience, and system design, that shape adaptation trajectories.

Together, these insights position adaptive performance as a valuable construct for advancing human-AI teaming research. By focusing on the mechanisms and trajectories underlying adaptation, rather than only end-state performance, adaptive performance offers a more nuanced lens for understanding and supporting effective human-AI collaboration.

Looking forward, the trajectory of adaptive performance as a metric should be one of refinement and differentiation. In particular, objective, process-oriented measures of adaptive performance warrant further expansion in human-AI research. The explicit modeling of transition adaptation, reacquisition adaptation, time-to-stability, and post-adaptation performance levels provides a structured framework for quantifying how human-AI fit evolves following perturbations. Expanding the use of such longitudinal and parameter-based approaches will enable researchers to move beyond static performance comparisons and toward a more precise characterization of adaptive dynamics.

At the same time, subjective indicators of adaptive performance require careful reevaluation and validation. Rather than relying predominantly on global, trait-like self-reports, future research should emphasize state-anchored, temporally sensitive measures that align with objectively modeled adaptation trajectories. Research testbeds in human-AI teaming offer a particularly valuable context for this validation effort. Conceptually, such testbeds should be designed to incorporate identifiable perturbation events, repeated behavioral measurements, and complementary self-reports, thereby enabling systematic examination of convergence and divergence between perceived and observed adaptation. Through such integrative designs, the field can clarify whether subjective and objective measures capture distinct facets of adaptation or reflect different temporal levels of the same process.

In this way, research testbeds function not merely as experimental platforms but as methodological proving grounds for refining adaptive performance metrics. By embedding longitudinal modeling, controlled perturbations, and multimethod assessment into these environments, scholars can iteratively evaluate, expand, and recalibrate the construct in light of empirical evidence. Such efforts will help ensure that adaptive performance remains a theoretically grounded and practically relevant metric for understanding human adjustment in increasingly complex AI-driven systems.

REFERENCES

Allworth, E., & Hesketh, B. (1999). Construct-oriented biodata: Capturing change-related and contextually relevant future performance. *International Journal of Selection and Assessment*, *7*, 97–111. https://doi.org/10.1111/1468-2389.00110

Baard, S. K., Rench, T. A., & Kozlowski, S. W. J. (2014). Performance adaptation: A theoretical integration and review. *Journal of Management*, *40*(1), 48–99. https://doi.org/10.1177/0149206313488210

Bansal, G., Nushi, B., Kamar, E., Lasecki, W. S., Weld, D. S., & Horvitz, E. (2019). Beyond accuracy: The role of mental models in human-AI team performance. *Proceedings of the AAAI Conference on Human Computation and Crowdsourcing*, *7*(1), 2–11. https://doi.org/10.1609/hcomp.v7i1.5285

Baschung Pfister, P., Tobler-Ammann, B., Knols, R. H., de Bruin, E. D., & de Bie, R. A. (2020). Usability and acceptance of an interactive tablet-based exercise application: A mixed methods study. *Frontiers in Digital Health*, *2*. https://doi.org/10.3389/fdgth.2020.578281

Beilock, S. L., & Carr, T. H. (2001). On the fragility of skilled performance: What governs choking under pressure? *Journal of Experimental Psychology: General*, *130*(4), 701–725. https://doi.org/10.1037/0096-3445.130.4.701

Brandtner, A., Liebherr, M., Schweig, S., Maas, N., Schramm, D., & Brand, M. (2019). Subjectively estimated vs. objectively measured adaptation to driving simulators – Effects of age, driving experience, and previous simulator adaptation. *Transportation Research Part F: Traffic Psychology and Behaviour*, *64*, 440–446. https://doi.org/10.1016/j.trf.2019.05.019

Burke, C., Pierce, L., & Salas, E. (2006). Understanding adaptability: a prerequisite for effective performance within complex environments. https://doi.org/10.1016/s1479-3601(05)06009-1

Campbell, J. P., McHenry, J. J., & Wise, L. L. (1990). Modeling job performance in a population of jobs. *Personnel Psychology*, *43*(2), 313–333. https://doi.org/10.1111/j.1744-6570.1990.tb01561.x

Cao, S., Gomez, C., & Huang, C.-M. (2023). How time pressure in different phases of decision-making influences human-AI collaboration. *Proceedings of the ACM on Human-Computer Interaction*, *7*(CSCW2), Article 277, 1–26. https://doi.org/10.1145/3610068

Charbonnier-Voirin, A., & Roussel, P. (2012). Adaptive performance: A new scale to measure individual performance in organizations. *Canadian Journal of Administrative Sciences-Revue Canadienne des Sciences de l'Administration*, *29*, 280–293. https://doi.org/10.1002/cjas.232

Cuppen, S., van Leunen, M., Henken, T., Goevaerts, M., Scherrenberg, M., Falter, M., Dendale, P., Kemps, H., & Kop, W. J. (2025). Association of technology-related skills and self-efficacy with willingness to participate in heart failure telemonitoring: Cross-sectional observational study. *JMIR Formative Research*, *9*, e68992. https://doi.org/10.2196/68992

Demir, M., McNeese, N. J., & Cooke, N. J. (2017). Team situation awareness within the context of human-autonomy teaming. *Cognitive Systems Research*, *46*, 3–12. https://doi.org/10.1016/j.cogsys.2016.11.003

Edmonds, J. (2016). Review of common human vulnerabilities and contributions to past accidents. In *Human factors in the chemical and process industries* (pp. 25–38). Elsevier. https://doi.org/10.1016/B978-0-12-803806-2.00003-0

Gorostiaga, A., Balluerka, N., & Rodríguez-López, A. (2022). Assessment of adaptive performance and its role as a moderator of the relationship between person–organization fit and organizational citizenship behaviors. *Psicothema*, *34*(1), 84–94. https://doi.org/10.7334/psicothema2021.310

Gößwein, E., & Liebherr, M. (2025). Harnessing the power of adaptation: A pre-registered systematic review on executive functions and the adaptation to new technologies. *Behaviour & Information Technology*, *44*(12), 2860–2875. https://doi.org/10.1080/0144929X.2024.2417381

Griffin, B., & Hesketh, B. (2003). Adaptable behaviours for successful work and career adjustment. *Australian Journal of Psychology*, *55*, 65–73. https://doi.org/10.1080/00049530412331312914

Howe, M. (2019). General mental ability and goal type as antecedents of recurrent adaptive task performance. *Journal of Applied Psychology*, *104*(6), 796–813. https://doi.org/10.1037/apl0000379

Köles, M. (2017). A review of pupillometry for human-computer interaction studies. *Periodica Polytechnica Electrical Engineering and Computer Science*, *61*(4), 320–326. https://doi.org/10.3311/PPee.10736

Kozlowski, S., Gully, S., Brown, K., Salas, E., Smith, E. M., & Nason, E. R. (2001). Effects of training goals and goal orientation traits on multidimensional training outcomes and performance adaptability. Organizational *Behavior and Human Decision Processes*, 85 (1), 1–31. https://doi.org/10.1006/OBHD.2000.2930

Krijgsheld, M., Tummers, L., & Scheepers, F. (2024). Adaptive performance scale: Translation and validation in English and Dutch. *Human Factors in Healthcare*, *6*, 100086. https://doi.org/10.1016/j.hfh.2024.100086

Lang, J. W., & Bliese, P. D. (2009). General mental ability and two types of adaptation to unforeseen change: Applying discontinuous growth models to the task-change paradigm. *Journal of Applied Psychology*, *94*(2), 411–428. https://doi.org/10.1037/a0013803

Leese, M. I., Dorociak, K. E., Noland, M., Gaugler, J. E., Mattek, N., & Hughes, A. (2021). Use of in-home activity monitoring technologies in older adult veterans with mild cognitive impairment: The impact of attitudes and cognition. *Gerontechnology*, *20*(2), 1–12. https://doi.org/10.4017/gt.2021.20.2.10.06

Liebherr, M., Enkel, E., Law, E. L. C., Mousavi, M. R., Sammartino, M., & Sieberg, P. (2026). Dynamic calibration of trust and trustworthiness in AI-enabled systems. *International Journal on Software Tools for Technology Transfer*. Advanced online publication. https://doi.org/10.1007/s10009-026-00840-6

Mitzner, T. L., Savla, J., Boot, W. R., Sharit, J., Charness, N., Czaja, S. J., & Rogers, W. A. (2019). Technology adoption by older adults: Findings from the PRISM trial. *The Gerontologist*, *59*(1), 34–44. https://doi.org/10.1093/geront/gny113

Motowidlo, S. J., & Kell, H. J. (2012). Job performance. In I. B. Weiner, N. W. Schmitt, & S. Highhouse (Eds.), *Handbook of psychology* (2nd ed.). Wiley. https://doi.org/10.1002/9781118133880.hop212005

Nalepka, P., Prants, M., Stening, H., Simpson, J., Kallen, R. W., Dras, M., Reichle, E. D., Hosking, S. G., Best, C., & Richardson, M. J. (2022). Assessing team effectiveness by how players structure their search in a first-person multiplayer video game. *Cognitive Science*, *46*(10), e13204. https://doi.org/10.1111/cogs.13204

Niessen, C., & Jimmieson, N. (2016). Threat of resource loss: The role of self-regulation in adaptive task performance. *Journal of Applied Psychology*, *101*(3), 450–462. https://doi.org/10.1037/apl0000049

Niessen, C., & Lang, J. W. B. (2021). Cognitive control strategies and adaptive performance in a complex work task. *Journal of Applied Psychology*, *106*(10), 1586–1599. https://doi.org/10.1037/apl0000830

Nissim, M., Shfir, O., & Ratzon, N. Z. (2024). Simulator driving abilities, executive functions, and adaptive behavior among adolescents with complex attention deficit hyperactivity disorder. *Journal of Attention Disorders*, *28*(3), 350–363. https://doi.org/10.1177/10870547231214975

Ployhart, R. E., & Bliese, P. D. (2006). Individual adaptability (I-ADAPT) theory: Conceptualizing the antecedents, consequences, and measurement of individual differences in adaptability. In C. Shawn Burke, L. G. Pierce, & E. Salas (Eds.), *Understanding adaptability: A prerequisite for effective performance within complex environments* (pp. 3–39). Emerald Group Publishing Limited. https://doi.org/10.1016/S1479-3601(05)06001-7

Pollak, A., Paliga, M., Pulopulos, M. M., Kozusznik, B., & Kozusznik, M. W. (2020). Stress in manual and autonomous modes of collaboration with a cobot. *Computers in Human Behavior*, *112*, 106469. https://doi.org/10.1016/j.chb.2020.106469

Pulakos, E., Arad, S., Donovan, M. A., & Plamondon, K. E. (2000). Adaptability in the workplace: Development of a taxonomy of adaptive performance. *Journal of Applied Psychology*, *85*(4), 612–624. https://doi.org/10.1037/0021-9010.85.4.612

Rieger, T., Onnasch, L., Roesler, E., & Manzey, D. (2025). Why highly reliable decision support systems often lead to suboptimal performance and what we can do about it. *IEEE Transactions on Human-Machine Systems*, *55*(5), 736–745. https://doi.org/10.1109/THMS.2025.3584662

Riley, M. A., Richardson, M. J., Shockley, K., & Ramenzoni, V. C. (2011). Interpersonal synergies. *Frontiers in Psychology*, *2*, 38.

Schoeffer, J., Jakubik, J., Voessing, M., Kuehl, N., & Satzger, G. (2023). AI reliance and decision quality: Fundamentals, interdependence, and the effects of interventions. *Journal of Artificial Intelligence Research*, *82*, 1235–1274. https://doi.org/10.1613/jair.1.15873

Seinsche, J., de Bruin, E. D., Saibene, E., Rizzo, F., Carpinella, I., Ferrarin, M., Moza, S., Ritter, T., & Giannouli, E. (2023). A newly developed exergame-based telerehabilitation system for older adults: Usability and technology acceptance study. *JMIR Human Factors*, *10*, e48845. https://doi.org/10.2196/48845

Stasielowicz, L. (2020). How important is cognitive ability when adapting to changes? A meta-analysis of the performance adaptation literature. *Personality and Individual Differences*, *166*, 110178. https://doi.org/10.1016/j.paid.2020.110178

Stokes, C., Schneider, T., & Lyons, J. (2010). Adaptive performance: A criterion problem. *Team Performance Management*, *16*, 212–230. https://doi.org/10.1108/13527591011053278

Tschandl, P., Rinner, C., Apalla, Z., Argenziano, G., Codella, N., Halpern, A., Janda, M., Lallas, A., Longo, C., Malvehy, J., Paoli, J., Puig, S., Rosendahl, C., Soyer, H. P., Zalaudek, I., & Kittler, H. (2020). Human–computer collaboration for skin cancer recognition. *Nature Medicine*, *26*(8), 1229–1234. https://doi.org/10.1038/s41591-020-0942-0

Vaccaro, M., Almaatouq, A., & Malone, T. W. (2024). When combinations of humans and AI are useful: A systematic review and meta-analysis. *Nature Human Behaviour*, *8*(12), 2293–2303. https://doi.org/10.1038/s41562-024-02024-1

van Dam, K. (2013). Employee adaptability to change at work: A multidimensional, resource-based framework. In S. Oreg, A. Michel, & R. T. By (Eds.), *The psychology of organizational change: Viewing change from the employee's perspective* (pp. 123–142). Cambridge University Press. https://doi.org/10.1017/CBO9781139096690.009

Vasconcelos, H., Jörke, M., Grunde-McLaughlin, M., Gerstenberg, T., Bernstein, M. S., & Krishna, R. (2023). Explanations can reduce overreliance on AI systems during decision-making. *Proceedings of the ACM on Human-Computer Interaction*, *7*(CSCW1), Article 129, 1–28. https://doi.org/10.1145/3579605

Venkatesh, V., Davis, F. D., & Zhu, Y. (2023). Competing roles of intention and habit in predicting behavior: A comprehensive literature review, synthesis, and longitudinal field study. *International Journal of Information Management*, *71*, 102644. https://doi.org/10.1016/j.ijinfomgt.2023.102644

Zhang, L. M., Aidman, E., Burns, B., & Kleitman, S. (2020). Integrating self-report and performance-based assessment of adaptability in a university context. *Journal of Research in Personality*, *88*, Article 103988. https://doi.org/10.1016/j.jrp.2020.103988

Section 3

Concepts

10 Goal-Based Control of Multiple Intelligent Autonomous Systems

A Paradigm Shift in Human-Autonomy Teaming

Cassandra Heffernan, Thomas Fahy, Kingsley Fletcher, and Joshua Charvetto

10.1 THE FUTURE OF AUTONOMOUS SYSTEMS

Machine systems have reached unprecedented levels of sensing, learning, reasoning, and communication (Werkhoven et al., 2018). With continuous advancements in artificial intelligence (AI), autonomous agents can perform a greater number of dynamic functions in both teamwork and taskwork than ever before (Seeber et al., 2020), and they are beginning to be viewed as teammates rather than tools by the human operator (Grimm et al., 2018; Lyons et al., 2018). The academic literature is starting to consider the implications of higher levels of intelligence in autonomous systems, particularly regarding human interaction with such systems. In the past, machine systems were often discussed as potential aids that were subservient to humans and as tools to support decision-making (O'Neil et al., 2020). With increases in automation, the role of the human changed from direct user of the system to supervisor of the system. Moving forward, it is expected that the construct will shift from automation to autonomy, further transforming the human operator's role from supervisor to teammate.

To discuss the implications of this change on human command and control (C2) of autonomous systems, and the likely human-autonomy teaming considerations, the distinction between automation and autonomy needs to be made. Automation has been defined as 'a device or system that accomplishes (partially or fully) a function that was previously, or conceivably could be, carried out (partially or fully) by a human operator' (Parasuraman et al., 2000, p. 286). Automation is viewed as executing preprogrammed tasks in which system designers have planned for routine inputs and responses from the system (Hancock et al., 2013). Automation performs well-defined tasks and can produce deterministic results based on a fixed set of rules and algorithms without AI technologies (Xu, 2021). Consequently, automation can only perform tasks that it was originally intended to perform and can function poorly on tasks beyond its original design parameters (Woods & Cook, 2006), requiring

DOI: 10.1201/9781003655923-13

human intervention in abnormal situations. Autonomy refers to the ability of an intelligent system to perform specific tasks independently (Xu, 2021). Autonomy may perform successfully in unanticipated situations, and the result of autonomous decision-making can be difficult to predict. Autonomy may therefore improve the performance of machine systems in dynamic and complex environments. However, the challenge changes from automation brittleness and the need for human intervention to understanding emergent system behaviour, particularly with autonomy that includes self-learning capability.

The potential for machine systems to work with increasing autonomy and intelligence opens interesting possible use cases for the Australian Defence Force (ADF). For example, one research team in Australia's Defence Science and Technology Group (DSTG) was tasked with unpacking use cases for the Royal Australian Navy (Navy) in particular. Future Navy surface combatants will be fitted with uninhabited aerial system (UAS) capability. While short-term capability may require multiple operators to manage UAS flight and payload capability, in the future the Navy is looking toward AI to augment and extend this capability. The Navy's Robotics, Autonomous Systems and Artificial Intelligence (RAS-AI) Strategy 2040 outlines the challenges and opportunities presented by these new technologies. The Navy is looking toward systems in the future that can independently carry out missions based on general information and automatically adapt to changing circumstances; that can understand the environment and think in higher levels of abstraction, judgement, and ideas-based reasoning; navigate in dense and dynamic environments; and work together as a coordinated swarm, with an understanding of intent and dynamic behaviour.

The move from short-term automated capability to future autonomous behaviour highlights not only a change in technology and how the human operator commands and controls UAS, but also the number of UAS a person may be expected to manage. Typically, a team of two or more crew members is required to manage a UAS, including interpreting and using information gathered from those vehicles. The Australian military is currently focusing on reducing crewing and finding ways in which a single operator can manage multiple UASs (Cummings et al., 2010). The limitation of this construct in the past has been the cognitive workload of an operator supervising and controlling multiple UASs. While it has been suggested that a significant amount of automation could enable a single operator to manage approximately four to five uninhabited vehicles at a time before performance degrades significantly (Cummings & Mitchell, 2008), the use of autonomy is thought to have the potential to overcome this limitation.

O'Neil et al. (2020) conducted a review and analysis of the empirical literature exploring human-autonomy teams and identified several themes for researchers to consider. Firstly, increasing levels of agent autonomy generally resulted in positive effects on outcomes. Autonomous agents with higher levels of autonomy are typically perceived as easier to work with (Johnson et al., 2012), resulting in higher communication efficiency (Wright et al., 2014), stronger coordination (Wright & Kaber, 2005), and better performance (Lewis et al., 2003). Participants also viewed agents with higher autonomy levels as easier to work with (Johnson et al., 2012) and as requiring lower workload (Wright et al., 2018). Humans also perceived better collaboration, stronger trust, and increased task understanding with greater levels of autonomy (Azhar & Sklar, 2017). However, other studies by Wright and colleagues have found that autonomous agents were better perceived when they provided a moderate level of decision-making input, compared to low or high levels (Wright & Kaber, 2005).

The control of multiple uninhabited vehicles (UxVs) poses a number of challenges for the human operator due to the behavioural autonomy, emergent behaviours, and distributed tasking of such systems. Operators cannot plan all system actions before or during missions, particularly in dynamic and complex scenarios. Novel approaches to the control of autonomous systems are necessary in order to realise their full capability potential. In order to explore how the control of multiple autonomous systems may function in the future, this chapter will describe current control concepts in the literature.

10.2 CURRENT CONTROL METHODS

The optimal approach to controlling autonomous systems is believed to be influenced by the level of system autonomy (Hua et al., 2017). While the previous section discusses the differences between autonomy and automation, these concepts are not all-or-none, and the level of autonomy or automation may change within one technological system depending on the task being completed. There is a range of models for defining human interaction with automated technology, such as Sheridan (1992), which describes levels of automation in robotics and introduces the concepts of trading control, sharing control, and supervisory control; and more recently, Mi and Yang's (2013) four levels of automation, which consist of manual control, human-oriented semi-autonomy, machine-oriented semi-autonomy, and fully autonomous. However, Parasuraman et al.'s (2000) model is the most widely cited in the literature. Their model examines human interaction in terms of 'levels of automation of decision and action selection', which is based on a 10-point scale ranging from the computer offering no assistance, and humans taking all decisions and actions, through to the computer deciding everything, acting autonomously, and ignoring the human.

While the literature presents a range of ways to define levels of automation, there appear to be fewer taxonomies for different approaches to control. Hua et al. (2017) highlight different levels of control from a systems architecture perspective, decomposing the control of autonomous systems into Mission control, Plan control, Task control, and Execution control. Mission control involves high-level mission planning, such as determining mission goals, dividing the mission into stages, and using the autonomous systems to meet mission goals at each stage. Plan control involves real-time perception and formation of mission instructions, including organising and collaborating throughout the mission to meet mission objectives. Task control is the real-time control of asset equipment and sensors, and execution control is the use of weapons or final effectors. Hua et al. (2017) considered how different information needs can be grouped or shared amongst different levels of a system. However, this paper did not explore the different human factors implications of controlling autonomous systems at different levels, or unpack which level may best support human-autonomy teaming. It is anticipated that more autonomy and intelligence are required in the autonomous systems for the plan and mission control levels to be supported, compared to the task and execution control levels, which are levels largely driven by human operator actions as opposed to autonomy decisions.

Some higher-level control approaches are being explored that align with the Plan control level described above. This research has largely been conducted under the Assistant Secretary of Defence for Research Engineering (ASD/R&E) sponsored research focusing on instantiating an 'Intelligence Multi-UxV Planner with Adaptive Collaborative Control Technologies (IMAPCT)' (Draper et al., 2018). This research aims to realise IMPACT by instantiating a 'play calling' paradigm for task delegation from a human operator to multiple UxVs. In this research environment, a range of interactions are supported by flexible operator-autonomy teaming options. This flexibility arises from the different types of control an operator can exact over the UxVs. At one level, the operator can manually control UxV movement or build up tasking from the ground up, specifying detailed parameters. This aligns with action-based control concepts seen to date and identified in Hua et al. (2017).

Alternatively, operators can quickly assign tasks to one or more UxVs by specifying a 'play' type and location, with all other task parameters captured within the task. Once the operator has designated the play type and location, the autonomy determines and recommends at least one play for the task (unless the operator's constraints cannot be satisfied). A 'play' is a high-level goal that consists of a detailed plan that calls all UxVs to reason over their availability and response time to fulfil that plan. The plays are 'called' with short commands, and the autonomous system responds by performing tasks that are pre-defined within the plays. For example, an operator may call a 'play' that states 'achieve air surveillance on a building'. The intelligent agent, built into the IMPACT system, then recommends one or more UxVs to use based on details such as estimated time en route, fuel use, and environmental conditions; an algorithm provides the shortest route to get to the building, considering operational constraints such as no-fly zones; and an autonomy framework monitors the play's ongoing status and provides the operator alerts regarding any issues encountered throughout the play. In sum, the research environment allows a variety of control methods, ranging from teleoperated (mouse/keyboard) control to high-level plays that define the actions of one or more UxVs.

Overall, the IMPACT research aimed to reduce operator workload by implementing automated vehicle path planning and sensor steering tasks that underlie each play. According to the 2018 research paper, a single IMPACT operator was able to task and manage a total of 12 heterogeneous UxVs in response to several unexpected events that arose during simulated base perimeter defence missions. This suggests that increasing the level of control from action-based to plan-based increases the number of heterogeneous autonomous systems one operator can manage.

Results from broader research evaluating the Playbook approach beyond the 2018 ASD/R&E paper have been mixed. Although the approach enables an operator to delegate plans to automation, it can also impose additional workload (Kaber & Prinzel, 2006). Another possible limitation of using pre-defined 'plays', or building up plays with detailed parameters in real time, is the reduced flexibility of autonomous systems to respond dynamically to the environment. When a play is called, the detailed parameters of the play have been specified, including expected flight paths and patterns, required sensors, and the time it is required to be completed by, and the centralised 'play manager' reasons over which UxVs have the capability to meet the play and if they are available to do so (i.e., not already completing a different play).

As such, the autonomy in this study does not refer to the capabilities of the UxVs, but rather to another piece of software that helps determine vehicle asset allocation, routing, and mission execution details, as well as a service that monitors the progress and success of the play. Furthermore, the play approach requires each plan to be executed in a linear fashion. If a new play is called, it will be achieved once an asset becomes available. The inability to take into account and action multiple goals simultaneously could decrease response time and reduce agility in the tactical environment.

There is also a risk that a quick change in plan or mission goal may be needed in the tactical environment, for which the cost of extra seconds or minutes to define parameters may mean the difference between mission success and failure. Kaber and Prinzel (2006) suggest that adaptable automation, such as the automation described in the ASD/R&E (2018) research where the operator can move between manual control and the issuing of plans, may be limited in what applications it can support. In particular, adaptable automation may only support fairly well-known task situations and pre-defined responses.

10.3 A NEW FRAMEWORK FOR AUTONOMOUS SYSTEM CONTROL

If the Navy is to use multiple, intelligent autonomous systems in dynamic, uncertain, and high-threat environments, a higher level of control beyond plan-based control may be necessary. Instead, the concept of goal-based control could be explored, where the human operator provides mission goals and constraints to autonomous systems, which then requires the autonomous systems to translate these mission goals and constraints into actions for implementation by each system element. Research being conducted in the Netherlands highlights the need for higher-level control concepts to fully realise the potential of intelligent, autonomous systems (Werkhoven et al., 2018). Their 2018 paper describes the difficult task of transferring decision-making to intelligent autonomous machine systems in a responsible and transparent way, leveraging the benefits of advanced AI while also ensuring the human can maintain meaningful control. Werkhoven et al. (2018) suggest that humans need to adopt a new role of providing intelligent autonomous systems with objectives or goals (including roles, norms, constraints, and moral values) to maintain meaningful control. It is then important that autonomous systems can break down the goals and constraints into actions in a way that is logical and transparent. To achieve this, consideration must be given to appropriate goal decomposition and how best this can be implemented. The following sections review the social and cognitive psychology theory around goals and how they influence human behaviour, and examine the implications for creating goals that can define and influence the behaviour of autonomous systems.

10.4 THE INFLUENCE OF GOALS ON HUMAN BEHAVIOUR

In the social psychology literature, a goal can be defined as a cognitive representation of a desired endpoint that impacts evaluations, emotions, and behaviours (Fishbach & Ferguson, 2007, p. 491). In the cognitive psychology domain, a goal has been defined as 'a mental representation of a state that a person wants to achieve'

(Neal et al., 2017). The end states of a goal can vary in their level of abstractness (Fishbach & Ferguson, 2007). For example, a goal may involve an end state that is tangible and perceptual in the world, such as having a cup of coffee, or one that is more abstract and conceptual, such as being regarded as successful in one's studies.

Goals entail more than just end states, however. They involve a range of behaviours, plans, and actions that enable one to reach that end state. For instance, the goal of obtaining a cup of coffee might entail temporally ordered, procedural information about first grinding coffee beans and then putting them into a coffee machine filter, and the goal of study success might include behaviours such as studying at the library and paying attention in class. The behaviours and actions associated with an end state can also vary in abstractness. For instance, the end state of achievement might include the specific behaviour of writing lecture notes as well as the relatively more general action of being punctual (Fishbach & Ferguson, 2007). Overall, a goal specifies an 'end state' that organises one's behaviour, and there is flexibility in the plans, behaviours, and tasks that can be employed to achieve the end state. The inability to use one means to achieve a goal does not imply that the end state is abandoned (Kruglanski et al., 2002). As stated in Fishbach and Ferguson (2007), goals may range from long-term personal projects or strivings, to mid-range aims or objectives, to task-specific aspirations, to action-specific intentions. As such, some goals may consist of pre-planned steps, while others may be achieved through a series of actions that do not require pre-planning.

For a goal or goals to motivate an individual into action requires more than a statement of an end state and the actions or plans needed for attaining it. Just because someone possesses knowledge about how to complete a task does not mean they want to do said task. As such, an important aspect of a goal is that the end state has to be desirable (Carver & Scheier, 1981). In social psychology, it is recognised that people are motivated to approach pleasure and avoid pain (Fishbach & Ferguson, 2007). Approach goals represent future states that are more desirable than current states and require actions that move someone closer to the desired state (Fishbach & Ferguson, 2007). In contrast, avoidance goals refer to impending future states that are less desirable than current states, warranting actions designed to prevent them (Fishbach & Ferguson, 2007).

The effect of a goal on human behaviour also changes over time. In the social psychology literature, it is said that the strength of a goal dissipates when the end state has been reached (Fishbach & Ferguson, 2007). Once a goal is perceived to be attained, or adequate goal progress is inferred to have occurred, it is implied that no future commitment of resources to that goal is necessary (Fishbach & Ferguson, 2007). However, there are competing theories regarding the strength of a goal when it is near completion. Lewin (1936) proposed that a goal remains active until the discrepancy between the actual and desired state is reduced, but the strength of the goal decreases as the discrepancy reduces. Others argue that goal strength increases over time until it is met (Atkinson & Birch, 1970; McClelland et al., 1953) or dissipates when the pursuit becomes too difficult to sustain (Brehm & Self, 1989; Wright, 1996). Within the cognitive psychology literature, it is also suggested that individuals monitor the discrepancy between the current state of a goal (i.e., the current level of goal achievement) and the reference level (i.e., the level by which a person would consider the goal achieved), and that the level of discrepancy influences the value of acting on that goal (Neal et al., 2017). The value of a goal is dynamic and signals the need for action. The value of a goal is dependent on the goal's importance (or how motivating it is in the social psychology literature) and the discrepancy (Neal et al., 2017). For an approach goal, value decreases as the variable gets closer to the desired level, indicating less need for action. For an avoidance goal, value increases as the variable gets closer to the undesired level, indicating a greater need to act.

Shah et al. (2009) state that goal pursuit does not exist in isolation but rather in a complex, dynamic system subject to constantly changing situational needs and affordances. The regulation of the pursuit of multiple goals requires constant management, and whether resources are allocated to a particular goal at a particular point in time depends on how motivating it is, or the level of discrepancy that is present at that moment.

10.5 TRANSLATING MISSION GOALS INTO AUTONOMOUS SYSTEM ACTIONS

The psychological literature provides a significant amount of research exploring the influence of goals on human behaviour. As outlined in the previous section, there are a number of concepts that surround goals, including the specification of end states, both to approach and avoid, as well as goal values, discrepancy, and the influence of pursuing multiple goals simultaneously. In order for a human to present an autonomous system with mission goals, these very human concepts need to be defined in an algorithmic way to influence the behaviour and actions of the autonomy.

One possible approach is the Multiple Goal Pursuit Model (MGPM). At any point, the human operator, or autonomous systems, may have several goals they are striving to achieve, some potentially conflicting with others. As a result, choices and trade-offs will need to be made. The environment may change rapidly or unpredictably, making it difficult to assess whether a goal can be achieved, whether a strategy for achieving a goal is still effective, or whether other goals need attention (Neal et al., 2017). The MGPM is based on human decision-making and aims to understand how people balance multiple goals. According to the MGPM, the choice between competing goals is determined by comparing the expected utilities of acting on each goal, and then selecting the goal with the highest expected utility (Neal et al., 2017). Expected utility represents the combination of expectancy and a sense of value, referred to as valence. Valence is the product of the discrepancy (i.e., need) and the degree of importance the goal has to the individual (i.e., gain). In simple terms, valence determines how motivating a goal is to pursue for an individual. Going forward, this chapter uses the term motivation to describe the sum of the discrepancy and importance of all goals an individual is pursuing. Expectancy represents a subjective sense of the resources required to achieve the goal compared to the resources an individual is willing or able to put forward. The subjective sense of resources required (e.g., time needed) is determined by the discrepancy and a belief regarding the rate at which the discrepancy can be reduced.

The MGPM accounts for the tendency of people to prioritise the goal with the greatest need when the expectancy of the goal is relatively high, and the goal with the best chance of being achieved when expectancy for one or more goals is relatively low (Neal et al., 2017). The model can explain some behaviour, such as the tendency to switch from prioritising the goal with the greatest need to the goal with the best chance of being achieved as a deadline approaches, or accounting for individual differences in the timing of this priority shift, as well as the effect of incentives on goal choice (Neal et al., 2017).

More recently, the MGPM was extended to incorporate Decision Field Theory, a sequential sampling model that describes how preferences for courses of action can develop when there are many possible consequences of the actions (Busemeyer & Townsend, 1993). According to the extended MGPM, people decide between actions by considering, one at a time, the possible consequences of the action with respect to the various goals they are pursuing. The person's preference for an action develops over time as they consider different consequences and accumulate more information about the effect an action may have on progress towards different goals. The updated MGPM is thought to better account for different types of goals, and can therefore explain why patterns of resource allocation change when people pursue different combinations of approach and avoidance goals (Neal et al., 2017). These combinations allow us to define five types of goals, adapted from Neal et al. (2019), which are outlined below and illustrated in Figure 10.1.

1. Approach Above Goal, where the discrepancy function is Δd = Reference state – Variable state for variable states less than the reference state, and $\Delta d = 0$ for variable states that are greater than or equal to the reference state. The goal is to decrease Δd. An example of this scenario is a factory manager striving to reach a production target.
2. Approach Below Goal, where the discrepancy function is $\Delta d = 0$ for variable states less than the reference state, and Δd = Reference state – Variable state for variable states that are greater than or equal to the reference state. The goal is to decrease Δd. An example of this scenario is a factory manager striving to reduce the defect rate below a specified level.
3. Avoid Above Goal, where the discrepancy function is Δd = Reference state – Variable state. The goal is to increase Δd. An example of this scenario is a factory manager striving to prevent inventory levels falling below a specified level.
4. Avoid Below Goal, where the discrepancy function is Δd = Variable state – Reference state. The goal is to increase Δd. An example of this scenario is a factory manager striving to prevent injury rates from reaching a critical threshold.
5. Approach Match Goal, where the discrepancy function is Δd = Reference state – Variable state for variable states less than the reference state, and Δd = Variable state – Reference state for variable states that are greater than the reference state. The goal is to decrease Δd. An example of this scenario is a factory manager trying to maintain an adequate level of staff.

For the MGPM to explain how people prioritise the allocation of time or resources when pursuing a set of competing goals, it assumes that people do this by monitoring

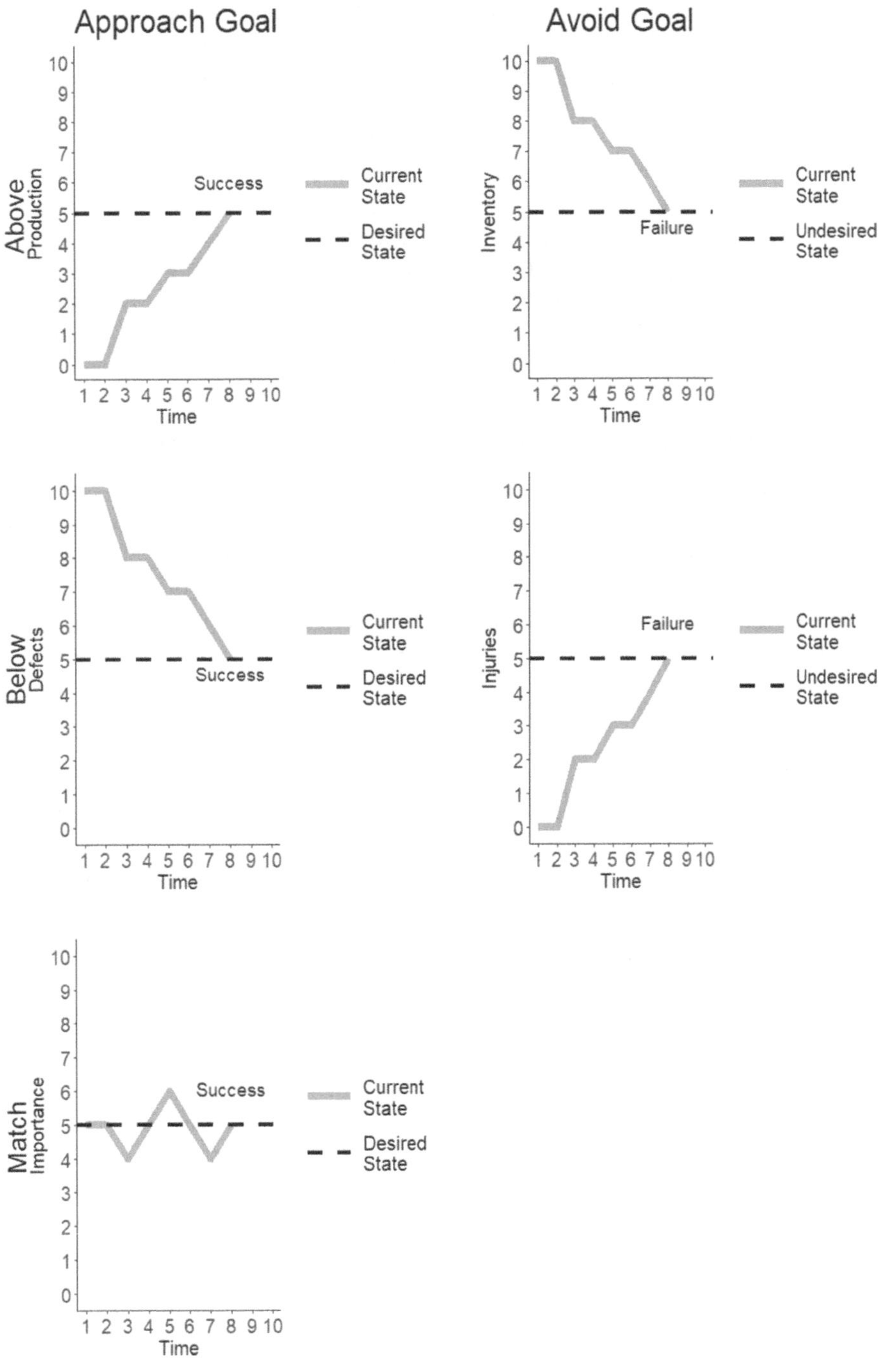

FIGURE 10.1 The five goal types. Approach Above (top left), Avoid Above (top right), Approach Below (middle left), Avoid Below (middle right), and Approach Match (bottom left).

the relevant variable for each goal and assessing the discrepancy between the current level of the variable and the reference level for that variable. Going beyond the use of the MGPM to understand individual decision-making, researchers at the University of Queensland found that it was possible to develop a computational model of distributed decision-making using the MGPM in a feasibility study (Farrell et al., 2019). This research used a future missile defence scenario in which a Maritime task force needed to defend itself against a series of incoming threats. This suggests that the MGPM can be used to facilitate autonomous system decision-making, and may be one way of allowing multiple autonomous systems to break down goals into actions and constraints, and allow the human to dynamically change and adjust mission goals. The use of the MGPM to support goal-based control may enable one human operator to manage and control multiple autonomous systems.

10.6 DEVELOPING A CONCEPTUAL CONTROL FRAMEWORK

In order to understand whether goal-based control supports human-autonomy teaming and C2, it is proposed that a new framework with different levels of control be defined and tested through experimentation. Based on existing control methods in the literature, and the need for higher-level control strategies, a three-level control framework was developed by DSTG. The framework is presented in Table 10.1. The levels of control framework assumes that the level of system autonomy is high and consistent across control levels.

10.7 HOW DOES GOAL-BASED CONTROL WORK?

Goal-based control is managed with the operator providing the aims and constraints for each goal for the UxVs to translate into actions. All UxVs are pursuing all defined goals simultaneously, and make decisions to maximise all goals at each time step. At each time step, each UxV has the option to pick one of a range of possible next

TABLE 10.1
Proposed Control Types

Proposed Levels of Autonomous System Control	Control Type Definition
Goal-based control	The human operator issues mission goals which require autonomous systems defined by autonomy functions to translate the goals into a series of actions for implementation by each system element
Plan-based control	The human operator issues a series of plans to each system element, which are expected to meet the mission goals, which requires each system element to translate the plan into a series of actions (e.g., a flight path) for implementation
Action-based control	The human operator controls each system element individually, issuing each system action (such as setting waypoints, slewing cameras, etc.) that are expected to meet mission plans and goals

actions. The waypoint with the highest overall motivation is selected as the next time-step location, where the motivation is defined as

$$M_j = \sum_{i=1}^{N} I_i \times \left(\tilde{V} * \Delta d\right)_j \qquad (10.1)$$

where for N goals, the value for a given waypoint j is calculated as the sum of all goals (from $i = 1$ to N), where each element is a combination of the relative goal importance I_i multiplied by the output of the function $(V * \Delta d)_j$, which provides the normalised value along the value function curve multiplied by the proposed change in discrepancy for the given waypoint. Δd is discrepancy, which is calculated as the difference between the current state of the UxV and a reference level determined by the goal. $\tilde{V}$ is the normalised position between 0 and 1, which represents the percentile along the importance value curve at the current state for that goal. This value is then normalised with respect to the maximum possible single-iteration increment for that goal. For approach goals, waypoints that reduce discrepancy (i.e., have a negative delta discrepancy) would have a positive motivation, and waypoints that increase discrepancy would have a negative motivation. For avoid goals, waypoints that increase discrepancy (i.e., have a positive delta discrepancy) would have a positive motivation, and waypoints that decrease discrepancy would have a negative motivation. Approach goals have zero discrepancy once the reference value is reached, while avoid goals always strive to increase the discrepancy between the reference value and variable state.

The value function is a logistic curve, which can have either a positive or a negative slope and associated steepness. Value functions with a positive slope produce low values for acting on a goal when the relevant environmental state is low, and high values when the relevant environmental state is high. Conversely, value functions with a negative slope produce high values for acting on a goal when the relevant environmental state is low, and low values when the relevant environmental state is high. The value function definition for each of the goal types is presented in Table 10.2.

The effectiveness of goal-based control is expected to be influenced by several factors, including task difficulty, the number of mission goals, the speed of actions, the range of possible actions, information availability, how dynamic or stable

TABLE 10.2
Proposed Goal Value Functions

Proposed Goal Types	Value Function Definition
Approach above	Value function is high for small variable states and low for large variable states
Approach below	Value function is low for small variable states and high for large variable states
Avoid above	Value function is low for small variable states and high for large variable states
Avoid below	Value function is high for small variable states and low for large variable states
Approach match	Value function is high for small and large variable states and low when the variable state is near the reference state

the environment is, and the level of uncertainty present in the environment. The appropriate level of control may vary depending on these factors, and therefore these control concepts need to be tested experimentally, manipulating these factors systematically. In order to do this, DSTG has developed an experimental test-bed, which will be described in the following section.

10.8 THE EXPERIMENTAL TEST-BED

The experimental test-bed was created by DSTG researchers in collaboration with software developers from Swordfish Computing to investigate and experiment with the three levels of control outlined in Table 10.1. In order to ground the test-bed within a realistic Defence context, a simplified maritime search and rescue scenario was created, where the objective is to search as much of the designated search area as possible within a designated timeframe using multiple UxVs. To successfully complete the mission, the UxVs must avoid colliding with each other, avoid entering any defined no-fly zones, and return to base before running out of battery. Graphs within the test-bed are defined by units relative to each specific goal. For example, both the do-not-collide and avoiding-no-fly-zone goals are in metres, where participants define minimum distances where it approaches the maximum value of acting on.

These goals represent a subset of possible goals, which are defined prior to the launch of the scenario through pre-defined frameworks within the configuration files. The configuration file also allows users to define the map area, the number and location of no-fly zones, and the number of UxVs. Figure 10.2 provides an example of the experimental

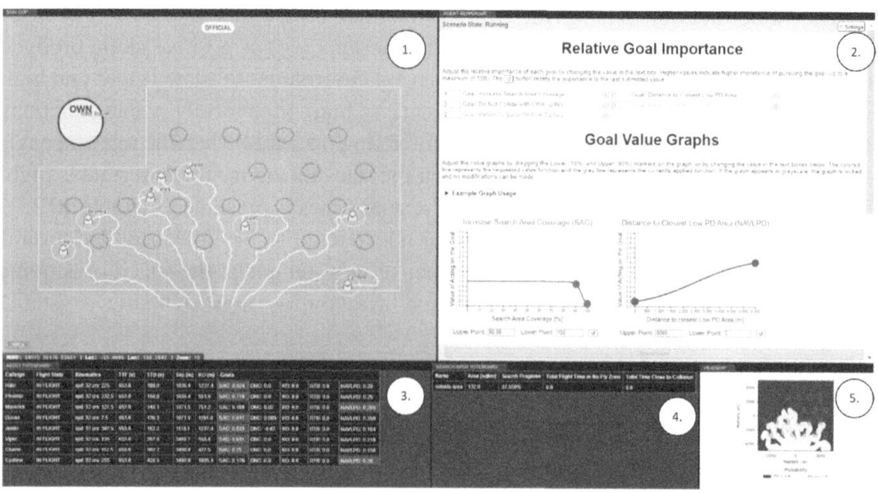

FIGURE 10.2 Goal-based control experimental test-bed configuration, which can be broken into five main components. Component 1 (the large area on the left) shows the live UxVs. Component 2 (top right) is where users interact with the goals. Component 3 (bottom left) shows real-time UxV information. Component 4 (bottom middle) shows key metrics related to the search area and goals. Component 5 (bottom right) contains the heatmap that shows what areas the UxVs have covered.

test-bed user interface, showing the goal-based control scenario, configuration, and components. Each component of the interface will be discussed in turn.

1. **Map**: The map component displays a map view of the experimental search area. It includes: an outside border indicating the search area's border, octagons marking no-fly zones, a large circular symbol in the top left corner representing the return-to-base location, small semi-elliptical symbols indicating the name and current location of each UxV, small dots showing the previous positions of the UxVs, and a ring around each of the UxVs representing the search area detection radius. All map parameters, such as the search area, size, number of no-fly zones, UxV base location and starting points, and number of UxVs, can be controlled within the configuration file prior to starting the experimental session. Individual waypoints and waypoint plans can also be created through the map component using a mouse during action-based or plan-based control scenarios.
2. **Goal Setting**: This component allows users to define and interact with the goals that control the decisions being made by the UxVs. The top part of this component is used for setting the relative importance of each goal, with users able to allocate an integer between zero and 100. The bottom part of this component allows users to set the value function for each goal, with the option to adjust the points on each graph or manually enter a value in the boxes below each graph. The upper point represents 90% of each goal's relative importance, while the lower point represents 10% of each goal's relative importance. The vertical axis of all plots is scaled relative to the highest relative importance value set in the top part of this component. Both the relative importance and value function sliders can be edited and re-submitted at any time, with the updated information being immediately passed to the UxVs, which adjust their decision-making accordingly.
3. **Asset Tote Board**: The asset tote board component displays real-time information for each UxV. Each row of the tote board corresponds to one UxV, with the columns displaying information including the UxV name, flight status, speed, course, remaining flight time, time to travel in a straight line back to the base location, distance to the nearest UxV, distance to the nearest no-fly zone, and the current motivation for each goal. The current implementation of goals includes searching the area (SAC), not colliding with another UxV (DNC), keeping out of the no-fly zones (KO), returning to base on low battery (RTB), and navigating to the nearest unsearched area (NAVLPD). The goal with the highest motivation value for each UxV is highlighted at each time step to indicate that it is having the highest influence on the UxVs' previous waypoint decision. The intent of this component is to provide users with a quick understanding of each UxVs' status and decision-making priorities.
4. **Search Area Tote Board**: The search area tote board provides an overview of the key metrics related to the UxVs' search area activities. This includes the percentage of the search area covered by the UxVs, the time spent collectively inside no-fly zones, and the time the UxVs spend collectively in

close proximity to each other. This data is updated in real time, giving users a view of the UxVs' search area performance and compliance with the avoidance goals.

5. **Heatmap**: This component displays a visualisation of the searched and unsearched areas. Outlines of the search area boundary (grey) and the no-fly zones (grey) are displayed, with the searched areas coloured light grey and the unsearched area remaining dark grey . Each UxV within the simulation shares a common heatmap.

Throughout the development of the goal-based control approach and associated test-bed, two user interaction studies have been conducted. Information about each of these studies will be outlined below.

10.9 USABILITY STUDY

The first experimental study (Summers et al., 2024) examined the usability of the novel control interface presented in this chapter. In this study, 10 participants from DSTG undertook the role of an operator in a simplified maritime search and rescue scenario and were asked to identify any human–machine interface features that hindered or enhanced the usability of the goal-based control concept. Prior to beginning the experimental task, participants were given approximately 30 minutes of training, which consisted of an initial PowerPoint slide deck, covering the background and non-specific examples of how to interact with the interface. They were then given a simplified version of the task to interact with, where they could ask questions of the researchers. Based on quantitative subjective assessments using the System Usability Scale (Brooke, 1996) and the User Experience Questionnaire (Laugwitz et al., 2008), the usability of the goal-based control system was considered below average when compared to industry benchmarks. Qualitative responses from participants indicated areas for further development. The usability study revealed a lack of transparency in the current interface. This led to the introduction of both the asset and search area tote boards to provide users with insight into the UxVs' decision-making and overall performance. Another theme that emerged from this study was the participants' interaction with the value function graphs. In the initial instantiation of the test-bed, users were only able to interact with the value functions by moving sliders underneath the graphs. This led to users stating that they felt they were unable to achieve specific values, while also noting that not being able to interact with the points on the graphs directly was unintuitive. Based on this feedback, the points became interactive and the sliders became manual inputs, which update when the points are moved.

10.10 PLAN-BASED vs. GOAL-BASED CONTROL COMPARATIVE STUDY

The second experimental study (O'Brien et al., 2025) examined the impact of increasing the number of UxVs being managed on task performance and workload when using goal-based control compared to the plan-based control paradigm. The study involved 24 university student participants and used a 2x2 mixed-methods design, with the control paradigm (action-based versus goal-based) serving as the between-participants factor and the number of UxVs being controlled (four versus eight) as a within-participants factor. Once again, the simplified maritime search task was used, and the instructions were identical for both control conditions, except that in action-based control, participants had to manually set all waypoints for each UxV. Once a UxV reached the end of a series of waypoints, it would loiter around that point until the participant set a new plan. As with the usability study, participants were given approximately 30 minutes of training, which consisted of an initial PowerPoint slide deck covering the background and non-specific examples of how to interact with the interface. They were then given a simplified version of the task (either action-based or goal-based) to interact with, where they could ask questions of the researchers.

This study found partial support for the goal-based control concept. Workload was higher overall when participants were controlling eight UxVs compared to four UxVs, and this increase in workload with additional UxVs was credibly lower when using goal-based control compared to action-based control. However, task performance was generally lower within the goal-based condition. The percentage of search area covered was higher when controlling eight UxVs compared to four, regardless of the control paradigm being used. However, less of the search area was covered when using goal-based control compared with plan-based control. The results indicated that participants may have struggled to fully understand the concepts involved within the goal-based control condition, or had some difficulty interacting with the test-bed interface. This possibility was supported by the analysis of the importance and value settings used by participants in goal-based control. This indicated that participants sometimes set goal value functions that produced UxV behaviours that were inconsistent with the desired goals. Sometimes, but not always, these settings were corrected during a scenario. In addition, the importance and value settings often differed between the four UxV and eight UxV conditions, which suggests that participants were exhibiting learning effects across the goal-based control scenarios.

These apparent learning effects make it difficult to draw strong conclusions about the impact of goal-based control on performance. Future studies should provide a greater level of training for participants, allowing them to become more familiar with both the concept of goal-based control and the test-bed interface, and include current UxV controllers as participants. Future studies should also consider possible improvements to the usability of the goal-based control interface. Participants in the GBC condition frequently reported finding the system difficult to use, and did not adequately understand the graphs or feel a sense of control. Participants often asked, 'why are they [the UxVs] doing that?', and a better interface design may address the apparent lack of connection between graphs, setting values, and the sense of control of UxVs. Future studies could also consider extending the complexity of the goals while exploring how to define and represent hierarchical goal structures (Whitney, 2024).

10.11 FUTURE DEVELOPMENT

Continual development of the test-bed is required to be able to translate the concept of GBC into both real and simulated UxVs. Through the use of MAVLink[1] protocols, the GBC paradigm has been able to be translated onto simulated UxVs within the Gazebo[2] simulation environment, with a plan to extend this to the control of physical UxVs. Further studies will also be required to help refine and develop the current concept and test-bed. The ability to run any further experimentation with more experienced and representative users will also allow for greater fine-tuning of the concept and hopefully avoid some of the errors associated with learning effects.

10.12 CONCLUSION

The shift from action-based and plan-based control to goal-based control represents a significant evolution in human–autonomy teaming, particularly in the context of managing multiple intelligent autonomous systems in dynamic and uncertain environments. This chapter has outlined the theoretical foundations, technical implementation, and experimental evaluation of goal-based control, demonstrating its potential to reduce operator workload and enhance system adaptability.

By leveraging psychological models such as the MGPM, goal-based control enables autonomous systems to interpret mission objectives, balance competing priorities, and make decisions that align with human intent. The experimental test-bed developed by DSTG has provided a valuable platform for exploring these concepts in a realistic Defence scenario, revealing both the promise and the challenges of this approach.

While initial studies suggest that goal-based control can support the management of more autonomous systems with lower cognitive demand, usability and interface design remain critical areas for improvement. Future research should focus on refining the user interaction in a goal-based control paradigm, enhancing transparency in autonomous decision-making, and expanding the complexity of goal structures to better reflect operational realities.

Goal-based control offers a potential pathway toward more effective and scalable human–autonomy teaming. As the ADF continues to integrate intelligent autonomous systems into operations, this paradigm may prove essential for maintaining meaningful human oversight, achieving mission success, and ensuring resilience in complex tactical environments.

NOTES

1 https://mavlink.io/en/
2 https://gazebosim.org/home

REFERENCES

Atkinson, J. W., & Birch, D. (1970). *The dynamics of action*. John Wiley.

Azhar, M. Q., & Sklar, E. I. (2017). A study measuring the impact of shared decision making in a human-robot team. *The International Journal of Robotics Research*, *36*, 461–482. https://doi.org/10.1177/0278364917710540

Busemeyer, J. R., & Townsend, J. T. (1993). Decision Field Theory: A dynamic-cognitive approach to decision making in an uncertain environment. *Psychological Review*, *100*, 432–459.

Brehm, J. W., & Self, E. A. (1989). The intensity of motivation. *Annual Review of Psychology, 40*, 109–131.

Brooke, J. (1996). SUS: A 'quick and dirty' usability scale. In P. W. Jordan, B. Thomas, B. A. Weerdmeester, & I. L. McClelland (Eds.), *Usability evaluation in industry* (pp. 189–194). Taylor & Francis.

Carver, C. S., & Scheier, M. F. (1981). *Attention and self-regulation: A control theory approach to human behaviour*. Springer. https://doi.org/10.1007/978-1-4612–5887-2

Cummings, M. L., Clare, A, & Hart, C. (2010). The role of human-automation consensus in multiple unmanned vehicle scheduling. *Human Factors, 52*(1), 17–27. https://doi.org/10.1177/0018720810368674

Cummings, M. L., & Mitchell, P. (2008). Predicting controller capacity in supervisory control of multiple UAVs. *IEEE Transactions on Systems, Man and Cybernetics – Part A: Systems and Humans, 38*(2), 451–460. https://doi.org/10.1109/TSMCA.2007.914757

Draper, M., Rowe, A., Douglass, S., Calhoun, S., Kingston, D., Humphrey, L., Patzek, M., Bartisk, J., Lange, D., Reeder, J., Croft, B., Chen, J., Barnes, M., Trafton, G., Apker, T., Behymer, K., Ruff, H., Rothwell, C., Evans, D., & Frost, E. (2018). Realising autonomy via intelligent adaptive hybrid control: adaptable autonomy for achieving UxV RSTA team decision superiority (also known as intelligent multi-UxV planner with adaptive collaborative/control technologies (IMPACT)) (Report No. AFRL-RH-WP-TR-2018-0005). Air Force Research Laboratory.

Farrell, S., Neal, A., Ballard, T., Palada, H., Fletcher, K., & Knight, D. (2019). Extension of the multiple goal pursuit model to multi-agent decision making. Australian Government, Department of Defence, Defence Science and Technology Group.

Fishbach, A., & Ferguson, M. J. (2007). The goal construct in social psychology. In A. W. Kruglanski & E. T. Higgins (Eds), *Social psychology: Handbook of basic principles* (pp. 490–515). The Guilford Press.

Grimm, D., Demir, M., Gorman, J. C., & Cooke, N. J. (2018). The complex dynamics of team situation awareness in human-autonomy teaming. *2018 IEEE Conference on Cognitive and Computational Aspects of Situation Management (CogSIMA),* Boston, USA, (pp. 103–109).

Hancock, P. A., Jagacinski, R. J., Parasuraman, R., Wickens, C. D., Wilson, G. F., & Kaber, D. B. (2013). Human-automation interaction research: Past, present and future. *Ergonomics in Design, 21*, 9–14.

Hua, H., Zhi-Fei, L., Wei-Ping, W., Yi-Fan, Z., Xiao-Bo, L. (2017). CB-MLC21 – A command and control architecture for unmanned combat SoS. *IEEE International Systems Engineering Symposium (ISSE),* Vienna, Austria, (pp. 1–6).

Johnson, M., Bradshaw, J. M., Feltovich, P. J., Jonker, C., van Riemsdijk, B., & Sierhuis, M. (2012). Autonomy and interdependence in human-agent-robot teams. *IEEE Intelligent Systems, 27*, 34–51.

Kaber, D. B. & Prinzel, L. J. (2006). Adaptive and adaptable automation design: a critical review of the literature and recommendations for future research. NASA Technical Report NASA TM-2006-214505. Langley Research Centre, VA: National Aeronautics and Space Administration.

Kruglanski, A. W., Shah, J. Y., Fishbach, A., Friedman, R., Chun, W. Y., Sleeth-Keppler, D. (2002). A theory of goal systems. *Advances in Experimental Social Psychology, 34*, 331–378. https://doi.org/10.1016/S0065-2601(02)80008-9

Laugwitz, B., Schrepp, M. & Held, T. (2008). Construction and evaluation of a user experience questionnaire. In: Holzinger, A. (Ed.): USAB 2008, LNCS 5298, pp. 63–67.

Lewin, K. (1936). *Principles of topological psychology* (F. Heider & G. M. Heider, Trans). McGraw-Hill. https://doi.org/10.1037/10019-000

Lewis, M., Sycara, K., & Payne, T. (2003). Agent roles in human teams. *AAMAS-03 Workshop on Humans and Multi-Agent Systems.* Melbourne, Australia.

Lyons, J. B., Mahoney, S., Wynne, K. T., & Roebke, M. A. (2018). Viewing machines as teammates: A qualitative study. *2017 AAAI Spring Symposium Series*. California, USA.

McClelland, D. C., Atkinson, J. W., Clark, R. A., & Lowell, E. L. (1953). Toward a theory of motivation. In D. C. McClelland, J. W., Atkinson, R. A., Clark, & E. L. Lowell (Eds.), *The achievement motive* (pp. 6–96). Appleton-Century Crofts. https://doi.org/10.1037/11144-002

Mi, Z., & Yang, Y. (2013). Human-robot interaction in UVs swarming: a survey. *International Journal of Computer Science Issues*, *10*(2), 273–280.

Neal, A., Ballard, T., Palada, H., Farrell, S., Fletcher, K., & Knight, D. (2019). Initial mapping of the multiple goal pursuit model onto a missile defence scenario. Australian Government, Department of Defence, Defence Science and Technology Group.

Neal, A., Ballard, T., & Vancouver, J. B. (2017). Dynamic self-regulation and multiple-goal pursuit. *Annual Review of Organizational Psychology and Organizational Behaviour*, *4*(1), 401–423. https://doi.org/10.1146/annurev-orgpsych-032516-113156

O'Brien, E., Fahy, T., & Fletcher, K. (2025). Operator workload and task performance under action-based and goal-based control of multiple uncrewed aerial vehicles. Australian Government, Department of Defence, Defence Science and Technology Group.

O'Neil, T., McNees, N., Barron, A., & Schelble, B. (2020). Human-autonomy teaming: A review and analysis of the empirical literature. *Human Factors*, 64(5), 904–938. https://journals.sagepub.com/doi/10.1177/0018720820960865

Parasuraman, R., Sheridan, T., & Wickens, C. (2000). A model for types and levels of human interaction with automation. *IEEE Transactions on Systems, Man, and Cybernetics – Part A: Systems and Humans*, *30*(3), 286–297. https://doi.org/10.1109/3468.844354

Royal Australian Navy. (2020). RAS-AI strategy 2040. Canberra: Defence Publishing Service.

Seeber, I., Bittner, E., Briggs, R. O., de Vreede, T., de Vreede, G., Elkins, A., Maier, R., Merz, A. B., Oeste-Reib, S., Randrup, N., Schwabe, G., & Soilner, M. (2020). Machines as teammates: A research agenda on AI in team collaboration. *Information & Management*, 57, 103174. https://doi.org/10.1016/j.im.2019.103174

Shah, J. Y., Hall, D., & Leander, N. P. (2009). Moments of motivation: Margins of opportunity in managing the efficacy, need, and transitions of striving. In G. B. Moskowitz, & H. Grant (Eds.), *The psychology of goals* (pp. 234–254). Guilford Press.

Sheridan, T. B. (1992). *Telerobotics automation, and human supervisory control*. Cambridge MIT Press.

Werkhoven, P., Neerincx, M., & Kester, L. (2018). Telling autonomous systems what to do. *European Conference on Cognitive Ergonomics 2018 Proceedings*. https://doi.org/10.1145/3232078.3232238

Summers, D., Fahy, T., & Fletcher, K. (2024). Usability testing of a goal-based control user interface prototype. (DSTG-TR-4111) Australian Government, Department of Defence, Defence Science and Technology Group.

Whitney, S. (2024). Goal decomposition in goal based control. (DSTG-GD-1266) Australian Government, Department of Defence, Defence Science and Technology Group.

Woods, D. D. M., & Cook, R. I. (2006). Incidents-markers of resilience or brittleness. In *Resilience Engineering: Concepts and Precepts* (pp. 69–76).

Wright, J. L., Chen, J. Y. C., & Barnes, M. J. (2018). Human-automation interaction for multiple robot control: The effect of varying automation assistance and individual differences on operator performance. *Ergonomics*, *61*, 1033–1045. https://doi.org/10.1080/00140139.2018.1441449

Wright, J. L., Quinn, S. A., Chen, J. Y., & Barnes, M. J. (2014). Individual differences in human-agent teaming: An analysis of workload and situation awareness through eye movements. *Proceedings of the Human Factors and Ergonomics Society Annual Meeting*, *57*, 1410–1414.

Wright, M. C., & Kaber, D. B. (2005). Effects of automation of information-processing functions on teamwork. *Human Factors: The Journal of Human Factors and Ergonomics Society*, *47*, 50–66. https://journals.sagepub.com/doi/10.1518/0018720053653776

Wright, R. A. (1996). Brehm's theory of motivation as a model of effort and cardiovascular response. In P. M. Gollwitzer & J. A. Bargh (Eds), *The psychology of action: Linking cognition and motivation to behaviour* (pp. 424–453). The Guilford Press.

Xu, W. (2021). From automation to autonomy and autonomous vehicles: Challenges and opportunities for human-computer interaction. *Interactions*, *28*(1), 49–53. https://doi.org/10.1145/3434580

11 Human-AI Leadership for Search and Rescue

Anirudh More, Tianzi Wang, and Nathan Lau

11.1 INTRODUCTION

The increasing capabilities and integration of AI (artificial intelligence) or autonomous agents into complex, high-stakes, dynamic environments such as search and rescue (SAR) are transforming team coordination and leadership in multi-agent systems. Operations in dynamic environments are often time-critical, geographically dispersed, and characterized by uncertainty, making (fully) centralized command and control inefficient (Kaber & Endsley, 2004; Roth et al., 2019). In such environments, the volume of data and rapid changes in conditions can overwhelm a single human leader, creating a bottleneck that hinders team performance and safety (Lyons et al., 2021; McNeese et al., 2018). Consequently, new interaction and coordination paradigms are required to maximize the potential of teams integrating human and AI agents.

Shared leadership, defined as the distribution or emergence of leadership responsibilities among team members based on expertise and situational demands, offers an alternative to traditional hierarchical structures (Carson et al., 2007; Pearce & Sims Jr, 2002). In human-agent teaming (HAT), this concept promotes fluid task management where agents negotiate leadership transfer to the most qualified agent, driven by factors such as workload redistribution or superior competence in unanticipated events. While organizational psychology research has demonstrated that shared leadership improves adaptability, resilience, and performance in human teams and organizations, particularly in dynamic environments (Zhu et al., 2018; Wang et al., 2014; Wu & Cormican, 2021), effective application of shared leadership in HAT presents a significant methodological challenge, as the mechanisms governing leadership negotiation and transfer between humans and AI are underexplored.

Experimental testbeds featuring AI and interaction capabilities are scarce, particularly those advancing shared leadership research in HAT. Existing testbeds have supported HAT research to improve team performance, effectiveness, and coordination (O'Neill et al., 2022). However, these testbeds primarily facilitate dyadic and/or one-to-one interaction between human and AI agents and may be insufficient for studying multi-agent systems, where decision-making authority must be negotiated among multiple, heterogeneous human and AI agents (Drescher et al., 2014). Research must also address many assumptions about human leadership negotiation behaviors and develop empirical models of human–AI leadership transfer.

To investigate shared leadership, a testbed must simulate representative tasks and capture nuanced human–agent decision-making for empirical modeling. We are

DOI: 10.1201/9781003655923-14

developing the Human-AI Leadership for Search And Rescue (HALSAR) testbed to simulate SAR missions and capture interaction data for modeling leadership emergence in multi-agent systems. This chapter presents the motivation, architecture, and simulation features of HALSAR, followed by the research questions and studies that the testbed addresses.

11.2 BACKGROUND AND MOTIVATION

Foundational research on human collaboration with automation, such as function allocation or distribution of tasks in a joint system (Hollnagel & Bye, 2000; Sheridan et al., 1978), prescribes static task assignments based on complementary capabilities. This approach typically allocates repetitive or data-intensive tasks to autonomous agents while leaving unstructured or ad-hoc tasks requiring ill-defined problem-solving to humans. While adequate for less intelligent and less complex systems, static assignments often produce suboptimal task distribution when the presumed differences in capabilities between humans and technology are less pronounced. Factors such as interdependence, contextual task demands, and flexibility play increasingly important roles in dynamic environments (Roth et al., 2019).

Advancing beyond static approaches, adaptive automation distributes tasks based on workload, system failures, and routines tailored to agent capabilities (Bernabei & Costantino, 2024; Brandt et al., 2018; Tokadli & Dorneich, 2019). While these approaches allow for function reallocation based on predefined triggers, their reliance on bounded decision spaces constrains responsiveness to unanticipated scenarios. Crucially, these approaches primarily focus on dyadic interactions between one human and one AI agent, precluding investigation of how leadership may shift among multiple agents with diverse competencies in a large team at any given moment (More & Lau, 2024).

Traditional approaches conceptualize human–automation relationships as static or hierarchical authority structures rather than dynamic, emergent leadership processes among agents. Because authority is predefined, traditional approaches neglect unscripted leadership transfers within a team. This limitation is particularly evident in dynamic environments such as SAR operations involving numerous aerial and ground robots (Cubber et al., 2017). For example, search robots leading initial reconnaissance may only recognize their own ineffectiveness on-site under local weather conditions, requiring flexible task reassignments and leadership transfers.

Recent research recognizes the need for dynamic transfer of leadership. These studies have investigated computational models, such as graph-based reinforcement learning and Markov decision processes, that enable autonomous agents to infer human goals or workload for re-assigning tasks (Byeon et al., 2023; Kwon et al., 2019). These models optimize function allocation and thus system performance but rely on simulated data, neglecting interactive negotiation between agents and embedding untested assumptions about human behaviors. Computational methods for dynamic function allocation need empirical support on interactive negotiation mechanisms and real-time interactions through which leadership emerges.

Shared leadership offers a promising framework for formulating new HAT. Instead of predefining authority, shared leadership facilitates an ongoing, interactive process where authority is distributed among team members based on the recognition of expertise and evolving situational demands (Carson et al., 2007; Pearce & Sims Jr, 2002). This approach enables flexible leadership transfer, reducing the risk of single-point failure. In human teams and organizations, shared leadership has demonstrated improved team performance and adaptability in dynamic environments by allowing the most qualified member to lead at the most appropriate time (Drescher et al., 2014; Imam & Zaheer, 2021; Klasmeier & Rowold, 2020, 2022). Supporting shared leadership in HAT is thus critical for enabling flexible leadership dynamics in real-world multi-agent systems.

The growing interest in dynamic allocation for HAT has not yet produced interaction designs that explicitly facilitate shared leadership. In many effective human teams, leadership is shared through a reciprocal process of assuming and relinquishing leadership roles, governed by the perception of individuals' context-dependent competence and trustworthiness (DeRue et al., 2015; Drescher et al., 2014; Muethel et al., 2012). The HAT literature has found that AI trustworthiness and competence influence human–AI collaboration (Freedy et al., 2007; Lee & See, 2004; Mioch et al., 2018; Shahrdar et al., 2018) but has not applied them to facilitate shared leadership. Reciprocal leadership dynamics can be operationalized in HAT through interactions of claiming and granting leadership. Claiming occurs when agents proactively assume leadership when their expertise is (or should be) most relevant to current situational demands, while granting involves voluntarily transferring leadership to others who can better manage the situation. Together, these behaviors enable flexible authority distribution or emergence between multiple agents, distinguishing shared leadership from hierarchical approaches.

Research on shared leadership in HAT presents significant challenges because adapting methods for studying shared leadership in human teams is insufficient. Questionnaire-based approaches provide standardized measurements of leadership behaviors and perceptions, supporting cross-study comparisons (Hoch & Morgeson, 2014; Muethel et al., 2012). Network analysis approaches provide density and centralization metrics, quantifying influence patterns and revealing emergent leadership structures (Carson et al., 2007; DeRue et al., 2015; Mayo et al., 2002). Observational approaches offer rich contextual insights through longitudinal case studies (Carte et al., 2006; Drescher et al., 2014). However, questionnaire methods may overlook real-time behavioral dynamics; network approaches may overlook individual motivations and nuanced interpersonal dynamics; and observational approaches face scalability challenges, potential observer effects, and difficulties in establishing causality. Critically, none of these methods support systematic manipulation of AI agent characteristics and leadership structures for advancing HAT research.

Shared leadership research in HAT requires a testbed that represents realistic operational environments and captures the nuanced interaction dynamics of leadership emergence. Such a testbed must support systematic manipulation of variables like agent capabilities, environmental conditions, and interaction mechanisms while measuring objective behaviors and subjective experience (e.g., trust, self-efficacy) over time. Dedicated experimental testbeds can provide the evidence base to translate shared leadership concepts into operational design principles. These reasons motivate the development of the HALSAR testbed.

11.3 HALSAR TESTBED: ARCHITECTURE, SIMULATION ENVIRONMENT, AND MEASUREMENT

HALSAR was grounded in human decision-making under dynamic environments. The SAR domain was chosen as an analog for dynamic operational environments where traditional, centralized command-and-control proves suboptimal. SAR operations involve inherent complexity and uncertainty, necessitating coordination of multiple distributed autonomous agents to be effective (Cubber et al., 2017). HALSAR simulates SAR missions where autonomous uncrewed aerial vehicles (UAVs) survey distinct geographical regions and detect objects, and the human collaborates with AI agents to identify detected objects and potentially reassign detections to optimal UAVs.

11.3.1 Architecture

HALSAR has a modular, three-tier architecture (Figure 11.1). The application layer serves as the testbed's core, consisting of a Django-based backend server. The Django

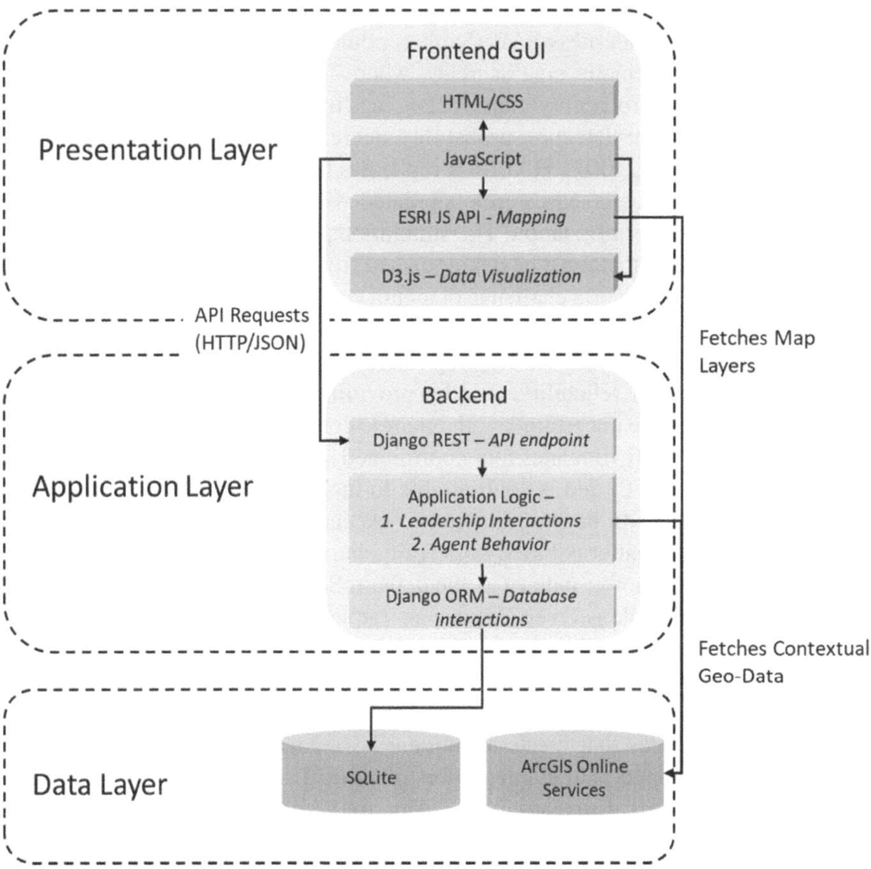

FIGURE 11.1 HALSAR architecture.

REST framework manages all experimental logic, including shared leadership interactions and agent behaviors, while the Django ORM handles database operations. The presentation layer comprises a GUI built with HTML, CSS, and JavaScript that handles all human–machine interactions, including UAV assignment, object detection notification and classification, and leadership claiming and granting inputs. This layer integrates the ESRI JavaScript API for interactive maps, and D3.js for area segmentation and visualization, with map layers fetched from ArcGIS Online Services via HTTP/JSON API requests. The data layer consists of a SQLite database for storing persistent application data (e.g., experiment status and user interactions), complemented by ArcGIS Online Services for contextual geospatial data including basemaps and terrain information. User inputs trigger asynchronous API calls, enabling real-time state changes and leadership negotiations between human and AI agents. The complete testbed implementation, including source code and documentation, is publicly available through GitLab at https://code.vt.edu/anirudh1/LEESH.

11.3.2 Simulation Environment

HALSAR simulates the time-critical and complex decision-making demands for the search phase of wilderness SAR (Statheropoulos et al., 2015). Traditional SAR missions involve systematic area searches where teams must prioritize segments based on lost person probability maps, balancing limited resources against expanding search areas where detection probability decreases significantly after the first 51 hours (Adams et al., 2007). HALSAR replicates these challenges by presenting a geographically distributed search area populated with multiple object types requiring classification and prioritization. The simulation presents participants with living and non-living targets that must be distinguished from living and non-living distractors, mirroring the real-world challenge of identifying clues amidst irrelevant objects. Detection events occur randomly across the search area under varying environmental conditions, including different lighting scenarios that affect sensor performance and image recognition reliability, thereby providing temporal and environmental variability to reflect the uncertainty and dynamic conditions of SAR operations.

Currently, HALSAR simulates four coordinated autonomous UAVs surveying the search area (Figure 11.2) and is configurable to include additional UAVs and other search robots. UAVs can be programmed to execute various search patterns such as spiral or adaptive patterns based on terrain characteristics and mission requirements, with the lawnmower pattern as the default. Search areas can be dynamically partitioned using flexible geometric divisions rather than fixed quadrants, allowing researchers to investigate different spatial allocation strategies and their effects on team coordination and leadership emergence (Williams et al., 2020).

HALSAR accommodates UAVs with diverse sensor configurations, simulating detection scenarios that create situational advantage for specific sensor types. The architecture supports the integration of optical, multispectral, or synthetic aperture radar systems. Currently, two UAVs are equipped with thermal imaging sensors, which excel at detecting living objects, while the remaining two feature near/short-wave infrared sensors for identifying non-living objects based on their spectral reflectance (Kaplan, 2007; Yang et al., 2021). Image recognition reliability

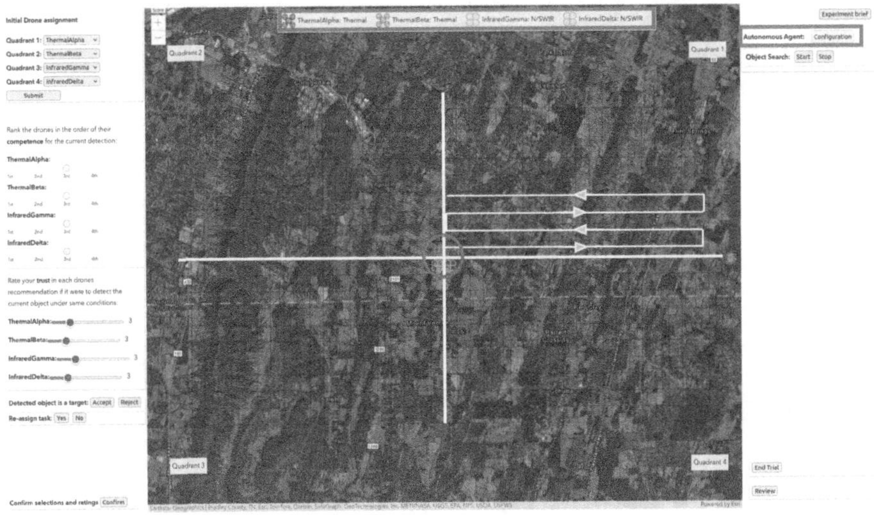

FIGURE 11.2 HALSAR platform with autonomous UAVs.

is systematically varied within each sensor type to create realistic performance differentials between UAVs, enabling investigation of how differentiated capabilities shape competence and trust perceptions that drive leadership delegation.

HALSAR enables autonomous assignment capabilities that can operate independently of human oversight for investigating the full spectrum of autonomy. Conflicts in leadership claims and grants between the AI agent and human operators are resolved through logic defining permitted authority structures, facilitating systematic examination of emergent coordination patterns. These autonomous assignment functions can be configured based on various criteria including sensor specialization, target proximity, remaining operational capacity, or composite performance metrics, providing flexibility for investigating different allocation strategies and their effects on team and leadership dynamics.

HALSAR integrates a lost person model providing probabilistic target distribution data to guide human and AI agents in decision-making. While full integration is under development, the platform currently displays lost person probability outputs as heatmaps and probability distributions accessible during mission execution (Wang, 2023; Wang & Lau, 2023). This integration enables investigation of how predictive modeling influences leadership patterns, with future enhancements planned to support real-time model updates based on detection events and environmental changes, creating dynamic probability landscapes that reflect evolving search conditions.

In HALSAR studies, human participants engage in critical decision-making beginning with initial UAV assignment to search different areas. During the search mission, participants receive notifications on classified objects after detection with information on the UAVs, an image of the object, environmental conditions, and image recognition/sensor reliability estimates. Depending on the leadership configuration, participants can reassign UAVs, transfer reassignment authority to an AI agent, or input subjective ratings on capabilities of UAV and AI agents (Figure 11.3). Performance feedback

FIGURE 11.3 HALSAR platform GUI for trust and competence ratings for the UAVs (left side) and AI agent (right side).

is provided through summary tables at experimenter-defined intervals. These interactions enable the study of how the human's subjective experience of AI agents, such as trust and competence perception, relates to leadership granting and claiming decisions across different operational and leadership configurations.

11.3.3 AI Algorithmic Capability

HALSAR incorporates multiple AI algorithmic components enabling autonomous decision-making and adaptive behaviors throughout the search mission. An image recognition algorithm uses static look-up tables to determine image recognition reliability based on sensor type, object type, and lighting scenarios. This image recognition algorithm can be substituted with computer vision models for more sophisticated object classification. A rule-based algorithm based on object type governs how the AI agent claims and grants leadership. This algorithm can be extended to incorporate probabilistic leadership decision-making based on environmental conditions and mission context. The AI agent relies on an algorithm to reassign detection tasks to UAVs based on their profiles of image recognition reliability for specific object types. Given the modular architecture, these components can be substituted to support investigations of how other AI design characteristics affect leadership dynamics amongst multiple human and AI agents.

11.3.4 Measurement Instruments

HALSAR is programmed to capture subjective perceptions and objective behaviors essential for modeling shared leadership (Table 11.1). These measures are collected throughout the mission for each detection event (i.e., a search task iteration), enabling

TABLE 11.1
Key Metrics Currently Computed in HALSAR

Data Type	Description	Collection Method/ Operationalization	HAT Research Applications
Trust ratings	Willingness to be vulnerable to an agent's actions	Embedded interval Likert scale for all agents	Temporal trust calibration, Influence on leadership delegation
Competence ratings	Assessment of an agent's capability to perform assigned tasks	Embedded ordinal ranking scale and interval Likert scale for AI agents (including UAVs)	Temporal competence modeling, Influence on leadership delegation
Leadership claims and grants	Timestamped record of leadership claims and grants between human and AI	Automated system logging (for each task iteration)	Network analysis of leadership emergence and distribution between human and AI
UAV reassignments	Timestamped record of UAV reassignments	Automated system logging	Network analysis of leadership distribution patterns from human or AI to UAV
Performance metrics	Mission effectiveness scores	Algorithmic score computation based on image classification and task assignment accuracy	Outcome assessment of different leadership structures

temporal analysis of evolving leadership dynamics. Trust and competence ratings are collected through embedded sliders at each search task iteration, providing continuous tracking of subjective perceptions. Trust ratings assess the human operator's willingness to be vulnerable to agent actions, indicating the operator's willingness to collaborate (Mioch et al., 2018; Xu & Dudek, 2015). Competence ratings assess perceived agent capability, indicating the operator's willingness to delegate leadership (Freedy et al., 2007; Shahrdar et al., 2018). The subjects (or content) of these ratings can be modified to examine other psychological constructs. This event-level data collection facilitates investigation of how trust and competence perceptions evolve throughout mission execution, capturing the dynamic calibration processes that govern leadership dynamics.

In addition to subjective ratings, HALSAR logs user interactions for behavioral process tracing. The system records every discrete interaction, including leadership claims and grants by human and AI agents, object classification decisions, and UAV assignments. This granular, objective behavioral data provides a detailed, chronological record of team decision-making and coordination, enabling analysis of emergent interaction patterns and providing insight into shared team cognition (Cooke et al., 2013).

The interaction logs support computation of network-based metrics for quantifying leadership distribution. Leadership density, an indicator of the proportion of agent interactions involving reciprocal leadership claims and grants, reflects the degree of collaborative leadership within the team. Leadership centralization, an indicator of leadership concentration within an agent, reflects how leadership distribution changes in response to agent performance. These metrics serve as the foundation for more advanced, graph-based learning methods. By generating network graphs for each search task iteration, this approach enables temporal tracking of leadership patterns and cross-referencing with trust and competence perceptions to identify optimal leadership patterns and emergent coordination strategies.

HALSAR computes mission performance metrics to assess different leadership structures and interaction designs. Classification accuracy measures the correctness of object identification decisions across different environmental conditions. Task assignment accuracy evaluates the appropriateness of UAV reassignments based on sensor–object compatibility. A composite performance score combining these metrics evaluates leadership structures holistically.

11.4 EMPIRICAL RESEARCH APPLICATIONS

Given the interaction features to operationalize shared leadership and extensive measurement capabilities to model collaboration dynamics, HALSAR supports unique study designs to build an empirical foundation for shared leadership in HAT.

11.4.1 Assessing Leadership Structures

HALSAR enables empirical comparison between leadership structures in terms of HAT processes and performance. HALSAR can be configured to accommodate hierarchical relationships (human- or AI-dominant control), collaborative partnerships (equal authority distribution), and supervisory configurations (human oversight with AI autonomy). This contrasts with existing HAT testbeds that primarily assess predefined human–AI authority or leadership schemas for function allocation based on environmental conditions, workload, or trust assessments (Aldridge & Bethel, 2023; Byeon et al., 2023). Currently, HALSAR enables researchers to configure leadership structures through (1) interaction mechanisms for granting and claiming leadership between agents and (2) logic governing these interactions to centralize or distribute leadership. For example, in a recently completed HALSAR study (More, 2025; More, Lau, & Foti, in review; More, Lau, Foti, et al., in press), the final UAV assignments in centralized leadership were always made by the human participant, whereas the final UAV assignment in shared leadership could be made by either the human or AI agent, depending on how both grant and claim leadership. When both simultaneously claim or grant leadership (i.e., conflicts in leadership), the human and AI agents alternate in making the final UAV assignments to maximize leadership distribution. More advanced logic for leadership resolution can be programmed to investigate other topics, such as AI decision explainability and iterative human–AI interactions. By examining mission performance scores across leadership structures, researchers can assess whether flexible authority allocation translates to measurable performance improvements in dynamic tasks, consistent with findings in human teams (Wang et al., 2014; Wu & Cormican, 2021).

11.4.2 Modeling Leadership Emergence

HALSAR supports empirical investigation of leadership distribution and emergence with respect to individualized interaction experience (or history) between human and AI agents, analogous to recent research on trust evolution in automation or AI (Duan et al., 2024; Humr et al., 2023). This approach moves beyond traditional HAT research on distributed trust that relies on predefined capability metrics or adaptive authority ladders to organize agents, which may inhibit organic leadership emergence (Schaefer et al., 2015; Schaefer & Cassenti, 2015). Specifically, HALSAR logs leadership claims and grants, subjective ratings, algorithmic AI capability, and optimal UAV assignments (based on normative decision-making) for every search task iteration to capture nuanced changes prior to leadership norm convergence. For example, a recent study of leadership emergence employed HALSAR to collect participant trust and competence ratings on four UAVs of different capabilities and reliabilities in object search and recognition, as well as UAV (re)assignment decisions for every search task iteration (More, 2025; More & Lau, in review). The study employed a Temporal Exponential Random Graph Model to model the empirical data as a sequence of human–agent networks, revealing that trust was more predictive (78.9%) than competence (7%–15%) in UAV assignment decisions. These graph models are critical for discovering how trust and competence perceptions shape stable leadership structures over time. Beyond analysis, these models can be embedded into an AI to predict human granting and claiming decisions driven by trust and competence.

11.4.3 Modeling Effects of Self-Directed/ Organic Interaction Experience

HALSAR enables investigation of how natural agent utilization and interaction exposure affect leadership emergence. This research advances existing HAT studies that often mandate human interaction and utilization of AI agents in a predefined manner (Brandt et al., 2018; Jay et al., 2016). Understanding natural utilization of agents is important because trust calibration can be governed by interaction exposure (Xu & Dudek, 2015, 2016), while human decisions are subject to availability heuristics (Mosier et al., 2017). Limited exposure to certain AI agents, whose capabilities may be irrelevant in prior tasks, can result in poorly calibrated trust and a bias toward selecting familiar agents as leaders for new tasks or situations. In one of the aforementioned HALSAR studies, the human participants were free to choose which UAV to re-assign in each task iteration, including the option for no reassignment (More & Lau, in review). This design aimed to support the human with a nuanced understanding of each UAV's capabilities and to examine how natural exposure shapes leadership emergence and agent utilization. Although yet to be investigated, HALSAR can be programmed to schedule predefined interaction exposure, potentially promoting recognition of complementary expertise, a critical component of shared leadership delegation in human teams (Drescher et al., 2014; Lyndon et al., 2022).

11.4.4 Understanding Team Composition

HALSAR facilitates examination of team composition effects on leadership emergence, complementing HAT research on dyadic interactions between a single human and a single AI agent (Bishop et al., 2020; Schaefer et al., 2019). HALSAR features a multi-agent architecture enabling multiple heterogeneous human and AI agents, including UAVs, to distribute leadership and tasks during a search mission. As mentioned earlier, one HALSAR study involved four UAVs equipped with distinct sensors, each exhibiting context-dependent object recognition accuracy. Results from that study suggest that humans learn to preferentially delegate leadership to UAVs whose sensor types align with object types, with reliability moderating this delegation (More & Lau, in review). Team composition and agent interactions likely influence leadership emergence, particularly in terms of the number of iterations required for agents to converge on a leader for a given situation. Further, HALSAR studies can readily scale to include more human and AI agents and varied human-to-AI ratios, as the granting and claiming interaction mechanisms are already programmed. However, data analysis and modeling complexity would increase due to combinatorial explosion with more agents. This research capability to study multi-agent leadership dynamics reflects empirical observation of shared leadership in real-world operations.

11.4.5 Beyond Leadership: Additional Research Applications

Designed primarily for shared leadership research, HALSAR's measurement capabilities and simulation features enable investigation of other HAT research questions:

- **Trust Dynamics**: Continuous trust rating collection enables studies of trust calibration, repair, and maintenance in multi-agent systems, examining how trust evolves across different AI capability and reliability conditions.
- **Expertise Recognition and Competence Assessment**: Interaction logs and performance data support investigation of how team members identify expertise among multiple agents and to optimize task delegation within heterogeneous teams.
- **Communication Patterns**: Logged user and AI interactions, and decision rationales enable investigation of information flow, coordination mechanisms, and communication effectiveness in human–AI teams.
- **Team Resilience**: Configurable system failures, environmental changes, or agent unavailability facilitate studies of how teams adapt and maintain performance under adverse conditions.
- **Agent Transparency and Explainability**: Graphical map overlays (e.g., lost person probability heatmaps) and text notifications support investigation of how AI transparency and explainability affect collaboration and leadership.

11.5 LIMITATIONS AND FUTURE WORK

HALSAR is a new testbed under continual development to advance shared leadership research in HAT. The testbed has been adopted only in a few studies given the novelty of the research questions and possesses limitations that must be addressed according to research outcomes.

One limitation is the sophistication of the AI agents in adapting interactions with human agents. The immediate development goal is to implement a new graph-learning module that predicts human decisions on leadership assignment based on trust and competence ratings (More & Lau, in review). This module will be integrated into the processing engine for adapting AI interactions with humans.

Future research will need to address the most effective interactions for AI agents in managing leadership decisions during agreement and conflict, including user interfaces providing transparency and explanation. These features and research studies are essential for addressing the lack of social intelligence and calibration in current autonomous systems in safety-critical domains (Farah & Dorneich, 2024), which stands in contrast with the sophisticated social calibration that characterizes effective human teams (Drescher et al., 2014). In general, HALSAR and other testbeds need to develop AI agents with advanced social intelligence, capable of learning from interaction history, inferring human intent, and dynamically adapting negotiation strategies, moving beyond current rule-based behaviors to become genuine partners.

HALSAR has not been deployed in studies involving more than one human participant even though the architecture and simulation environment were designed to support such study designs. Presumably, the detailed methodological design for multi-human studies would reveal other essential features.

HALSAR discretizes a search mission into stochastic multifactorial detection events, simulating a search scenario lasting about 20 minutes. This structure may not fully capture the continuous coordination demands and extended timeframes of real-world SAR operations. Although HALSAR can be modified to approximate the frequency of detection events and mission duration based on empirical data, such extended experimental studies would be impractical. Moreover, leadership transitions in SAR could occur outside of discrete detection events and often involve psychological stress given life-or-death consequences. Thus, the inherent complexities and ill-defined aspects of SAR are not fully captured in HALSAR's experimental tasks, though they are relevant to leadership emergence in multi-agent operations. These limitations are common to simulation-based research involving human participants and reflect inherent tradeoffs between experimental control and ecological validity.

As a new testbed, HALSAR has not yet matured to become "plug-and-play" like other human factors testbeds such as MAT-B II (Ngo et al., 2022, 2024). The publicly available source code provides the core functionalities for simulating search missions with real-world geographical information, collecting subjective ratings, logging interactions, automating UAV deployment, and displaying search outcomes with decision feedback. However, the most intriguing investigations, such as user interface designs that accelerate leadership convergence among agents or algorithms resolving leadership conflicts, require careful methodological consideration, innovative human factors design, and moderate coding investment. While a "plug-and-play" nature undoubtedly facilitates study execution and publication, there are risks pertaining to the abuse of the testbed for superficial research circumventing essential technical understanding and innovation (as observed in some human factors studies in highly technology-focused domains, such as cybersecurity). From this perspective, HALSAR could better serve the research community by adopting a modular design that can support efficient programming and crowdsourcing of new platform features.

11.6 CONCLUSION

HALSAR represents a significant methodological advancement for investigating shared leadership principles for HAT. As the first experimental testbed specifically designed to operationalize shared leadership, HALSAR contributes to scientific advances in HAT by enabling studies of leadership negotiation and transfer between human and AI agents, thereby enhancing adaptability and performance in complex operational environments. The capability of the testbed to systematically manipulate leadership structures while capturing both behavioral and self-reported data empowers researchers to investigate how shared leadership principles translate from human–human to human–AI contexts.

HALSAR addresses fundamental questions about authority distribution and emergence in multi-agent systems. Unlike traditional function allocation approaches or rigid organizational structures that prescribe static role assignments, shared leadership enables real-time authority reallocation based on situational demands and agent competencies. This dynamic approach has demonstrated effectiveness in human teams, yet its application to HAT requires careful consideration of the unique characteristics of human–AI interaction, including trust calibration, competence assessment, and the lack of inherent social intelligence in autonomous systems.

Beyond its immediate research applications, HALSAR lays a foundation for developing evidence-based approaches to human–AI collaboration. Future research directions include examining shared leadership principles across different team compositions, investigating the role of AI behavior in shared leadership, and exploring how interaction experience influences leadership negotiation and patterns. The platform's extensibility supports adaptation to other collaborative domains beyond SAR, enabling broader investigation of shared leadership principles across diverse contexts.

FUNDING

This research is supported by the National Science Foundation for the awarded projects entitled "Leadership Emergence for Synchronous Human-Autonomy Teaming" (#2212386) and "Balancing Collaboration and Autonomy for Multi-Robot Multi-Human Search and Rescue" (#1830414).

REFERENCES

Adams, A. L., Schmidt, T. A., Newgard, C. D., Federiuk, C. S., Christie, M., Scorvo, S., & DeFreest, M. (2007). Search is a time-critical event: When search and rescue missions may become futile. *Wilderness & Environmental Medicine*, *18*(2), 95–101.

Aldridge, A. L., & Bethel, C. L. (2023). M-OAT shared meta-model framework for effective collaborative human-autonomy teaming. *Companion of the 2023 ACM/IEEE International Conference on Human-Robot Interaction*, Stockholm, Sweden (pp. 663–666).

Bernabei, M., & Costantino, F. (2024). Adaptive automation: Status of research and future challenges. *Robotics and Computer-Integrated Manufacturing*, *88*, 102724.

Bishop, J., Burgess, J., Ramos, C., Driggs, J. B., Williams, T., Tossell, C. C., Phillips, E., Shaw, T. H., & Visser, E. J. D. (2020). CHAOPT: A testbed for evaluating human-autonomy team collaboration using the video game overcooked! 2. *2020 Systems and Information Engineering Design Symposium (SIEDS)*, Charlottesville, Virginia, USA (pp. 1–6). IEEE.

Brandt, S. L., Lachter, J., Russell, R., & Shively, R. J. (2018). A human-autonomy teaming approach for a flight-following task. *International Conference on Applied Human Factors and Ergonomics*, Orlando, Florida, USA (pp. 12–22). Springer International Publishing.

Byeon, S., Choi, J., & Hwang, I. (2023). A computational framework for optimal adaptive function allocation in a human-autonomy teaming scenario. *IEEE Open Journal of Control Systems*, *3*, 32–44.

Carson, J. B., Tesluk, P. E., & Marrone, J. A. (2007). Shared leadership in teams: An investigation of antecedent conditions and performance. *Academy of Management Journal*, *50*(5), 1217–1234.

Carte, T. A., Chidambaram, L., & Becker, A. (2006). Emergent leadership in self-managed virtual teams: A longitudinal study of concentrated and shared leadership behaviors. *Group Decision and Negotiation*, *15*(4), 323–343.

Cooke, N. J., Gorman, J. C., Myers, C. W., & Duran, J. L. (2013). Interactive team cognition. *Cognitive Science*, *37*(2), 255–285.

Cubber, G. D., Doroftei, D., Rudin, K., Berns, K., Matos, A., Serrano, D., Sanchez, J., Govindaraj, S., Bedkowski, J., & Roda, R. (2017). Introduction to the use of robotic tools for search and rescue. In *Search and rescue robotics-from theory to practice* (pp. 1–17). IntechOpen.

DeRue, D. S., Nahrgang, J. D., & Ashford, S. J. (2015). Interpersonal perceptions and the emergence of leadership structures in groups: A network perspective. *Organization Science*, *26*(4), 1192–1209.

Drescher, M. A., Korsgaard, M. A., Welpe, I. M., Picot, A., & Wigand, R. T. (2014). The dynamics of shared leadership: Building trust and enhancing performance. *Journal of Applied Psychology*, *99*(5), 771–784.

Duan, W., Zhou, S., Scalia, M. J., Yin, X., Weng, N., Zhang, R., Freeman, G., McNeese, N., Gorman, J., & Tolston, M. (2024). Understanding the evolvement of trust over time within Human-AI teams. *Proceedings of the ACM on Human-Computer Interaction*, *8*(CSCW2), 1–31.

Farah, Y. A., & Dorneich, M. C. (2024). Human-Autonomy teaming in a cooperative gamified testbed: How can AI teammates support teamwork processes? *Proceedings of the Human Factors and Ergonomics Society Annual Meeting*, *68*(1), 386–392.

Freedy, A., DeVisser, E., Weltman, G., & Coeyman, N. (2007). Measurement of trust in human-robot collaboration. *Proceedings of the 2007 International Conference on Collaboration Technologies and Systems* (pp. 106–114), Orlando, Florida, USA. IEEE.

Hoch, J. E., & Morgeson, F. P. (2014). Vertical and shared leadership processes: Exploring team leadership dynamics. *Academy of Management Proceedings*, *2014*(1), 16977.

Hollnagel, E., & Bye, A. (2000). Principles for modelling function allocation. *International Journal of Human-Computer Studies*, *52*(2), 253–265.

Humr, S. A., Canan, M., & Demir, M. (2023). Temporal evolution of trust in artificial intelligence-supported decision-making. *Proceedings of the Human Factors and Ergonomics Society Annual Meeting*, *67*(1), 1936–1941.

Imam, H., & Zaheer, M. K. (2021). Shared leadership and project success: The roles of knowledge sharing, cohesion and trust in the team. *International Journal of Project Management*, *39*(5), 463–473.

Jay, S. R., Brandt, S. L., Lachter, J., Matessa, M., Sadler, G., & Battiste, H. (2016). Application of human-autonomy teaming (HAT) patterns to reduced crew operations (RCO). *Proceedings of the 13th International Conference on Engineering Psychology and Cognitive Ergonomics*, Toronto, Ontario, Canada (Vol. 9736, pp. 244–255). Springer.

Kaber, D. B., & Endsley, M. R. (2004). The effects of level of automation and adaptive automation on human performance, situation awareness and workload in a dynamic control task. *Theoretical Issues in Ergonomics Science*, *5*(2), 113–153.

Kaplan, H. (2007). *Practical applications of infrared thermal sensing and imaging equipment* (Vol. 75). SPIE Press.

Klasmeier, K. N., & Rowold, J. (2020). A multilevel investigation of predictors and outcomes of shared leadership. *Journal of Organizational Behavior*, *41*(9), 915–930.

Klasmeier, K. N., & Rowold, J. (2022). A diary study on shared leadership, team work engagement, and goal attainment. *Journal of Occupational and Organizational Psychology*, *95*(1), 36–59.

Kwon, M., Li, M., Bucquet, A., & Sadigh, D. (2019). Influencing leading and following in human-robot teams. *Autonomous Robots*, *45*(7), 959–978.

Lee, J., & See, K. (2004). Trust in automation: Designing for appropriate reliance. *Human Factors*, *46*(1), 50–80.

Lyndon, S., Pandey, A., & Navare, A. (2022). Emergence and outcomes of shared leadership: Unraveling the role of transactive memory system and team mindfulness using mixed-methods approach. *Leadership & Organization Development Journal*, *43*(2), 196–210.

Lyons, J. B., Sycara, K., Lewis, M., & Capiola, A. (2021). Human–autonomy teaming: Definitions, debates, and directions. *Frontiers in Psychology*, *12*, 589585.

Mayo, M., Meindl, J. R., & Pastor, J. C. (2002). Shared leadership in work teams: A social network approach. In C. L. Pearce & J. A. Conger (Eds.), *Shared leadership: Reframing the hows and whys of leadership* (pp. 193–214). SAGE Publications, Inc.

McNeese, N. J., Demir, M., Cooke, N. J., & Myers, C. (2018). Teaming with a synthetic teammate: Insights into human-autonomy teaming. *Human Factors*, *60*(2), 262–273.

Mioch, T., Peeters, M. M., & Neerincx, M. A. (2018). Improving adaptive human-robot cooperation through work agreements. In *2018 27th IEEE International Symposium on Robot and Human Interactive Communication (RO-MAN)*, Nanjing, China (pp. 1105–1110). IEEE.

More, A. (2025). *Engineering shared leadership for human and autonomy collaboration in multi-agent systems* (Doctoral dissertation, Virginia Polytechnic Institute and State University). Vtechworks.

More, A., & Lau, N. (in review). Trust and competence calibration in leadership assignment for multi-agent systems. *International Journal of Human-Computer Studies*.

More, A., & Lau, N. (2024). Exploratory literature review on shared leadership to aid collaboration and function allocation in human-autonomy teams. *Proceedings of the Human Factors and Ergonomics Society Annual Meeting*, *68*(1), 1491–1495.

More, A., Lau, N., & Foti, R. (in press). Development and evaluation of the human agent leadership organization scale. *Computers in Human Behavior Reports*.

More, A., Lau, N., Foti, R., & Williams, R. K. (in review). Shared versus centralized leadership on task allocation decisions in multiagent systems. *Proceedings of the Human Factors and Ergonomics Society Annual Meeting.*

Mosier, K. L., Fischer, U., Burian, B. K., & Kochan, J. A. (2017). *Autonomous, context-sensitive, task management systems and decision support tools I: Human-autonomy teaming fundamentals and state of the art.* NASA STI Program.

Muethel, M., Gehrlein, S., & Hoegl, M. (2012). Socio-demographic factors and shared leadership behaviors in dispersed teams: Implications for human resource management. *Human Resource Management, 51*(4), 525–548.

Ngo, M., Rauffet, P., & Banks, S. (2022). Exploring multitasking performance and fatigue with the MATB-II: A narrative. *Human Error, Reliability, Resilience, and Performance, 33*, 22–35.

Ngo, M., Rauffet, P., Coussens, S., Banks, S., & Yates, C. (2024). Multiple uses: A pilot study exploring the impact of multitasking performance with a switching automation aid using the Multi-Attribute Task Battery II (MAT-B II). *Proceedings of Australasian Experimental Psychology Conference*, Adelaide, Australia.

O'Neill, T., McNeese, N., Barron, A., & Schelble, B. (2022). Human–autonomy teaming: A review and analysis of the empirical literature. *Human Factors, 64*(5), 904–938.

Pearce, C. L., & Sims Jr, H. P. (2002). Vertical versus shared leadership as predictors of the effectiveness of change management teams: An examination of aversive, directive, transactional, transformational, and empowering leader behaviors. *Group Dynamics: Theory, Research, and Practice, 6*(2), 172.

Roth, E. M., Sushereba, C., Militello, L. G., Diiulio, J., & Ernst, K. (2019). Function allocation considerations in the era of human autonomy teaming. *Journal of Cognitive Engineering and Decision Making, 13*(4), 199–220.

Schaefer, K. E., Baker, A. L., Brewer, R. W., Patton, D., Canady, J., & Metcalfe, J. S. (2019). Assessing multi-agent human-autonomy teams: US Army robotic wingman gunnery operations. *Micro-and Nanotechnology Sensors, Systems, and Applications XI* (Vol. 10982, pp. 353–372), Baltimore, Maryland, USA. SPIE.

Schaefer, K. E., & Cassenti, D. N. (2015). A network science approach to future human-robot interaction. *2015 IEEE International Multi-Disciplinary Conference on Cognitive Methods in Situation Awareness and Decision* (pp. 132–136), Orlando, Florida, USA. IEEE.

Schaefer, K. E., Evans III, A. W., & Hill, S. G. (2015). Command and control in network-centric operations: Trust and robot autonomy. *Proceedings of the 20th International Command and Control Research and Technology Symposium*, Baltimore, Maryland, USA.

Shahrdar, S., Menezes, L., & Nojoumian, M. (2018). A survey on trust in autonomous systems. *Science and Information Conference*, London, UK (pp. 368–386). Springer International Publishing.

Sheridan, T. B., Verplank, W. L., & Brooks, T. (1978). Human/computer control of undersea teleoperators. *Proceedings of the 14th Annual Conference on Manual Control, Moffett Field,* California, USA. NASA Ames Research Center..

Statheropoulos, M., Agapiou, A., Pallis, G. C., Mikedi, K., Karma, S., Vamvakari, J., Dandoulaki, M., Andritsos, F., & Thomas, C. P. (2015). Factors that affect rescue time in urban search and rescue (USAR) operations. *Natural Hazards, 75*(1), 57–69.

Tokadli, G., & Dorneich, M. C. (2019). Interaction paradigms: From human-human teaming to human-autonomy teaming. *2019 IEEE/AIAA 38th Digital Avionics Systems Conference (DASC)*, San Diego, California, USA (pp. 1–8). IEEE.

Wang, D., Waldman, D. A., & Zhang, Z. (2014). A meta-analysis of shared leadership and team effectiveness. *Journal of Applied Psychology, 99*(2), 181–198.

Wang, T. (2023). *Transparency, trust, and level of detail in user interface design for human autonomy teaming* (Doctoral dissertation, Virginia Polytechnic Institute and State University). Vtechworks.

Wang, T., & Lau, N. (2023). Level of detail in visualization for human autonomy teaming: Speed, accuracy, and workload effects. *Proceedings of the Human Factors and Ergonomics Society Annual Meeting*, *67*(1), 549–555.

Williams, R. K., Abaid, N., McClure, J., Lau, N., Heintzman, L., Hashimoto, A., Wang, T., Patnayak, C., & Kumar, A. (2020). Collaborative multi-robot multi-human teams in search and rescue. *Proceedings of the international ISCRAM conference*, Blacksburg, Virginia, USA (Vol 1, pp. 973–983).

Wu, Q., & Cormican, K. (2021). Shared leadership and team effectiveness: An investigation of whether and when in engineering design teams. *Frontiers in Psychology*, *11*, 569198.

Xu, A., & Dudek, G. (2015). OPTIMo: Online probabilistic trust inference model for asymmetric human-robot collaborations. *Proceedings of the Tenth Annual ACM/IEEE International Conference on Human-Robot Interaction*, Portland, Oregon, USA (pp. 221–228).

Xu, A., & Dudek, G. (2016). Towards modeling real-time trust in asymmetric human–robot collaborations. *Robotics Research: The 16th International Symposium ISRR*, Singapore (pp. 113–129). Springer International Publishing.

Yang, J., Zhang, X., Zhang, X., Wang, L., Feng, W., & Li, Q. (2021). Beyond the visible: Bioinspired infrared adaptive materials. *Advanced Materials*, *33*(14), 2004754.

Zhu, J., Liao, Z., Yam, K. C., & Johnson, R. E. (2018). Shared leadership: A state-of-the-art review and future research agenda. *Journal of Organizational Behavior*, 39(7), 834–852.

12 Joint Activity Testing

A Foundation for Human-AI Evaluation Testbeds

Mohammadreza Jalaeian, Dane A. Morey, Amy Ryder, Jennifer Winner, Mimi Cai, Liana Algarín, and Michael F. Rayo

12.1 INTRODUCTION

Amid the near-deafening hype of artificial intelligence (AI), it is clear that AI-infused technologies have the potential to both enhance and diminish overall system performance. AI is transforming the execution and expectations of defense operations and other safety-critical domains. AI algorithms can provide increased speed in sensemaking, targeting, and enhancing decision-making efficiency, especially on benchmark tasks. However, the performance of those same algorithms can drop precipitously as they encounter situations outside of their training sets (Zech et al., 2018). AI decision-support tools can improve human performance in some cases where the AI performs well, while degrading human performance on other occasions where the system failed to surface its errors (Morey, Rayo, et al., 2025). Improving algorithm performance alone does not fully mitigate these risks. In healthcare, AI has achieved expert-level performance in areas such as radiology, pathology, and medical imaging, but has also created new forms of risk when not rigorously evaluated (Kiani et al., 2020; Liu et al., 2019; Rajpurkar et al., 2022). These examples highlight an important challenge: we cannot assess the effectiveness of a human–AI system by evaluating the algorithms or operators in isolation.

Joint activity testing (JAT) addresses many of the shortcomings that make many of the test and evaluation (T&E) methods insufficient for the complexity of current joint cognitive systems (JCSs). JAT addresses these shortcomings by directly measuring performance, providing broad macrocognition coverage, utilizing continuous, scientifically grounded testing variables, and focusing on the JCS.

Direct Measurement of Performance: JAT joins a number of other methods (see Gray, 2002; Howie & Vicente, 1998) explicitly designed to measure direct task performance as well as a number of other performance-related measures including situational awareness, user workload, and trust. Direct performance measures are necessary, as performance-related measures have been shown in multiple settings to not be good proxies (Morey et al., in press).

Broad Macrocognition Coverage: JAT performance and challenge measures are explicitly associated with all macrocognitive functions and processes, including event detection, sensemaking, replanning, and action-taking (Klein et al., 2003).

DOI: 10.1201/9781003655923-15

By contrast, usability testing was devised to evaluate products that only support action-taking, and lacks specific mechanisms to evaluate support for the other macrocognitive functions (Rayo, 2017).

Scientifically Grounded Testing Variables: Although JAT can utilize any performance and challenge measures, it is meant to be used with scientifically grounded measures drawn directly from research findings. A catalog of measures has been created (https://go.osu.edu/jat-measures) that draws from basic performance tradeoffs including efficiency/thoroughness and acute/chronic goal attainment (Hoffman & Woods, 2011; Hollnagel, 2017), facets of work complexity including data overload and ambiguous signals (Patterson, Roth, et al., 2010), and facets of task complexity including the number of interdependent components and the degree of uncertainty (Wood, 1986).

Continuous Measures to Reveal Brittleness Regions: Empirical evaluations that either do not vary the difficulty of their test scenarios or model challenge as binary (e.g., "easy" and "hard") or categorical (e.g., "low," "medium," or "high" difficulty) will obscure the more subtle and non-linear ways that team performance collapses as challenges exceed system capacity (Damacharla et al., 2018). This relationship between performance and challenge is critical to understanding a system's potential for resilient performance: the system's ability to sustain critical functions as demands increase. The extent to which systems demonstrate resilient performance depends upon the capacity of the team to stretch beyond what was planned (Woods, 2018b). In situations of increasing demand, teams demonstrate adaptive capacity by changing roles, reprioritizing cues, or shifting approaches while keeping performance steady. Brittleness emerges as a sharp drop in system performance, the point where the boundary of stretching capacity has been reached. These dynamics become visible only when testing actually approaches these boundary conditions. Woods (2015, 2018a) notes that resilience is only revealed when the system is pushed to its limits or near limits, as both adaptive capacity and brittleness become visible. However, most current T&E practices are not designed to assess the transition which occurs across these boundary conditions, so they are limited in the extent to which they reveal early warnings of strain, or how a system might collapse under pressures which fall outside the testing envelope.

Focus on Joint Human–Machine System: JAT explicitly focuses on the performance of the joint human–machine system (which will be referred to as the joint cognitive system, or JCS), comprised of a number of people and a number of machines working together in some fashion. By contrast, most T&E approaches primarily emphasize the verification and validation of algorithmic accuracy.

Another variant of this shortcoming is the goal to evaluate and optimize humans and machines separately rather than examining how their behaviors combine in practice. The separate optimization of humans and machines can produce the paradox of a reduced collective effectiveness in human–machine systems (Bansal et al., 2021; Sorkin & Woods, 1985). For example, Morey, Rayo, et al. (2025) found that the interaction of people and AI algorithms introduced new vulnerabilities which were not evident in isolated evaluations of either, despite these isolated evaluations suggesting that AI algorithms were vastly superior to current technologies and that people were competent in areas of AI weaknesses. This is because the *joint activity architecture*, which is the way work and roles are distributed across people and machines, often evolves differently in practice than the designers envision. Without methods to inspect system brittleness, there is a high risk of drifting into fragile configurations that appear robust in limited tests but collapse with increased challenge. Dekker (2011) describes this phenomenon as a "drift into failure": a gradual erosion of safety hidden behind reassuring evaluation results.

To avoid the above-mentioned limitations, the focus of human–machine systems evaluation should move from individual components to joint performance, which is the interaction of humans and machines working together as a whole. This perspective is rooted in cognitive systems engineering (CSE) and resilience engineering, in which the focus is on how adaptive capacity emerges from the coordination of people and machines in joint activity architectures (Dekker & Woods, 2002; Klein et al., 2005; Patterson, Rogers, et al., 2010). Focusing on adaptive capacity, which has also been described as adaptive performance (Baard et al., 2014; Lang & Bliese, 2009), reveals periods of performance decline due to new challenges, followed by either recovery or rapid decline. Building on this foundation, Morey et al. (2020) proposed joint activity testing (JAT) through continuous measures of both performance and challenge rather than purely categorical or averaged scores. This type of data makes it possible to determine performance strain, sudden collapses, insights into the types of adaptive performance being observed (Lang & Bliese, 2009), and when adaptive capacity allows the JCS to extend performance beyond its expected limits.

This chapter introduces JAT as a human–machine systems evaluation approach to address this gap. It evaluates human–machine JCSs by examining how performance changes in relation to increasing demand on the team (Morey et al., 2022). The primary output of JAT is the joint performance graph (JPG), which plots performance in relation to challenge such that patterns of resilience and brittleness can be directly observed. This method enables evaluators to map breakdown points, compare alternative human–machine JCS designs, and have a clear vision that extends beyond small test sets.

We proceed in five parts. First, we present the JAT foundations in cognitive systems and resilience engineering. Next, we define the core constructs of JAT, including continuous performance and challenge measures, and explain how JPGs are constructed and interpreted. We then provide a walkthrough of the method in practice, with a focus on the test design and analysis phases. We conclude with an exploration of JAT in relation to existing approaches, addressing limitations and future directions. In each section, we reference examples from published JAT studies to illustrate core concepts. Collectively, we describe JAT as a process for evaluating joint performance, rather than operator or algorithm performance independently, and a foundation for creating resilient human–AI systems.

12.2 JAT ORIGINS AND FOUNDATIONS

JAT emerged from decades of research evaluating human–machine systems. While automation can mitigate some errors, it is likely to create new ones, especially when human involvement comes late. Classic *ironies of automation* (Bainbridge, 1983) revealed that when automated systems fail, people are left doing the most difficult work while under time pressure and with incomplete context. That shift in human–machine coordination introduces brittle spots, not due to the negligence of the operators, but because the role of the human operator changes at the worst possible moment. This observation has shaped many of the subsequent studies on human–machine systems, all confirming this paradox. Wickens and Dixon (2007), for example, noted that operators often lean more heavily on automation under high workload conditions, even when automation is imperfect. This reliance can create new risks if the system fails, especially when human intervention is needed. More recently, Endsley (2023) argues that advanced modern AI systems may increase such risk by disturbing situation awareness, encouraging overtrust, and slowing down recovery following failures. The recognition of such challenges gives legitimate reason for evaluations to consider not just the accuracy of algorithms or workload alone but joint human–machine activity, particularly when performance collapses or fails under pressure and increased challenges to the system.

Over time, researchers in the field have begun to recognize that the challenge of evaluating human–machine systems is not only about operator workload or technology usability. In such systems, the automation does not simply substitute a task, but it fundamentally alters the architecture of joint activity between human cognition and machines. A study on decision-making with respect to expert monitoring (Klein, 1993) showed how the role of expertise would prompt the operator to engage recognition-primed decision-making, which senses relevant cues in context using prior experience rather than the use of abstract rules. In practice, people and tools coordinate attention, cues, and roles. Therefore, any intervention, whether it be a checklist or an AI decision aid, will always change the dynamics of the interaction.

CSE, introduced by Hollnagel and Woods (1982), emerged to deal with this complexity: emphasizing that performance is always relative to the larger system of roles, signals, and common goals (Patterson, Rogers, et al., 2010). It reframed human–machine systems as JCSs. In this perspective, humans and machines behave as adaptive partners, with the two sharing an internal model of each other to facilitate effective coordination. Shortly after, resilience engineering introduced a complementary perspective. Hollnagel (2014) proposed a shift from the "Safety-I" paradigm, which assessed only the absence of "failure" in action, to "Safety-II," which emphasizes the system's ability to adapt and continue performing under increased challenges without collapsing. From these views emerged the concept of Graceful Extensibility, or the idea that systems can stretch beyond designed capacity when demand is greater than expected (Woods, 2018b, 2020; Woods & Branlat, 2011a, 2011b). This notion served as a foundation for assessing brittleness and resilience in complex situations. While engineering approaches have traditionally focused on optimizing for stability, graceful extensibility argues for the need for flexibility at the boundaries of performance where brittleness occurs most often. JAT describes an empirical method that observes how human–AI teams extend or collapse as demands escalate, transforming graceful extensibility into a practical evaluation.

12.3 JOINT PERFORMANCE GRAPH (JPG): PERFORMANCE vs. CHALLENGE

The ideas from CSE and resilience theory point to a central challenge: how do we measure brittleness and adaptation in practice? This is where JAT introduces its core analytic tool: the JPG. JPGs (see Figure 12.1) plot measures of JCS performance

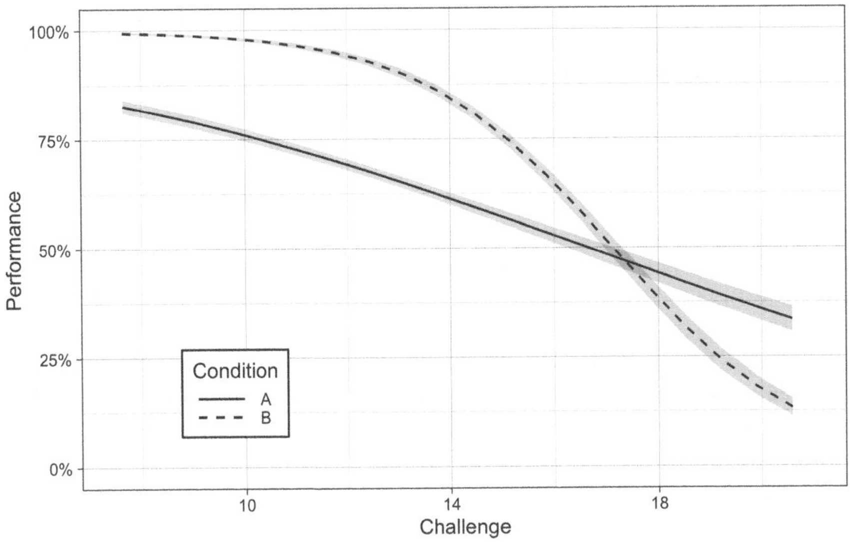

FIGURE 12.1 Example JPG: Performance plotted against challenge, by condition. Reprinted with permission (Morey, Ryder, et al., 2025).

against increasing levels of challenge, turning previously mentioned concepts of brittleness, adaptive capacity, and graceful extensibility into measurable patterns. Figure 12.1 illustrates a notional example of a JPG, with performance on the vertical (Y) axis and challenge on the horizontal (X) axis. In this figure, a comparison between two evaluation conditions, A and B, is represented. The lines depict the aggregated performance of multiple operators across a range of sampled challenge levels. In the figure, challenge increases as you move from left to right on the horizontal axis. In this example, the JPG shows that performance drops as the challenge increases, as noted by the sharp decrease in the values along the vertical (Y) axis. Similar JPGs constructed from real-world JAT studies are published elsewhere (Morey, Rayo, et al., 2025).

Once the JPGs have been developed, evaluators can compare them to identify which joint activity architectures sustain performance longer, recover more effectively, or collapse more rapidly as challenges are increased (Morey et al., 2020). Performance can take many forms depending on the domain, including general counts of detected events, classification accuracy, or task completion times. Performance measures typically align with measures of correctness, accuracy, or efficiency. Challenge can be altered along dimensions such as dataset size, algorithm quality, time pressure, or signal uncertainty. Challenge measures typically reflect known patterns of what makes cognitive work difficult (Patterson, Roth, et al., 2010). In this case, both performance and challenge are continuously measured so that the interaction between the two can be plotted. For example, Morey, Rayo, et al. (2025) plotted the correctness of nurses' decisions (performance) against the magnitude of AI algorithm error (challenge), both of which were computed by comparing to ground truth about individual patients.

The derived curves provide insights into the system's resilience. A relatively flat slope implies stable performance, and a steep decline implies brittle behavior (Morey et al., 2020; Woods, 2018b). A system may demonstrate high performance under relatively low challenge conditions, only to collapse suddenly under higher challenge conditions, exposing vulnerabilities not readily apparent when comparing average performance. As shown in Figure 12.1, one system performed better under lower challenge but lost stability as demands increased, revealing a sharp drop in performance at more challenging levels. This pattern illustrates how adaptive systems often fail gradually at first and then abruptly once their capacity limits are reached (Woods & Branlat, 2011a). By plotting performance versus challenge, JPGs enable organizations to determine if a tool is enhancing resilience or creating brittleness in its operation. JPGs can also initiate meaningful conversations about how the costs and benefits are distributed across system configurations and provide a structured way to predict how performance may decrease in future scenarios as challenges increase (Morey, Ryder, et al., 2025). For example, Morey, Rayo, et al. (2025) revealed how the addition of an AI-infused decision aid made some situations easier for nurses to handle, but made others much more difficult. JPGs exemplify the principles of JAT: they visualize how a JCS extends or collapses under increased challenge and evaluate performance as joint performance rather than the performance of humans and machines as isolated measures. In addition to understanding where a given solution's performance collapses, JPGs also allow visual comparisons between multiple solutions for a given problem or task. This allows quick comparisons of display design strategies or even changes in the architecture of how human and machine roles are working together (i.e., the joint activity architecture). This can include evaluations of people unaided by the proposed technology as well as the performance of AI algorithms in isolation. These comparisons are easily made through the entire range of challenge represented in the evaluation. It also allows for model-aided extrapolations of each solution's performance outside of the bounds of the test (for example, predicting performance for cases more challenging than encountered in the evaluation).

The value of JPGs becomes apparent in the application of real-world context. JAT has already been utilized in multiple domains, each demonstrating how the method can illuminate brittleness and resilience in ways that traditional evaluations cannot. In healthcare, Morey, Rayo, et. al. (2025) demonstrated how AI decision-support could in some cases develop or degrade clinician performance, depending on the accuracy of the AI-guided decision-support. For technologies including aviation and nuclear control commands, automation surprises have long represented safety threats. JAT provides a disciplined way to surface vulnerabilities that often lie below the surface (Dekker & Woods, 2002; Woods, 1985). JAT is also being used in multiple US Department of Defense settings to evaluate new AI-enabled systems, identify potential brittleness, and guide acquisition decisions (Morey et al., 2022).

In all these contexts, the purpose of JAT is uniform: to reveal patterns of performance, including resilience and brittleness, before implementation so organizations can understand vulnerabilities and rework system designs so they can adapt when situations are beyond expectations. Furthermore, JAT is timely and relevant to the evolving discourse regarding ethical AI. For instance, policy frameworks such as the EU AI Act and the NIST AI Risk Management Framework include human oversight but do not indicate how AI will alter the performance of human partners. JAT addresses this limitation by incorporating joint activity and resilience explicitly in the process of evaluation. Similarly, it responds to recent calls from the AI ethics community to provide evaluation strategies that go beyond high-level principles (Floridi et al., 2018; Hutchinson et al., 2022; Mikalef et al., 2022).

In this section we introduced the theoretical foundations of JAT and JPGs as its methodological core. It is equally important to understand how these ideas are enacted in practice. In the next section of the chapter, we turn from concepts to application, describing the phases of a JAT evaluation as they unfold in real-world studies.

12.4 JAT METHODOLOGY IN PRACTICE

In this section we discuss the JAT methodology and how we recommend JAT to be integrated into standard evaluation processes (see Table 12.1). For this purpose, we simplify evaluations into five phases (Morey, Ryder, et al., 2025): (1) frame the evaluation, (2) plan and design the evaluation, (3) collect the data, (4) analyze the data, and (5) report the evaluation results. Although each phase is presented sequentially,

TABLE 12.1
Key Steps for Joint Activity Testing (JAT) with Evaluation Phases

Phase	Key Steps	JAT Goals
Phase I: Frame the evaluation	Conduct a task analysis to identify target tasks and what makes them difficult to perform	Understand the goals of work and what makes work difficult to validate and substantiate performance and challenge measures.
Phase II: Plan and design the evaluation	Define JAT performance measures	Define continuous measures of joint system performance to reflect multiple aspects or trade-offs in determining joint system (or mission) success or failure.
	Define JAT challenge measures	Define continuous measures of joint system challenge to reflect multiple (hypothesized but substantiated) aspects of what makes work difficult.
	Design practical data collection methods for both performance and challenge measures	Operationalize performance and challenge measures for a specific setting and study.
	Design a set of testing cases	Select and counterbalance a set of testing cases covering a wide range of challenges to maximize the potential for observing a brittle collapse of performance.
	Prepare "ground truth" or "answer keys"	Establish a benchmark for understanding and comparing participant behaviors in each testing case.
Phase III: Collect the data	None	N/A

(Continued)

TABLE 12.1 (*Continued*)
Key Steps for Joint Activity Testing (JAT) with Evaluation Phases

Phase	Key Steps	JAT Goals
Phase IV: Analyze the data	Produce JPGs	Model and analyze the relationship(s) between measures of performance and measures of challenge. Visualize these results by constructing JPGs for each significant combination of performance and challenge measures.
	Analyze qualitative data to give context for JPGs	Perform qualitative analyses of observation notes and participant interviews to understand the "why's" behind the JPGs "what's"
Phase V: Report evaluation results	Discuss the potential JCS benefits and risks	Discuss the potential for brittleness (or resilience) and ranges of acceptable (or unacceptable) performance to understand potential benefits and risks.

each evaluation may dictate a different sequence or revisiting the same step multiple times as an evaluation team iterates toward a solution.

This section briefly reviews all five phases but focuses most on the framing and analysis stages, which represent the majority of unique JAT contributions. Planning and data collection are discussed in outline form, since their general principles overlap with conventional research practices, though important adaptations for JAT will be noted.

12.4.1 Phase I: Frame the Evaluation

The first phase of a JAT evaluation is about framing. This step sets the foundation for everything that follows. At its core, framing means defining the scope of the study, identifying the evaluation questions, and making sure the right people, resources, and ethical protections are in place before planning begins in detail. Unlike many traditional T&E efforts, JAT uses this phase to look ahead to mission-level challenges and task demands, so the evaluation can surface not only where systems perform well but also where vulnerabilities are most likely to appear. Establishing the scope and timeline is one of the first framing activities. Scope includes milestones, deliverables, and dependencies, along with a plan that can be tracked over the course of the project, working backward from the start of data collection to accommodate for pilot runs, adjustments, and reporting. Teams often leverage structured means such as project trackers or milestone spreadsheets to keep this plan in view and allow for everyone to understand the order of activities.

The next critical step is defining evaluation questions that direct the evaluation toward the most important aspects of system performance. The evaluation questions should come from careful consideration of the tasks and what makes the tasks challenging. This is where approaches like cognitive task analysis (CTA) become important. These approaches focus on more than basic descriptions of a task; they reveal the decision-making, judgments, and experts' thinking processes along with the contextual factors that shape those processes (Clark et al., 2007; Crandall et al., 2006; Hoffman & Militello, 2008; Schraagen et al., 2000). CTA is a recognized and widely used approach within the human factors and cognitive psychology literature. It uses a set of structured yet flexible methods to elicit how experts think and behave when they face difficult and ambiguous problems in real-world contexts. Some useful steps for obtaining high-quality data include engaging stakeholders to narrow the focus, reviewing the documentation, interviewing subject matter experts and frontline workers, and observing subject matter experts and frontline workers in real-world settings. Through these activities, the evaluators are able to identify "critical tasks," usually judged on criteria such as how central the task is to the mission, how often it is performed, and what the potential consequences of poor task performance are.

By using CTA methods in the framing stage, evaluators are able to form questions that are based on actual cognitive demands of the work rather than the functional specifications of the system or simple workflows. They connect abstract evaluation objectives with the operator's lived experience, providing a grounding to see the evaluation as more than just a test of technology in isolation, but rather a study of how the technology interacts with the actual complexity of work in the real world. This first step is important to ground later steps in the realities of the work that is intended to be supported.

After drafting the evaluation questions during the framing phase, the focus switches to the technologies that will be included. Assessing human readiness levels and technology readiness levels may help to identify technologies that are too immature for a full JAT evaluation and may fit better in a lighter method, such as a Joint Activity Design heuristic review (Jalaeian et al., 2025). For systems that are more mature, JAT can be applied directly, often with comparisons to human-only or machine-only baselines. This is the step to ensure the evaluations are at the appropriate stage of development, and that resources are utilized effectively (Morey, Ryder, et al., 2025).

Given the involvement of human participants in JAT studies and, in some cases, classified or otherwise sensitive information, both ethical review and security must be incorporated from the very beginning. These reviews also require a larger and more diverse team than is typical for usability evaluations. The roles needed for JAT studies range from cognitive analysts and domain experts to facilitators and program managers. Scope and roles are addressed early to help maintain coordination and make sure that ethical protections, security checks, and technical perspectives are all addressed appropriately. The prioritization of protections and roles sets the scene so that later phases can reveal brittleness and resilience rather than becoming overwhelmed with preventable barriers.

All of these steps together show why framing is not just a first step but the foundation of JAT. Establishing scope, formulating evaluative questions using CTA, assessing levels of readiness, raising ethical and security issues, and forming a competent team all help ensure that the evaluation will produce meaningful insights. Most importantly, the framing stage in JAT is designed with brittleness in mind. The framing phase process is not just aimed at seeing how the system performs under straightforward circumstances, but setting the stage for understanding how the JCS performs when conditions become challenging and where it becomes brittle. For example, Morey, Rayo, et al. (2025) deliberately included rare but severe AI errors as part of the scope, rather than typical AI performance alone.

12.4.2 Phase II: Plan and Design the Evaluation

The ideas generated during the framing phase are developed into a concrete evaluation plan in phase II. Anyone who has been involved in experimental design will recognize much of what takes place in this phase: participant selection, protocol preparation, and data capture setup. However, JAT introduces a unique component in that its measures and scenarios are tied back to the cognitive analyses and mission challenges identified during phase I. This anchors the evaluation in real-world complexity and guards against oversimplified or abstract tests.

The central design tasks in this phase are the definition of performance measures and challenge measures. JAT's emphasis on multi-dimensional performance is based on the foundational understanding that a system is simultaneously trying to accomplish multiple goals, many of which conflict with each other (Woods et al., 1994). Each measure of performance contributes an important but incomplete picture of performance: only by utilizing multiple measures, including task performance, decision-making quality, and coordination, can a more complete picture of performance emerge (Campbell et al., 1990). Building on this view, JAT preserves multiple continuous measures of performance, which allows evaluators to observe how different aspects of performance change as challenges increase rather than falling into the trap of reducing everything into a single score. This matches real-world settings in which performance can never be fully reduced to one dimension. An intelligence analyst, for example, is expected to work accurately, prioritize the most critical items, and use time efficiently. These dimensions are often at odds with one another. A faster pace increases the possibility of missing important details, while an excessively thorough approach may sacrifice the timeliness of reporting. For this reason, JAT emphasizes defining multiple continuous measures that can capture how these dimensions are traded off against one another. Rather than collapsing all the dimensions into a single index prematurely, JAT retains them separately to allow an evaluator to determine where a system improves one aspect of performance while degrading another. This keeps the measures tied to the mission value of the work and not just to what is easy to record. Finally, JAT is more effective when performance variables are measured on a continuous ratio scale, as opposed to a categorical or binary scale. This increased resolution of performance, where appropriate, allows for much more competent modeling of results.

Determining appropriate challenge variables is central to JAT. Challenge measures describe what makes work difficult. Rather than simply labeling tasks as being either "easy" or "hard," JAT translates challenge into something more granular and scalable. For an analyst, for example, "challenge" might refer to working with a larger-than-usual dataset, signal ambiguity, increased time pressure, or needing to coordinate across multiple streams of information. The challenge on each of these levels can be carefully controlled so that evaluators can follow along with how performance maintains its level, begins to strain, or gives way entirely to the increased demand challenge. Each challenge measure isn't guaranteed to capture brittleness by itself. Sometimes what seems like a non-challenge is in fact a stumbling block, and vice versa. This is why JAT supports relying on several complementary challenge measures that span enough levels of challenge to identify at what points joint performance bends and eventually breaks. Like performance measures, continuous challenge measures are preferred as they increase the resolution of differences between the test cases and support much more powerful analyses.

Morey, Rayo, et al. (2025) illustrate how performance measures and challenge measures can be operationalized in the context of a study. For performance, the authors measured nurses' concern or perceived urgency for a patient on a continuous sliding scale, rather than collecting a binary decision. This reflected the continuum of responses nurses may have for a patient, including those that fall in between the extremes of responsiveness (e.g., increased monitoring). Furthermore, the authors separately measured nurses' performance on emergency patients and control patients to reflect the dual goals of being responsive to urgent patients without being overly responsive to non-urgent patients. For challenge, the authors measured the magnitude of AI error for individual patient recommendations, rather than categorizing AI recommendations (e.g., hit, miss, false alarms, correct rejection), which provided additional granularity to insights.

After identifying performance and challenge measures, the next step is to design cases or scenarios that operationalize those measures. As an empirical method, which is based on observing performance in a scenario or across a set of scenarios, the strength of the JAT method depends on how realistic those scenarios are to the work domain. Thus, to capture brittleness, the scenarios should include a wide range of levels for the challenge measure. Evaluators must avoid clustering too many challenge levels at either the easy or moderate range. Figures 12.2 and 12.3 compare strong and weak distributions of challenge levels in a scenario. Poor distribution of

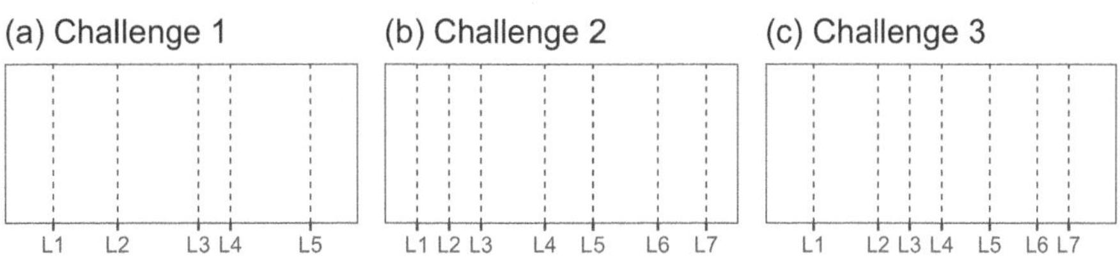

FIGURE 12.2 (a–c) Recommended distribution of challenge levels (L#) across challenge measures.

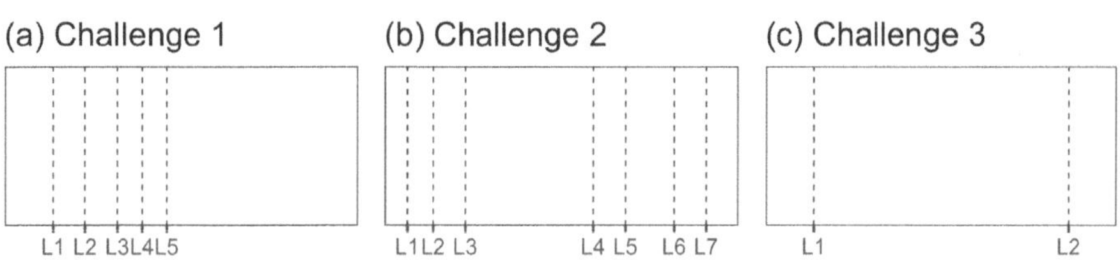

FIGURE 12.3 (a–c) Not recommended distributions of challenge levels (L#) across challenge measures.

challenge levels risks missing the sharp drops in performance, whereas distributing the challenge levels more evenly in a scenario can help evaluators see how performance holds up, begins to drop, and ultimately collapses when work demands rise. For example, Morey, Rayo, et al. (2025) tested a wide range of AI performance with cases that were (approximately) evenly spaced across the range, which facilitated modeling the relationship between performance and challenge. As noted earlier, CTA may be used to identify points where brittleness is most likely to occur, and scenarios should include those conditions purposely. Practically, this equates to selecting datasets that vary in signal clarity, time pressure, or competing demands and ensuring those "in-between" conditions are sampled because those are the conditions where errors are most likely. Designing scenarios in this way links abstract challenge measures to concrete testing and ensures that evaluations yield evidence regarding resilience and brittleness prior to deployment of systems.

JAT studies have often utilized ground truth, especially in calculating performance measures. As such, before the evaluation begins, the research team typically must produce (or identify) answer keys or ground truth to define the success associated with the selected performance measures. This could be a simple list of correct responses or a more complicated graded set of possible interpretations in ambiguous conditions. For example, Morey, Rayo, et al. (2025) utilized historical records of emergency events as ground truth about the urgency of a patient. Truthing does not necessarily eliminate uncertainty. If the best standard in a domain is to indicate a scenario is indeterminate, then indeterminacy is still a legitimate ground truth. What is important is that the truthing happens early enough to guide the choice of scenarios and to avoid any bias being introduced once data collection from participants has begun.

All this preparatory work is what makes phase II the place to sharpen the evaluation before the testing phase. The details are, of course, important. How is performance measured? What is the challenge? How are scenarios built? How are the answers truthed? Each of these details matters in shaping what the evaluation will tell us. What sets this phase apart from a more routine planning phase is the way in which JAT pairs performance and challenge. By treating both as continuous and grounded in the real challenges of analytic work, the evaluation is designed to show not simply whether the system works, but how performance changes jointly as demands increase. This is the foundation for how JAT makes brittleness visible prior to the deployment of systems.

12.4.3 Phase III: Collect the Data

In the data collection phase of a study, JAT adopts many of the standards or best practices of experimental research. These practices include having clear protocols, organized workstations, participant immersion, reference materials, training, proficiency checks, data logging, structured or unstructured observations, debriefs, and clear documentation throughout. These activities are not unique to JAT but are nevertheless important for effective evaluations.

What is unique to JAT is not necessarily *what* is done in data collection, but rather *how* data collection focuses on the relationship between performance and challenge.

It is not only about collecting accurate data, but about understanding how the JCS behaves under real conditions. Data are collected in multiple forms and integrated to form a complementary picture of how people adapt to sustain performance in the face of growing challenges. For example, data collection may include real-time observation, unobtrusive behavior capture (via interface interactions), training notes, survey instruments, and debrief sessions. The iterative synthesis of these varied data sources generates a deeper understanding of both the system being evaluated and the evaluation process itself.

12.4.4 Phase IV: Analyze the Data

Much of the data analysis process, such as checking for errors, cleaning datasets, and synthesizing observers' field notes, is standard. However, what makes JAT distinct is the way in which the results are connected back to performance and challenge. The goal is not only to summarize what happened but also to show how performance changes as tasks become harder. Thus, the first step is to check to see if ground truth has been fully specified for each scenario. This work may be completed in advance, but sometimes with a large or complex study, the truthing set must be expanded post-data collection. If truthing data is being generated by subject matter experts, they should not be exposed to the participants' responses to avoid truthing bias. With the truthing established, the team begins with quantitative data preparation. Quantitative preparation includes cleaning files, verifying that values make sense, and confirming that the correct measures are present for analysis. Concurrently, the qualitative data organization process begins. Observer notes, participant comments, and daily debriefs are reviewed, coded, and sorted into themes. These two processes then comprise JAT's main output: the JPGs (discussed in detail in the previous section of this chapter). The JPGs illustrate the changes in performance across different challenge levels. This is where system brittleness becomes visible. The numbers, by themselves, indicate the performance drop, but do not explain the reason. Notes, debriefs, and other forms of qualitative data help to understand why participants picked a specific path, why some errors were repeated, and why adaptation worked in certain scenarios and not others.

The last step is comparing the quantitative and qualitative findings. The quantitative results show where performance dropped, and the qualitative results explain how and why it happened. Together, they provide a perspective not possible by either measure alone. Ultimately, the end product is not merely data, but an understanding of how a JCS adapts when challenges increase and where it is most likely to fail.

12.4.5 Phase V: Report Evaluation Results

When delivering results, JAT facilitates the transparent disclosure of strengths and weaknesses. Both performance improvements and brittleness points are reported. Emphasizing both sides of the findings assists decision-makers in seeing not only what the system can do but also where a new risks may be created. This information is then pulled together in the final report. The final report contains all the standard sections of studies (introduction, methods, results, and discussion) while adding recommendations from after-action reviews and recommendations for future evaluation. Slide briefings are prepared alongside or to accompany the final report, focusing on less text and more visuals, staying within timing constraints, and leaving time for questions. In some situations, parts of the findings may be published broadly to share results with the wider community working on JCSs. Therefore, a report does not simply end the evaluation; it also acts as a source for a larger cycle of learning and improvement.

12.5 ROLE OF JAT IN THE FUTURE OF HUMAN–AI EVALUATION

Current efforts to improve JAT focus on making the method easier and quicker to implement and more sensitive to non-linear performance. To accelerate study setup, we have compiled a set of potential performance and challenge measures that can be tailored or used as inspiration for subsequent projects at https://go.osu.edu/jat-measures. We are working to make the data collection of participant behaviors more unobtrusive and automated so that a research team can test substantially more participants. Efforts are also underway to introduce automation into qualitative data analysis (Asabere et al., 2025), as we know that structured data collections can distort important details of participants' thoughts and actions (Morey, Cai, et al., 2025). For analysis, efforts are underway to provide guidance and potential semi-automation for model-based analysis, including linear mixed effects models and automation of JPG creation. In this way, we hope to make JAT a more pragmatic method that is appropriate for more contexts and earlier stages of the solution development lifecycle.

REFERENCES

Asabere, D., Sridhar, S., Ramnath, R., Parthasarathy, S., Modali, A., Hasenstab, K. A., Moore, C. O., Hood, D. B., & Rayo, M. F. (2025). Computational ethnography: Scaling community-based participatory research and planning. *Proceedings of the Human Factors and Ergonomics Society Annual Meeting*, *69*(1), 2080–2085. https://doi.org/10.1177/10711813251358782

Baard, S. K., Rench, T. A., & Kozlowski, S. W. J. (2014). Performance adaptation: A theoretical integration and review. *Journal of Management*, *40*(1), 48–99. https://doi.org/10.1177/0149206313488210

Bainbridge, L. (1983). Ironies of automation. *Automatica*, *19*(6), 775–779.

Bansal, G., Nushi, B., Kamar, E., Horvitz, E., & Weld, D. S. (2021). Is the most accurate AI the best teammate? Optimizing AI for teamwork. *Proceedings of the AAAI Conference on Artificial Intelligence*, *35*(13), Article 13. https://doi.org/10.1609/aaai.v35i13.17359

Campbell, J. P., McCloy, R. A., Oppler, S. H., & Sager, C. E. (1990). A theory of performance. In N. Schmitt & W. C. Borman (Eds.), *Personnel selection in organizations* (pp. 35–70). Jossey-Bass.

Clark, R. E., Feldon, D. F., Van Merriënboer, J., Yates, K. A., & Early, S. (2007). Cognitive task analysis. In J. M. Spector, M. D. Merrill, J. J. G. Van Merriënboer, & M. P. Driscoll (Eds.), *Handbook of research on educational communications and technology* (3rd ed.). Springer Science+Business Media.

Crandall, B., Klein, G., & Hoffman, R. R. (2006). *Working minds: A practitioner's guide to cognitive task analysis*. Mit Press.

Damacharla, P., Javaid, A. Y., Gallimore, J. J., & Devabhaktuni, V. K. (2018). Common metrics to benchmark human-machine teams (HMT): A review. *IEEE Access*, *6*, 38637–38655. https://doi.org/10.1109/ACCESS.2018.2853560

Dekker, S. (2011). The criminalization of human error in aviation and healthcare: A review. *Safety Science*, *49*(2), 121–127. https://doi.org/10.1016/j.ssci.2010.09.010

Dekker, S., & Woods, D. D. (2002). MABA-MABA or abracadabra? Progress on human–automation co-ordination. *Cognition, Technology & Work*, *4*(4), 240–244. https://doi.org/10.1007/s101110200022

Endsley, M. R. (2023). Supporting human-AI teams: Transparency, explainability, and situation awareness. *Computers in Human Behavior*, *140*, 107574. https://doi.org/10.1016/j.chb.2022.107574

Floridi, L., Cowls, J., Beltrametti, M., Chatila, R., Chazerand, P., Dignum, V., Luetge, C., Madelin, R., Pagallo, U., Rossi, F., Schafer, B., Valcke, P., & Vayena, E. (2018). AI4People—An ethical framework for a good AI society: Opportunities, risks, principles, and recommendations. *Minds and Machines*, *28*(4), 689–707. https://doi.org/10.1007/s11023-018-9482-5

Gray, W. D. (2002). Simulated task environments: The role of high-fidelity simulations, scaled worlds, synthetic environments, and laboratory tasks in basic and applied cognitive research. *Cognitive Science Quarterly*, *2*(2), 205–227.

Hoffman, R. R., & Militello, L. G. (2008). *Perspectives on cognitive task analysis* (1st ed.). Psychology Press. https://doi.org/10.4324/9780203809877

Hoffman, R. R., & Woods, D. D. (2011). Beyond Simon's slice: Five fundamental trade-offs that bound the performance of macrocognitive work systems. *IEEE Intelligent Systems*, *26*(6), 67–71. IEEE Intelligent Systems. https://doi.org/10.1109/MIS.2011.97

Hollnagel, E. (2014). *Safety-I and safety-II*. Ashgate Publishing, Ltd.

Hollnagel, E. (2017). *The ETTO principle: Efficiency-thoroughness trade-off*. CRC Press.

Hollnagel, E., & Woods, D. D. (1982). *Cognitive systems engineering: New wine on new bottles* (Risø-M No. 2330). Risø National Laboratory. https://erikhollnagel.com/onewebmedia/CSE_NWINB.pdf

Howie, D. E., & Vicente, K. J. (1998). Measures of operator performance in complex, dynamic microworlds: Advancing the state of the art. *Ergonomics*, *41*(4), 485–500. https://doi.org/10.1080/782155459

Hutchinson, B., Dekker, S., & Rae, A. (2022). Writing plans instead of eliminating risks: How can written safety artefacts reduce safety? *Safety Science*, *151*, 105738. https://doi.org/10.1016/j.ssci.2022.105738

Jalaeian, M., Morey, D. A., & Rayo, M. F. (2025). *Joint activity design heuristics for enhancing human-machine collaboration* (arXiv:2512.08036). arXiv. https://doi.org/10.48550/arXiv.2512.08036

Kiani, A., Uyumazturk, B., Rajpurkar, P., Wang, A., Gao, R., Jones, E., Yu, Y., Langlotz, C. P., Ball, R. L., Montine, T. J., Martin, B. A., Berry, G. J., Ozawa, M. G., Hazard, F. K., Brown, R. A., Chen, S. B., Wood, M., Allard, L. S., Ylagan, L., … Shen, J. (2020). Impact of a deep learning assistant on the histopathologic classification of liver cancer. *Npj Digital Medicine*, *3*(1), 1–8. https://doi.org/10.1038/s41746-020-0232-8

Klein, G. (1993). A recognition-primed decision (RPD) model of rapid decision making. In G. A. Klein, J. Orasanu, R. Calderwood, & C. E. Zsambok (Eds.), *Decision making in action: Models and methods* (4th ed., Vol. 5, pp. 138–147). Ablex Publishing.

Klein, G., Feltovich, P. J., Bradshaw, J. M., & Woods, D. D. (2005). Common ground and coordination in joint activity. In W. B. Rouse & K. R. Boff (Eds.), *Organizational simulation* (pp. 139–184). John Wiley & Sons, Inc. https://doi.org/10.1002/0471739448.ch6

Klein, G., Ross, K. G., Moon, B. M., Klein, D. E., Hoffman, R. R., & Hollnagel, E. (2003). Macrocognition. *IEEE Intelligent Systems*, *18*(3), 81–85. IEEE Intelligent Systems. https://doi.org/10.1109/MIS.2003.1200735

Lang, J. W. B., & Bliese, P. D. (2009). General mental ability and two types of adaptation to unforeseen change: Applying discontinuous growth models to the task-change paradigm. *Journal of Applied Psychology*, *94*(2), 411–428. https://doi.org/10.1037/a0013803

Liu, X., Faes, L., Kale, A. U., Wagner, S. K., Fu, D. J., Bruynseels, A., Mahendiran, T., Moraes, G., Shamdas, M., Kern, C., Ledsam, J. R., Schmid, M. K., Balaskas, K., Topol, E. J., Bachmann, L. M., Keane, P. A., & Denniston, A. K. (2019). A comparison of deep learning performance against health-care professionals in detecting diseases from medical imaging: A systematic review and meta-analysis. *The Lancet Digital Health*, *1*(6), e271–e297. https://doi.org/10.1016/S2589-7500(19)30123-2

Mikalef, P., Conboy, K., Lundström, J. E., & Popovič, A. (2022). Thinking responsibly about responsible AI and 'the dark side' of AI. *European Journal of Information Systems*, *31*(3), 257–268. https://doi.org/10.1080/0960085X.2022.2026621

Morey, D. A., Cai, M., Hrebenach, J., Li, M., Walli, P., & Rayo, M. F. (2025). Toward scalable research: Comparing structured and unstructured data collection mechanisms. *Proceedings of the Human Factors and Ergonomics Society Annual Meeting*, 10711813251358234. https://doi.org/10.1177/10711813251358234

Morey, D. A., Della Vella, D., Rayo, M. F., Zelik, D. J., & Murphy, T. B. (2022). Joint activity testing: Towards a multi-dimensional, high-resolution evaluation method for human-machine teaming. *Proceedings of the Human Factors and Ergonomics Society Annual Meeting*, *66*, 2214–2219.

Morey, D. A., Jalaeian, M., Reynolds, M. E., McGeorge, N. M., Oldham, H., & Rayo, M. F. (in press). Four aspirations for evaluating AI-infused technologies: A narrative review. *Journal of Cognitive Engineering and Decision Making*.

Morey, D. A., Marquisee, J. M., Gifford, R. C., Fitzgerald, M. C., & Rayo, M. F. (2020). Predicting graceful extensibility of human-machine systems: A new analysis method for evaluating extensibility plots to anticipate distributed system performance. *Proceedings of the Human Factors and Ergonomics Society Annual Meeting*, *64*, 313–318. https://doi.org/10.1177/1071181320641072

Morey, D. A., Rayo, M. F., & Woods, D. D. (2025). Empirically derived evaluation requirements for responsible deployments of AI in safety-critical settings. *Npj Digital Medicine*, *8*(1), 374. https://doi.org/10.1038/s41746-025-01784-y

Morey, D. A., Ryder, A., Winner, J. L., Cai, M., Algarin, L., Rayo, M. F., Parker, C., Tewani, P., Kopren, K., Green, J., & Kinney, P. (2025). *Joint activity testing implementation guide* (NGA-U-2025-01382). The Ohio State University.

Patterson, E. S., Rogers, M. L., Tomolo, A. M., Wears, R. L., & Tsevat, J. (2010). Comparison of extent of use, information accuracy, and functions for manual and electronic patient status boards. *International Journal of Medical Informatics*, *79*(12), 817–823. https://doi.org/10.1016/j.ijmedinf.2010.08.002

Patterson, E. S., Roth, E. M., & Woods, D. D. (2010). Facets of complexity in situated work. In E. S. Patterson & J. E. Miller (Eds.), *Macrocognition metrics and scenarios: Design and evaluation for real-world teams* (1st ed., pp. 221–251). Taylor & Francis Group. https://www.worldcat.org/title/macrocognition-metrics-and-scenarios-design-and-evaluation-for-real-world-teams/oclc/754793942

Rajpurkar, P., Chen, E., Banerjee, O., & Topol, E. J. (2022). AI in health and medicine. *Nature Medicine*, *28*(1), Article 1. https://doi.org/10.1038/s41591-021-01614-0

Rayo, M. F. (2017). Designing for collaborative autonomy: Updating user-centered design heuristics and evaluation methods. *Proceedings of the Human Factors and Ergonomics Society Annual Meeting*, *61*, 1091–1095. https://doi.org/10.1177/1541931213601877

Schraagen, J. M., Chipman, S. F., & Shalin, V. L. (Eds.). (2000). *Cognitive task analysis* (1st ed.). Psychology Press. https://doi.org/10.4324/9781410605795

Sorkin, R. D., & Woods, D. D. (1985). Systems with human monitors: A signal detection analysis. *Human–Computer Interaction*, *1*(1), 49–75. https://doi.org/10.1207/s15327051hci0101_2

Wickens, C. D., & Dixon, S. R. (2007). The benefits of imperfect diagnostic automation: A synthesis of the literature. *Theoretical Issues in Ergonomics Science*, *8*(3), 201–212. https://doi.org/10.1080/14639220500370105

Wood, R. E. (1986). Task complexity: Definition of the construct. *Organizational Behavior and Human Decision Processes*, *37*(1), 60–82. https://doi.org/10.1016/0749-5978(86)90044-0

Woods, D. D. (1985). Cognitive technologies: The design of joint human-machine cognitive systems. *AI Magazine*, *6*(4), 1–7. https://doi.org/10.1609/aimag.v6i4.511

Woods, D. D. (2015). Four concepts for resilience and the implications for the future of resilience engineering. *Reliability Engineering & System Safety*, *141*, 5–9. https://doi.org/10.1016/j.ress.2015.03.018

Woods, D. D. (2018a). Resilience is a verb. In B. D. Trump, M.- V. Florin, & I. Linkov (Eds.), *IRGC resource guide on resilience: Domains of resilience for complex interconnected systems* (Vol. 2). EPFL International Risk Governance Center (IRGC). igrc.epfl.ch

Woods, D. D. (2018b). The theory of graceful extensibility: Basic rules that govern adaptive systems. *Environment Systems and Decisions*, *38*(4), 433–457. https://doi.org/10.1007/s10669-018-9708-3

Woods, D. D. (2020). *Building adaptive capacity in this series of beyond-surge-capacity outbreaks*. https://doi.org/10.5281/ZENODO.3748069

Woods, D. D., & Branlat, M. (2011a). Basic patterns in how adaptive systems fail. In E. Hollnagel, J. Pariès, D. D. Woods, & J. Wreathall (Eds.), *Resilience engineering in practice: A guidebook* (1st ed., pp. 127–143). Ashgate. https://doi.org/10.1201/9781317065265-10

Woods, D. D., & Branlat, M. (2011b). How human adaptive systems balance fundamental trade-offs: Implications for polycentric governance architectures. In E. Hollnagel, É. Rigaud, & D. Besnard (Eds.), *Proceedings of the 4th Resilience Engineering Symposium* (pp. 277–283). Presses des Mines. https://doi.org/10.4000/books.pressesmines.1125

Woods, D. D., Johannesen, L. J., Cook, R. I., & Sarter, N. B. (1994). *Behind human error: Cognitive systems, computers, and hindsight* (State-of-the-Art Report CSERIAC SOAR 94-01). The Ohio State University.

Zech, J. R., Badgeley, M. A., Liu, M., Costa, A. B., Titano, J. J., & Oermann, E. K. (2018). Variable generalization performance of a deep learning model to detect pneumonia in chest radiographs: A cross-sectional study. *PLoS Medicine*, *15*(11), e1002683. https://doi.org/10.1371/journal.pmed.1002683

13 The ABC–DEFs of HATs
A Cyber-Social Paradigm for Human-AI Teaming

Mukhunth Raghavan, Daniel Griffin, Stephanie A. Boettcher, Max Brunell, and Joshua Pearman

13.1 INTRODUCTION

Advances in artificial intelligence (AI) have fundamentally reshaped how humans interact with computational systems at work. Early conceptualizations of human–machine collaboration largely treated algorithms as tools or automated processes that humans supervise (Seeber et al., 2020). However, modern AI systems increasingly demonstrate autonomy, context sensitivity, and adaptive behavior, enabling them to participate in decision-making and coordination in ways that resemble traditional human teammates (Dennis et al., 2023; Landers & Behrend, 2024). This shift has accelerated the emergence of human–AI teams (HATs; Iftikhar et al., 2023; O'Neill et al., 2022), in which human and algorithmic partners work interdependently toward shared goals (Larson & DeChurch, 2020). As AI systems assume more proactive and socially embedded roles in work teams, existing models of teamwork that were developed to explain and measure interactions between human teammates are no longer sufficient for understanding or optimizing performance in these hybrid work groups.

Research within all-human teams and HATs alike has found that affective emergent states encompassing trust, shared motivation, and cohesion play a critical role in enabling and shaping the collective behaviors and processes of teamwork (Bell et al., 2018; Marks et al., 2001). Similarly, emergent team cognition, encompassing concepts such as shared mental models, reflects to what extent team members have a common understanding of goals and actions (Mathieu et al., 2000; Mohammed et al., 2021). Within HATs, humans routinely ascribe competence, intention, and even social presence to algorithmic partners, particularly when those systems exhibit autonomous and contextually responsive behavior (Glikson & Woolley, 2020).

Yet, despite these changing expectations, current theorizing about HATs remains asymmetrically human-centered. Research overwhelmingly focuses on human trust of AI (Dennis et al., 2023; Duan et al., 2024), decision-making (Sawant et al., 2022), and other human reactions, while giving comparatively little attention to what the AI agent internally represents, how it evaluates situations, and how its internal processes shape the joint functioning of the team. This imbalance obscures the mechanisms by

DOI: 10.1201/9781003655923-16

which HATs function effectively and limits the development of AI systems that can integrate smoothly into human social and cognitive structures. In this chapter, we propose the cyber-social framework as a conceptual paradigm for understanding the unique context that emerges from the collaboration between human and AI teammates, and expand on this to demonstrate its potential as an empirical tool.

13.2 WHY HATs NEED A CYBER-SOCIAL PARADIGM

Although humans frequently perceive and treat advanced AI as teammates, they do so in a context where the agent's internal "psychology" is opaque or neglected. Existing HAT research tends to describe human affect, behavior, and cognition (the ABCs of teamwork; Bell et al., 2018) without offering a parallel account of AI processes that can influence team functioning. However, as noted by Andrews and colleagues (2023), AI team members will have their own "mental" representations of humans, as well as their own actions and behaviors that shape their contributions to interdependence and coordination. When only the human components of teamwork are theorized, half of the team's causal system remains underspecified.

A growing barrier to effective human–AI coordination is the lack of understanding of the shared representations of the task and context, a foundational element of human teamwork long attributed to Theory of Mind (ToM; Premack & Woodruff, 1978). Teams that develop a strong shared understanding can coordinate and collaborate more effectively, often without the need for explicit communication (Mathieu et al., 2000; Müller & Antoni, 2022). ToM suggests that humans continuously infer the thoughts, feelings, and intentions of others based on their beliefs about those individuals' mental states (Li et al., 2022). When collective ToM is explored within teams, it has been shown to strengthen coordination and foster strong relational ties (Shteynberg et al., 2023). Without such mutual inference, collaboration devolves into misaligned expectations, over- or under-reliance on automation, and breakdowns in coordinated action (Bansal et al., 2019; Carter-Browne et al., 2021).

As a result, collaboration in HATs is hindered by significant challenges (e.g., coordinating interdependent tasks, recovering from breakdowns at the team level; Bansal et al., 2019; Iftikhar et al., 2023; G. Zhang et al., 2021), and humans continue to distrust their AI counterparts (Georganta & Ulfert, 2024), making coordination difficult. Although AI agents are efficient at accomplishing the specific tasks they are designed to achieve, they are often poor at integrating within a human team's social and decision-making structure (Bansal et al., 2019), making them suboptimal team members. For example, human teammates tend to disregard AI decision-making in favor of human decision-making due to a variety of biases toward AI competence in team scenarios, as reported in specific contexts (see Ashktorab et al., 2021; Ishowo-Oloko et al., 2019). The absence of a unified conceptual framework for modeling both human and algorithmic teammates limits theoretical integration and hinders empirical progress. A new paradigm is needed to articulate the cyber-social context of HATs in which computational and human processes interact to produce emergent team phenomena.

13.2.1 Machine Theory of Mind (MToM) is the Cyber-Social Keystone

Machine Theory of Mind (MToM; Rabinowitz et al., 2018) provides the conceptual bridge necessary for integrating humans and AI within a single cyber-social framework. Formally, MToM is defined as the ability of an AI agent to update posterior probabilities based on the mental states and latent characteristics of other agents, allowing them to more deeply understand other-agent mindsets and adjust their actions in ways that are sensitive to their human and AI teammates' preferences, knowledge, and intentions (Rabinowitz et al., 2018). MToM is a key computational mechanism that enables shared understanding, anticipation, and proactive coordination, which are all foundational for effective teamwork in all teams.

Empirical work demonstrates that AI agents equipped with even basic levels of ToM capabilities infer trustworthiness (Vinanzi et al., 2019), predict human tendencies in shared workspaces (S. Zhang et al., 2024), and dynamically adapt communication and action strategies to support their human partners. These findings underscore that effective MToM is not merely an optional enhancement but a necessary condition for treating AI agents as teammates rather than tools, whereby AI agents can participate in proactive coordination and develop shared mental models that facilitate team performance (Langley et al., 2022).

Although MToM-equipped systems are currently nascent, the trajectory of AI development suggests increasing integration of such capabilities. This framing allows us to systematically map the parallels between human and AI processes, aligning ABC in humans alongside their algorithmic analogs in AI agents. To study these interactions meaningfully and to move beyond our reliance on perceptions of AI agents as teammates, we introduce the *cyber-social framework*, which offers a shared lens for evaluating interdependence, coordination, and emergent states in HATs.

13.3 DEFINING THE CYBER-SOCIAL CONTEXT

Human teamwork is often studied based on the key components of Affect, Behavior, and Cognition (the ABCs of teamwork; Bell et al., 2018). The Affective component refers to team members' emotions, attitudes, and motivations, such as trust, psychological safety, and collective efficacy. Positive affective states foster open communication, greater team cohesion, and better collaboration, directly enhancing team performance. The Behavioral component refers to the actions team members engage in, including task work, communication, coordination, and conflict management. Effective behavioral processes are linked to higher productivity and better performance outcomes. The Cognitive component refers to individual and shared mental models, understanding, and knowledge among team members. Strong cognitive alignment is associated with better anticipation of team needs and proactive behaviors, which are critical for complex and high-stakes scenarios. The ABC constructs predict key team outcomes across domains but apply only to human teammates.

Affective states such as team trust, cohesion, shared positive affect, and collective efficacy have all been identified as critical antecedents of team outcomes, including effectiveness (Black et al., 2019; Grossman et al., 2022; Kaplan et al., 2013; Meneghel et al., 2016; Ramzaninezhad et al., 2009), adaptability (Dervitsiotis, 2006; Elms, 2025;

Hoandră, 2017; Marques-Quinteiro et al., 2015), creativity (Emich & Lu, 2021), and well-being (Conti et al., 2018; Krug et al., 2021). Similarly, team-centric behaviors such as planning (Mehta et al., 2009; Weingart, 1992), coordination (Janicik & Bartel, 2003; Rico et al., 2018, 2019), and communication (Marlow et al., 2018; Valls et al., 2016), along with collective cognition (i.e., shared mental models and transactive memory systems; Bachrach et al., 2019; Lim & Klein, 2006; Mohammed et al., 2021), have been identified as critical factors shaping teams and their outcomes.

Although AI agents are not designed to feel, act, or think in the same way as humans, they perform computational functions that are structurally analogous to these human processes. Modern AI systems are capable of continuously appraising options in a given situation, executing functions that shape team outcomes, and maintaining internal representations of the state, as well as memories of past events or conversations (Wu et al., 2025; Zhong et al., 2024). To capture this symmetry and enable rigorous testing of the cyber-social paradigm, we introduce a complementary triad for AI team members—Determination, Execution, and Framing—as the DEFs representing algorithmic counterparts to the human ABCs.

13.3.1 Determination

Affective states serve as heuristic evaluative tools that shape actions taken by the team (Bell et al., 2018). In much the same way, autonomous AI agents in HATs must repeatedly perform mission-critical evaluations using analogous functions that assess risks, prioritize objectives, and adjust expectations. For example, just as humans must make judgment calls regarding whom to trust (i.e., be vulnerable to), AI agents will need to make judgment calls regarding the reliability of working with specific humans. Just as humans make affective judgment calls regarding the probability of loss and reward in the form of positive-negative affect (Neville et al., 2021), AI algorithms generally must evaluate risk and reward probabilities. In fact, risk-averse algorithms could be described as having a high degree of negative affect (Alessandro et al., 2025). Similarly, AI agents that tend to ignore risk and focus on desired objectives could be described in terms of positive affect (e.g., socially supportive AI; Hu et al., 2025).

AI agents requiring robust MToM evaluative capabilities could be programmed to detect a human's frustration or confusion (e.g., through tone of communication or hesitation), for instance, and determine the appropriate anticipatory response with an explanation or reassurance, helping to build trust over time. On the other hand, when using a pre-built AI agent, the determination quality could be measured by analyzing the AI agent's logs to uncover whether it correctly identified cues and provided appropriate, anticipatory responses.

It is important to note that while determinations used by AI agents are not an exact representation of the "affect" experienced by AI team members, they do present some conceptual parallels. We suggest that in order to fully understand the cyber-social context of work done in HATs, it is critical to understand the evaluation process performed by the algorithms. In the near term, progress on capturing AI determinations may depend on qualitative insights into how agents evaluate goals, analyze risk, and enforce security protocols. With time, however, research should aim to develop systematic methods for capturing these evaluative processes directly, extracting metrics that can be compared alongside measures of human teammates' affect to illuminate the interaction and relationship between human affect and agent determination.

13.3.2 Execution

Humans engage in task-relevant behaviors such as action and transition processes (Marks et al., 2001). AI agents are similarly designed to execute one or more functions that may include action and transition processes (e.g., planning team behaviors; Bezrucav & Corves, 2020). While the way in which humans and AI agents engage in their respective behaviors and executions is different, the fundamental idea is the same: AI executions correspond directly to human team behaviors and processes.

One advantage when studying behaviors is that they naturally lend themselves to objective measures. Objective measures enable researchers to evaluate an individual or team while minimizing the effect of individual perceptions. Realistically, objective measures are not only valuable in studying any team (Salas et al., 2008) but are necessary for HATs, where individual perceptions are not clearly defined. For humans, team behavior encompasses actions such as communication, planning, task execution, and even physical movement. These are generally observable and objectively quantifiable, yet the dominant approaches are self-reports (Maynard et al., 2021). For instance, researchers have demonstrated the value of objective measures based on text data as a source for studying team processes (Mathieu et al., 2022). Similarly, AI agents' functions include the execution of tasks, decision outputs, and communication with humans and potentially other AI agents, which can all be objectively recorded by reviewing data logs, providing objective metrics of the AI agents' behaviors.

When building AI agents for teamwork, programmers can ensure that the agents show proactive behavior by anticipating their teammates' needs. For example, an agent with high MToM capacity could predict that a teammate is about to make a mistake based on incomplete or incorrect information, intervening with a warning or providing the correct response instead. The agent's ability to accurately infer the human's mental model and take aligned actions can be measured as an index of the AI agent's execution capacity. These behaviors can be easily measured and assessed; greater capacity to successfully execute proactive actions will be indicative of an advanced AI agent with MToM capabilities and foster stronger team cohesion.

13.3.3 Framing

Teams research has consistently highlighted the role of shared cognition in human teams (DeChurch & Mesmer-Magnus, 2010; Marks et al., 2001; Mohammed et al., 2021). Just as team cognition involves the development of internal mental models of a team's work, AI algorithms must internally describe and frame the environment and task domain they work in. Although this computerized framing is different from human team cognition, there is significant overlap in the purpose and general content that these internalized representations must include.

When employing conversational AI as a teammate (such as for decision-making support; e.g., Zercher et al., 2025), the AI agent's ability to frame a shared mental model can be simply assessed by asking the agent to explain its understanding of the team's goals, team member roles, and task execution plans, which can then be rated for accuracy based on a predetermined rubric. Similarly, for embedded AI systems, metrics such as accurate predictions of human teammates' actions and needs in a simulated environment based on the team's past actions and current state would provide evidence for a strong internal description of the team environment and task.

By properly assessing a computer's internalized description of a situation, we suggest that AI framing could be meaningfully studied as a parallel construct to human team cognition. Such efforts would require substantial expertise to explicitly program tools, such as composite measures of data logs and technical metrics or conversational prompts, to assess an AI agent's internalized description. Despite this complexity, we suggest that it may be less problematic than studying team cognition in humans. Measurement of human team cognition generally uses questionnaires that better measure perceptions of shared cognition than shared cognition itself (e.g., Bell et al., 2018), or requires sophisticated measurement techniques (Ilgen et al., 2005) that are fraught with their own challenges (e.g., potential bias) (DeChurch & Mesmer-Magnus, 2010; Salas et al., 2015).

In contrast, it is quite possible to develop tools that directly and objectively measure the framing used by an AI agent. Utilizing metadata that reports an AI agent's internal dialogue through the use of decision trees, utility models, etc. (Zhao et al., 2022), researchers can keep track of AI framing and determinations that ultimately lead to executions being made in the team setting. Oftentimes, researchers will develop AI cognition themselves, allowing for direct measures of the Ds and Fs to be pulled using metadata with ease. Even without such efforts, it is reasonable that subject matter experts could provide meaningful qualitative assessments of the computerized description, giving a pulse on the sharedness of the internal models of humans and AI agents.

Measuring AI agents' DEFs is not simply about applying psychometric constructs available for studying human teams; it requires defining and tracking AI-specific metrics that reflect how these systems internally represent, evaluate, and act within a team context. Mapping these AI systems to the ABCs of teamwork highlights the similarities and differences between human and AI processes, and underscores the need to develop tools and methodologies that allow for meaningful comparison and study of human-AI team dynamics. Table 13.1 details the DEF triad, outlining how each must be designed and measured to ensure AI agents integrate effectively with human ABCs.

If HAT researchers do not consider the DEFs of AI team processes, they will miss out on potentially rich data demonstrating an AI agent's contributions to interdependent team goals. By focusing solely on human processes, researchers will only understand half of the picture. This is even more relevant to understand for the teams themselves. If human teammates disregard AI insight, the AI agent will learn to avoid interdependent goals with the humans and focus solely on team goals it can achieve by itself. This collapse of trust may further exacerbate human wariness of AI, ultimately destroying team chemistry.

TABLE 13.1
Definitions, Example Constructs, and Their Operationalization for ABC-DEFs of HATs

		Affect	**Determination**
Definition		Feelings and attitudes toward one's team and their goals.	Utilizing utility equations to weigh the value of risks, rewards, and resources.
Example constructs and operationalization	1	*Trust* Self-reported cognitive and affective trust in teammates (McAllister, 1995)	*Risk-Assessment* Assessors calculate risk-assessments by identifying potential risks, estimating likelihood and severity bands, and explicitly documenting evidence, underlying assumptions, and analyses at appropriate granularities (Wisakanto et al., 2025).
	2	*Cohesion* Self-reported attraction and integration toward a group's social and task processes (GEQ, Brawley et al., 1987)	*Utility equation* AI neural utility functions based in utility theory that dissect item relationships and reason about trade-offs to make more accurate recommendations (Jenkins et al., 2021).
		Behavior	**Execution**
Definition		Interdependent actions that translate inputs into outputs.	Exertion of physical, social, or other behavioral processes to accomplish team goals.
Example constructs and operationalization	1	*Communication* Self-reported quality and frequency of communication (Marlow et al., 2018)	*Trajectory Planning* Integration of real-time human motion data to respond to environmental changes during physical collaboration (Shen et al., 2024)
	2	*Coordination* Team-member cooperative movements in an interdependent simulation (Boettcher et al., 2025)	*Backup Behaviors* Agents equipped with long short term memory and a deep neural network estimate driver intentions and can intervene to control a vehicle in real-time (Janwani et al., 2024).
		Cognition	**Framing**
Definition		Structure of knowledge and collective perception.	Pseudo-cognitive representations of the team system and role responsibilities.
Example constructs and operationalization	1	*Team Mental Model* Degree of sharedness of ideas (subject overlap, agreement), which represents similarity in team-member mental models (Jonker et al., 2011)	*Goal Hierarchies* Hierarchical Goal Network (HGN): combining heuristics and reasoning from classical planning with the scalability of hierarchical task networks to conduct advanced planning in ambiguous environments (Shivashankar, 2015).
	2	*Transactive Memory System*	*Task Dependency Mapping*

(*Continued*)

TABLE 13.1 (*Continued*)
Definitions, Example Constructs, and Their Operationalization for ABC-DEFs of HATs

	Affect	Determination
	Knowledge mapping: ask team members who knows what and cross-compare for accuracy (Lewis, 2003).	Topological task sorting to determine task dependencies and optimize task execution order (Wang et al., 2025).

13.3.4 The Need for Tools to Study the Cyber-Social Context

Current research examining human cognition and affect is troubled by a reliance on self-report methodology (Podsakoff et al., 2024). To advance our understanding of HATs, it is critical to develop tools and methodologies that enable researchers to capture both human ABC constructs and AI DEF constructs within a unified measurement architecture. Traditional methods, which are typically reliant on self-report measures, are insufficient in cyber-social contexts because they illuminate only the human side of the team. AI systems cannot generate subjective reports, yet they produce abundant behavioral and computational trace data that reflect DEF processes. Without methodologies that translate these data into interpretable indicators aligned with human constructs, researchers cannot fully characterize interdependence, coordination, or breakdowns in HAT performance. Therefore, new tools must support the extraction, synchronization, and analysis of parallel human and AI indicators in real time, enabling high-resolution assessment of emergent team states and processes.

Recent reviews highlight the need for real-time measurement approaches in HATs to capture the dynamic interplay of human and AI teammates (Duan et al., 2025). Likewise, foundational team science has emphasized that objective behavioral indicators are essential for advancing beyond static or perception-only models of teamwork (Salas et al., 2008; Wespi et al., 2023). Simulation- and game-based approaches present a potentially powerful way to address the significant challenges to studying HATs from a cyber-social perspective. High-fidelity simulated environments allow researchers to embed both human and AI teammates within dynamic, adversarial, and cognitively demanding tasks that resemble those experienced in real situations (Wiernik et al., 2022), while automatically collecting granular behavioral trace data (e.g., movement and coordination, decision-making timing, and situational awareness). For example, the rapid growth of esports (i.e., competitive video games) has brought a new wealth of available data and research to draw upon (Reitman et al., 2020), with the potential to augment teams research (Wagner, 2006).

Operationalizing the cyber-social framework requires moving beyond self-report measures, which dominate traditional team research but are ill-suited for AI agents that are not conversational AI. Instead, we advocate for objective, high-resolution behavioral metrics that can serve as proxies for both ABC and DEF constructs.

Recognizing the value of simulations for examining the cyber-social dynamics of human-AI teams, we describe a few novel approaches that demonstrate how ABC-DEF constructs can be objectively captured from both human and AI teammates through behavioral trace data.

13.3.4.1 Determination

It is valuable to illustrate modalities by which AI agents make determinations in a team setting. To demonstrate this, a task wherein AI agents make assumptions of human competence is helpful. Hanabi is a validated team-based task that has been widely used to investigate human coordination alongside AI algorithms (Eger et al., 2017, 2021). Hanabi is a cooperative 2–5 player card game where humans and AI collaborate by giving hints about cards held by their teammates, with the goal of playing cards accurately and maximizing a shared team score. The scarcity, and sometimes relative ambiguity, of hints given in Hanabi leads players to rely on set strategies and inferences about their teammates' lines of thought and intentionality. If properly equipped with ToM, AI agents who play Hanabi can make more informed inferences of human actions, and layer their strategy to align with a human teammate's train of thought.

The intentional AI developed by Eger and colleagues (2017) using this paradigm is explicitly designed to simulate potential actions of its human teammates, infer how the potential actions change the player's mental state, and then predict what the human's most probable action is. The AI in this task then assigns scores to the humans' varying potential moves and calculates the move that generates the highest possible score. Like a utility equation, this AI makes a determination of a human's potential logical progressions and chooses the move that has the highest predicted utility payoff. This AI is programmed to make determinations of human thought and predict which potential human move is beneficial for the team's collective goal.

AI integration into Hanabi teams is a telling example of the capabilities of AI agents that are endowed with MToM. AI agents that make inferences of human thought can more accurately estimate perceived utility, and resultingly, make cognitively aware decisions in interdependent team settings. These determinations provide AI with human-like reasoning and advanced risk assessment competence.

13.3.4.2 Execution

Boettcher et al. (2025) developed a methodology using a high-fidelity virtual combat simulation built on the popular gaming platform *Counter-Strike 2 (CS2)*. In their study, Boettcher and colleagues (2025) identified game-based metrics of team coordination behavior based on archival data of more than 1,200 rounds of gameplay. This methodology has the capability to generate granular, objective behavioral trace data which can be transformed into indicators of spatial coordination and team performance for all-human teams.

Building upon this methodology for identifying coordination metrics in human teams, we sought to extend their approach to evaluate analogous metrics for AI agents functioning as teammates. To demonstrate this, we replicated the study using human-AI teams, illustrating one example of how DEF processes could be extracted from AI trace data, in a manner comparable to the human behavioral metrics.

We present the following as a brief overview of our approach, intended as a proof-of-concept for how meaningful, parallel teamwork metrics can be obtained for both human and AI teammates.

Participants in our study engaged in a total of 112 rounds of *Counter-Strike 2* gameplay, where each player played alongside a human partner and an expert-level AI teammate, against a team of AI-controlled opponents. The overall goal of the simulation is for teams to coordinate to prevent the opposing team from successfully planting and defending a bomb, either by preventing the plant, defusing the bomb if planted, or eliminating all opposing players. Defensive objectives involved monitoring and controlling multiple approach paths to prevent the opposing team from successfully planting the bomb, including holding and guarding bomb sites to prevent entry and delay advances. In contrast, offensive objectives involved coordinated engagements such as retaking a bomb site or neutralizing opponents after a plant. Across rounds, teams alternated between such defensive objectives and offensive objectives, requiring coordination around both prevention and recovery goals while controlling designated objective locations (i.e., bomb sites). This allowed for the measurement of real-time coordination of HATs during a dynamic, high-tempo, adversarial environment in which teammates collaborated toward shared goals.

The simulation generated continuous behavioral trace data processed into one-second intervals to assess team spatial coordination. Team spatial coordination is commonly operationalized using positional indicators that capture how teammates distribute proximity, attention, and movement relative to one another during task completion. Accordingly, prior work highlights the use of positional tracking data in team sports, highlighting that inter-player proximity, collective coverage of space, and movement toward shared targets are core components of spatial coordination (Bourbousson et al., 2010; Rico-González et al., 2022). Three objective indicators of such coordination were derived to represent human teammates' behaviors and the AI teammate's execution: Colocation, representing how close teammates are to each other; View Coverage, reflecting the total coverage of the teammates' visual orientations; and Convergence, indicating the rate at which team members moved toward each other and toward a common point. These measures were designed to operationalize how teams allocate space, attention, and movement in response to changing mission demands, while enabling the direct comparison between similar constructs for human and AI team members without relying on subjective perceptions.

Our analyses revealed that coordination states were strongly predictive of performance outcomes for both humans and AI agents in consistent ways. For defensive objectives, greater View Coverage was significantly associated with lower damage taken and defeats, while Colocation was associated with reduced damage taken. For offensive objectives, Convergence emerged as a strong predictor of both damage to opponents and opponent neutralization, while lower View Coverage (i.e., greater collective team focus on the target) was associated with higher offensive effectiveness. Importantly, these relationships held consistently across human and AI teammates, suggesting that our objective coordination metrics provide a common framework for evaluating both entities within HATs.

These results demonstrate that AI agents, like human teammates, exhibit measurable patterns of coordination that can be systematically studied. More broadly, this work illustrates how the cyber-social paradigm can move beyond human-centric models to incorporate AI as a full-fledged team member, paving the way for richer theories, more precise measurement, and targeted interventions to optimize human-AI collaboration.

This study presented a specific opportunity to use a simulation approach to measure one of the DEF processes AI agents engage in within coordinated team settings. The simulation enabled the simultaneous measurement of AI teammates' execution and human teammates' behavior through spatial movement and coordination strategies used by team members. Many HAT paradigms are troubled by distinct processes and behaviors that underpin human and AI team dynamics, often shrouding AI processes behind the "black box" (see: Neerincx et al., 2018). By identifying operationally similar metrics for human and AI teammates, this study allows us to peek behind the "black box" and provides a replicable method for evaluating coordination and decision-making in HATs. The cyber-social paradigm of teamwork in HATs offers a unified framework for assessing both human and AI contributions within the same operational context, reducing ambiguity and enhancing transparency in team dynamics.

13.3.4.3 Framing

AI must accurately understand the roles, social structure, and goal-setting of their newly formed teams to functionally assimilate and become team members. The means by which they frame this understanding of the team's structure is paramount for this task, and is critical for AI algorithms that must integrate human teammate advice into their own task-relevant estimations. Goal optimization is a key method by which AI frame the goals and roles of their team to then make determinations and executions in team settings. One aspect of this framing that has been reliably demonstrated across empirical HAT studies is the goal-valuation and optimization of goal hierarchies conducted by AI. Autonomous agents who work alongside humans may encounter varying sub-goals and challenges that impede the overall objective attainment of the AI agent. To adequately rank these goals in order of importance, AI agents must assign values of relevance to each goal to mathematically designate their sequence.

Demonstrating the utility of goal hierarchies in independent HAT settings, Ma and colleagues (2023) developed a Human-AI Goal Optimization algorithm (HAI-GO). HAI-GO synthesizes human encoding of knowledge related to goals and its own optimization of goal sequencing to determine optimal goal-attainment structure. By framing the human's relevant knowledge and integrating its own advanced processing, HAI-GO makes accurate estimations of the team's needs prior to executing actions. HAI-GO was tested on its ability to navigate complex maze environments by ordering performance-relevant sub-goals. This algorithm proficiently encoded human input and optimized sub-goals to navigate complex environments in an efficient manner.

By capturing dynamic, objective, and task-oriented data, approaches like the three provided allow researchers to overcome key limitations of traditional self-report measures and enable the study of AI agents' role in team coordination through constructs comparable to their human counterparts. While external validity remains to be tested, this work illustrates a methodological foundation and a proof-of-concept for future research on the ABC-DEFs of teamwork in HATs, paving the way for deeper insights into the cyber-social dynamics of human-AI collaboration.

13.4 IMPLEMENTING THE CYBER-SOCIAL PARADIGM IN RESEARCH AND PRACTICE

The cyber-social paradigm offers a structural lens for integrating human and AI teammates by explicitly pairing human ABC with AI parallels of DEF. This symmetric framework advances the theoretical precision of HAT research by specifying the full set of mechanisms that drive emergent team phenomena.

Decades of teams research have clearly identified affective emergent states, cognitive emergent states, and team behavioral processes as key factors shaping collective outcomes (Rapp et al., 2021) and even representing them as core human needs in a collective (Pincus, 2025). These findings remain fully relevant in HATs, but represent only half the story. In contemporary work settings, people increasingly treat sufficiently autonomous and competent AI systems as teammates, attributing trust, intention, and even relational qualities to AI teammates (Duan et al., 2025; Georganta & Ulfert, 2024). Therefore, limiting conceptualizations and analysis of teamwork in HATs to human psychology leaves the broader cyber-social system underspecified. To truly understand the totality of systems underlying HAT functioning, researchers and practitioners must account not only for the human ABCs of teamwork, but also for the AI agents' task representations, decision processes, and performance characteristics, which are all elements that often remain opaque and unmeasured.

Crucially, we advocate for the implementation and assessment of MToM as a capability built into the functioning of the AI agent. MToM is critical for supporting shared understanding and enabling interdependence between human and AI teammates. Without MToM, AI agents remain limited to providing reactive assistance and would not be able to anticipate the team's needs, coordinate with team members, and act proactively toward team goals.

Our cyber-social paradigm addresses these requirements by pairing human ABC with AI DEF, enabling conceptually parallel and empirically measurable assessments of both sides of the team. Treating advanced AI agents as teammates when they coordinate and make complex decisions is therefore not just a conceptual refinement but a measurement and design imperative for advancing HAT science. The symmetric ABC-DEF framework advances the theoretical precision of HAT research by specifying the full set of mechanisms that drive emergent team phenomena in HATs.

13.4.1 Implications for Research

The cyber-social framework opens novel avenues for empirical research by providing a common language and measurable constructs for studying HATs. Researchers are encouraged to design studies that capture both the human ABC and AI DEF indicators, enabling high-fidelity examination of team dynamics to examine how human cognition aligns with AI framing, how human affect interacts with AI determinations under duress, and how coordinated human behavior and AI execution jointly influence team performance.

Existing research on human perceptions of their AI teammates could also be enhanced by integrating the cyber-social paradigm into the construction and assessment of the AI agent's capabilities. For example, Farah and Dorneich (2024) used a game-based scenario to demonstrate that proactive AI agent teammates were rated higher on perceived AI performance and teammate likeness, specifying the extent to which human teammates perceive AI agents as true teammates as opposed to tools (Wynne & Lyons, 2018). This is likely due to features consistent with MToM, such as anticipatory communication and planning. However, without assessing the AI agent's factors by recording its determinations, actions, and context framing, only half of the team's dynamics can be realistically understood. Future research that extends this work by including the entire cyber-social network, combining human and AI measures, would allow for richer causal modeling of HAT processes.

Future research should also explore more methods for extracting DEF indicators from AI behaviors and data logs to validate whether these constructs predict coordination and performance outcomes in a range of different situations. While we presented experimental test cases reflective of the cyber-social paradigm based on simulated environments, future research on business applications of HATs in real-world scenarios, such as healthcare and emergency response, would provide rich insights into the interplay between the ABC-DEF constructs and team performance across diverse contexts and AI agent architectures.

13.4.2 Implications for Practice

A core contribution of the cyber-social framework is its ability to bridge the gap between theory and application. By defining constructs that are both theoretically grounded and practically measurable, DEF provides practitioners with guidance for AI design, team composition, and performance monitoring.

Organizations should treat DEFs as design requirements, ensuring that sufficiently capable AI teammates are engineered with robust Determination and Framing capabilities that are observable through the agents' Execution to foster humans' trust in the agent and its accountability, increasing team cohesion and performance. Roles requiring anticipatory or interdependent coordination, for example, should be assigned only to AI agents with strong MToM and Determination capabilities. DEF-derived metrics can further support team diagnostics and continuous improvement, enabling organizations to identify breakdowns and optimize human–AI coordination.

AI systems built with these design principles can more effectively augment teamwork in HATs by explicitly enabling the AI agents to internalize, conceptualize, and respond to human team members' ABC. Encoding DEF mechanisms directly into the AI agents' architecture also promotes interpretability, offering the potential to further aid humans in developing clearer shared cognition and healthy affective responses to AI agent team members, ultimately enhancing the effectiveness of these integrated teams.

13.4.3 Challenges and Future Considerations

While the cyber-social paradigm offers a promising roadmap, it is important to acknowledge the current challenges across the HAT landscape. First, AI agents' internal representations and evaluative processes are rarely accessible or interpretable in most current systems. Unlike humans, we cannot simply ask an AI agent what it "thinks" or "feels," and even when the AI agent is able to provide answers to those questions (such as a conversational AI), the responses may not reflect the true underlying mechanisms. DEF mitigates this by emphasizing observable trace data, but continued refinement and validation of DEF indicators are essential, especially for fostering trust and dependence on AI agents as capable teammates. For instance, Wildman et al. (2024) propose a list of research questions to explore emergent forms of trust in HATs that are largely only applicable to sufficiently advanced AI agents.

Second, the utility of the DEF framework depends on the observability of the AI agent's processes. Proprietary models from AI vendors may not easily reveal internal states or provide trace data. Some of the AI agent's determinations and framing could potentially be inferred through observable behavior only, but this will limit diagnostic power accordingly, and may obscure additional factors influencing the behaviors.

Third, to be considered a teammate and hence fit into the cyber-social paradigm, the AI agent will have to be sufficiently advanced to possess MToM. This requires the engineering of algorithms that possess the ability to infer information based on context and act proactively to support the team's goals. While technology is rapidly advancing, such advanced AI agents are not yet commonplace. Current limitations in many commercial AI systems, including low-effort "AI workslop" (Niederhoffer et al., 2025), can produce confusion and degrade team performance. Thoughtful integration of AI solutions, clearer norms, and better guidance on task-aware deployment of AI are necessary precursors to implementing effective DEF-capable agents.

Lastly, mapping trace data to meaningful DEF constructs requires interpretive frameworks and validation. Future research should strive to examine and highlight common agentic traits defined by the DEF constructs and how these interact with human ABC indicators to explain and augment team effectiveness. Developing composite DEF indicators for different types of agents deployed in different contexts is a critical next step to advance the cyber-social paradigm from a conceptual framework to an implementation architecture.

To begin addressing some of these limitations and guide future needs for studying and designing AI agents for HATs, we outline several specific challenges and priorities mapped to each DEF dimension in Table 13.2. These challenges span issues discussed above, such as the difficulty of accessing internal AI states, the need for observable and interpretable outputs, the importance of modeling teammate reliability, and the role of MToM in enabling proactive, adaptable, and context-aware behavior. The issues raised here offer particular considerations for applying the cyber-social framework so that researchers and practitioners can better evaluate and design AI agents that function as effective teammates.

TABLE 13.2
Challenges, Key Considerations, and Potential Directions for Applying DEF Dimensions of the Cyber-Social Framework for Studying and Designing Effective AI Agents for HATs

Challenges	Key Considerations	Relevant DEF Dimensions	Potential Research and Design Directions
1. Observability of AI internal processes	Commercial and proprietary AI systems rarely expose their underlying representations, limiting accessibility.	Determination, framing	Develop standardized trace-logging protocols[a], advocate for AI transparency, and create DEF-aligned telemetry to capture goals, evaluations, and responsive task functioning.
2. Mapping trace data to DEF constructs	Behavioral logs alone may obscure internal decision processes. Composite metrics require interpretive frameworks and validation of underlying mechanisms.	Determination, execution, framing	Design DEF extraction and coding protocols, validate construct–outcome relationships in controlled studies, and test automated DEF extraction across diverse contexts.
3. Machine Theory of Mind (MToM) capabilities	True teamwork requires contextual inference, anticipation of human needs, and proactive coordination, which are not yet common in most current AI systems.	Determination, framing	Develop and benchmark MToM-enabled AI models and evaluate how agent inferences about human goals influence HAT coordination.
4. Reliability and trust calibration between human and AI teammates	Humans vary in accuracy, bias, and workload. AI agents must be able to model these factors to coordinate effectively and mitigate risk. Overtrust or undertrust could result from poor transparency, degrading team performance.	Determination	Model human reliability as inputs potentially by using utility equations and predictive modeling as pseudo-trust constructs[b]. Examine how AI transparency features influence trust alignment with HATs.
5. Information sharing and explainability	Effective coordination requires communication that avoids information overload and supports shared mental models across various communication and decision-making styles. AI explainability is crucial for building trust, coordinating integration, and reducing confusion.	Execution	Design human-centered and explainable communication protocols. Study and calibrate information-sharing equilibrium[c] under varying time pressure and uncertainty.
6. Role clarity, leadership, and teammate adaptability	Leadership structures such as flat and hierarchical affect team performance in different ways. AI agents must understand their role in the team and adapt to various leadership behaviors based on context.	Determination, framing	Study how leadership coordination affects AI outcomes and how AI agents appraise their roles in the team and the overall leadership hierarchy. Develop adaptive role-framing algorithms[d] for shifting task and team demands.

(Continued)

TABLE 13.2 (*Continued*)
Challenges, Key Considerations, and Potential Directions for Applying DEF Dimensions of the Cyber-Social Framework for Studying and Designing Effective AI Agents for HATs

Challenges	Key Considerations	Relevant DEF Dimensions	Potential Research and Design Directions
7. AI teammate characteristics and fit	AI that is overly assertive, hyper-intelligent, or unhelpfully verbose can impair team functioning. Perceptions of AI agents' competence, trustworthiness, usefulness, and cohesion are critical for effective integration.	Execution, framing	Identify optimal AI teammate profiles for different tasks and situations by testing personality-fit models[e] for AI and tune Execution behaviors to match team and organizational norms.
8. Standardization across contexts and agent architectures	DEF evaluation must remain robust across AI types (e.g., LLMs, multi-agent systems) and diverse operational domains (e.g., healthcare, finance, and emergency response).	Determination, execution, framing	Establish cross-domain DEF assessment frameworks to develop benchmarks for comparing DEF constructs across agent types and contexts and testing scalability of DEF indicators.
9. Ethical and organizational risks	Poorly designed agents may lead to lack of accountability, overtrust, or degraded human skills. AI workslop (Niederhoffer et al., 2025) illustrates the risk of low-quality outputs.	Execution, framing	Establish safeguards to clarify levels of responsibility when using AI, evaluate conditions leading to overreliance, and design Execution checks to limit unhelpful, misleading, or superficial outputs.
10. Methodological and validation needs	Implementing DEFs for research and practice requires careful consideration for measuring and integrating behavioral data, cognitive markers, and model introspection to identify key indicators and validate measurement approaches.	Determination, execution, framing	Develop multimethod data collection procedures to test DEF–ABC interactions experimentally and validate DEF predictors across a variety of simulated and true operational environments.

References:

[a] Collecting and logging digital trace data (Koc et al., 2025; Raetze et al., 2025; Y. Zhang et al., 2023);
[b] Assessing utility of AI teammates as an explainable reason for trusting human-AI collaboration (Davis et al., 2020; Lee & See, 2004);
[c] Determine level of information sharing required for effective coordination (Schelble et al., 2025);
[d] Determine and maintain distinct roles to improve role-clarity and responsibilities in HATs (Liao & Sundar, 2021; Tarafdar et al., 2023);
[e] Manage perceptions of AI personality to improve human-AI collaboration and fit within teams (Ju & Aral, 2025; Kovbasiuk et al., 2026).

13.5 CONCLUSION

The cyber-social paradigm offers a structural approach for integrating AI agents as true teammates rather than just passive tools, bridging the gap between human-centric and machine-centric perspectives in traditional teams research. By introducing the DEF triad as algorithmic counterparts to human ABC, this framework enables both measurement symmetry and design guidance for future HAT research and practice. The DEF framework provides a roadmap for building and evaluating AI agents capable of shared understanding and proactive coordination. As human–AI collaboration becomes increasingly prevalent across domains, adopting the cyber-social framework will be essential for creating and monitoring teams that are not only efficient but also adaptive, transparent, and collaborative.

We invite researchers and practitioners alike to approach HAT research from the proposed cyber-social framework, identifying meaningful metrics that can be applied to understand both human and AI team members. As we continue to navigate the uncharted waters of human-AI collaboration, the synthesis of diverse perspectives will be essential.

REFERENCES

Alessandro, G., Dimitri, O., Cristina, B., & Anna, M. (2025). The emotional impact of generative AI: Negative emotions and perception of threat. *Behaviour & Information Technology*, *44*(4), 676–693. https://doi.org/10.1080/0144929X.2024.2333933

Andrews, R. W., Lilly, J. M., Srivastava, D., & Feigh, K. M. (2023). The role of shared mental models in human-AI teams: A theoretical review. *Theoretical Issues in Ergonomics Science*, *24*(2), 129–175. https://doi.org/10.1080/1463922X.2022.2061080

Ashktorab, Z., Dugan, C., Johnson, J., Pan, Q., Zhang, W., Kumaravel, S., & Campbell, M. (2021). Effects of communication directionality and AI Agent differences in Human-AI interaction. *Proceedings of the 2021 CHI Conference on Human Factors in Computing Systems* (pp. 1–15). https://doi.org/10.1145/3411764.3445256

Bachrach, D. G., Lewis, K., Kim, Y., Patel, P. C., Campion, M. C., & Thatcher, S. M. B. (2019). Transactive memory systems in context: A meta-analytic examination of contextual factors in transactive memory systems development and team performance. *Journal of Applied Psychology*, *104*(3), 464–493. https://doi.org/10.1037/apl0000329

Bansal, G., Nushi, B., Kamar, E., Lasecki, W. S., Weld, D. S., & Horvitz, E. (2019). Beyond accuracy: The role of mental models in Human-AI team performance. *Proceedings of the AAAI Conference on Human Computation and Crowdsourcing*, *7*, 2–11. https://doi.org/10.1609/hcomp.v7i1.5285

Bell, S. T., Brown, S. G., Colaneri, A., & Outland, N. (2018). Team composition and the ABCs of teamwork. *American Psychologist*, *73*(4), 349–362. https://doi.org/10.1037/amp0000305

Bezrucav, S. O., & Corves, B. (2020, October). Improved AI planning for cooperating teams of humans and robots. In *Proceedings of the Planning and Robotics (PlanRob) Workshop—ICAPS* (Vol. 176). https://doi.org/10.5281/zenodo.4286242

Black, J., Kim, K., Rhee, S., Wang, K., & Sakchutchawan, S. (2019). Self-efficacy and emotional intelligence: Influencing team cohesion to enhance team performance. *Team Performance Management*, *25*(1–2), 100–119. https://doi.org/10.1108/TPM-01-2018-0005

Boettcher, S. A., Brunell, M. J., Griffin, D. J., Pearman, J. J., & Katz, A. L. (2025). Cultivating teamwork in esports: A framework for assessment and training. *Team Performance Management*, *31*(5–6), 390–407. https://doi.org/10.1108/TPM-10-2024-0124

Bourbousson, J., Sève, C., & McGarry, T. (2010). Space–time coordination dynamics in basketball: Part 2. The interaction between the two teams. *Journal of Sports Sciences*, *28*(3), 349–358. https://doi.org/10.1080/02640410903503640

Brawley, L. R., Carron, A. V., & Widmeyer, W. N. (1987). Assessing the cohesion of teams: Validity of the Group Environment Questionnaire. *Journal of Sport and Exercise Psychology*, *9*(3), 275–294. https://doi.org/10.1123/jsp.9.3.275

Carter-Browne, B. M., Paletz, S. B. F., Campbell, S. G., Carraway, M. J., Vahlkamp, S. H., Schwartz, J., & O'Rourke, P. (2021). *There is no "AI" in teams: A multidisciplinary framework for AIs to work in human teams* (ARLIS Report). Applied Research Laboratory for Intelligence and Security, University of Maryland. https://www.arlis.umd.edu/sites/default/files/2022-03/No_AI_In_Teams_FinalReport%20%281%29.pdf

Conti, M. E., Arcuri, M., & Simone, C. (2018). Wellbeing, conflict and teamworking: The social role of the team leader - an overview. *International Journal of Environment and Health*, *9*(2), 151. https://doi.org/10.1504/IJENVH.2018.092805

Davis, B., Glenski, M., Sealy, W., & Arendt, D. (2020). *Measure utility, gain trust: Practical advice for XAI researcher* (arXiv:2009.12924). arXiv. https://doi.org/10.48550/arXiv.2009.12924

DeChurch, L. A., & Mesmer-Magnus, J. R. (2010). The cognitive underpinnings of effective teamwork: A meta-analysis. *Journal of Applied Psychology*, *95*(1), 32–53. https://doi.org/10.1037/a0017328

Dennis, A. R., Lakhiwal, A., & Sachdeva, A. (2023). AI Agents as team members: Effects on satisfaction, conflict, trustworthiness, and willingness to work with. *Journal of Management Information Systems*, *40*(2), 307–337. https://doi.org/10.1080/07421222.2023.2196773

Dervitsiotis, K. N. (2006). Building trust for excellence in performance and adaptation to change. *Total Quality Management & Business Excellence*, *17*(7), 795–810. https://doi.org/10.1080/14783360600595153

Duan, W., Flathmann, C., McNeese, N., Scalia, M. J., Zhang, R., Gorman, J., Freeman, G., Zhou, S., Hauptman, A. I., & Yin, X. (2025). Trusting autonomous teammates in Human-AI Teams—A literature review. *Proceedings of the 2025 CHI Conference on Human Factors in Computing Systems* (pp. 1–23). https://doi.org/10.1145/3706598.3713527

Duan, W., Zhou, S., Scalia, M. J., Yin, X., Weng, N., Zhang, R., Freeman, G., McNeese, N., Gorman, J., & Tolston, M. (2024). Understanding the evolvement of trust over time within Human-AI teams. *Proceedings of the ACM on Human-Computer Interaction*, *8*(CSCW2). https://doi.org/10.1145/3687060

Eger, M., Martens, C., Chacón, P. S., Córdoba, M. A., & Hidalgo-Cespedes, J. (2021). Operationalizing intentionality to play *Hanabi* with human players. *IEEE Transactions on Games*, *13*(4), 388–397. https://doi.org/10.1109/TG.2020.3009359

Eger, M., Martens, C., & Cordoba, M. A. (2017). An intentional AI for Hanabi. *2017 IEEE Conference on Computational Intelligence and Games (CIG)* (pp. 68–75). https://doi.org/10.1109/CIG.2017.8080417

Elms, A. (2025). *The power of trust: How felt trust enhances employee empowerment and adaptive performance* [University of Guelph]. https://hdl.handle.net/10214/28792

Emich, K. J., & Lu, L. (2021). Mixed feelings: How shared and unshared affect impact team creative success. In *Creative success in teams* (pp. 145–166). Elsevier Academic Press. https://doi.org/10.1016/B978-0-12-819993-0.00008-4

Farah, Y. A., & Dorneich, M. C. (2024). Human-Autonomy teaming in a cooperative gamified testbed: How can AI teammates support teamwork processes? *Proceedings of the Human Factors and Ergonomics Society Annual Meeting*, *68*(1), 386–392. https://doi.org/10.1177/10711813241269252

Georganta, E., & Ulfert, A.-S. (2024). Would you trust an AI team member? Team trust in Human–AI teams. *Journal of Occupational and Organizational Psychology*, *97*(3), 1212–1241. https://doi.org/10.1111/joop.12504

Glikson, E., & Woolley, A. W. (2020). Human trust in artificial intelligence: Review of empirical research. *Academy of Management Annals*, *14*(2), 627–660. https://doi.org/10.5465/annals.2018.0057

Grossman, R., Nolan, K., Rosch, Z., Mazer, D., & Salas, E. (2022). The team cohesion-performance relationship: A meta-analysis exploring measurement approaches and the changing team landscape. *Organizational Psychology Review*, *12*(2), 181–238. https://doi.org/10.1177/20413866211041157

Hoandră, M. G. (2017). The mediating role of task conflict and cross-understanding in the relation between transactional leadership and team adaptive performance. *Psihologia Resurselor Umane Revista Asociaţiei de Psihologie Industrială Şi Organizaţională*, *15*(1), 56–68.

Hu, M., Chua, X. C. W., Diong, S. F., Kasturiratna, K. T. A. S., Majeed, N. M., & Hartanto, A. (2025). AI as your ally: The effects of AI-assisted venting on negative affect and perceived social support. *Applied Psychology: Health and Well-Being*, *17*(1), e12621. https://doi.org/10.1111/aphw.12621

Iftikhar, R., Chiu, Y.-T., Khan, M. S., & Caudwell, C. (2023). Human–Agent team dynamics: A review and future research opportunities. *IEEE Transactions on Engineering Management*, 1–16. https://doi.org/10.1109/TEM.2023.3331369

Ilgen, D. R., Hollenbeck, J. R., Johnson, M., & Jundt, D. (2005). Teams in organizations: From input-process-output models to IMOI models. *Annual Review of Psychology*, *56*, 517–543. https://doi.org/10.1146/annurev.psych.56.091103.070250

Ishowo-Oloko, F., Bonnefon, J.-F., Soroye, Z., Crandall, J., Rahwan, I., & Rahwan, T. (2019). Behavioural evidence for a transparency–efficiency tradeoff in human–machine cooperation. *Nature Machine Intelligence*, *1*(11), 517–521. https://doi.org/10.1038/s42256-019-0113-5

Janicik, G. A., & Bartel, C. A. (2003). Talking about time: Effects of temporal planning and time awareness norms on group coordination and performance. *Group Dynamics: Theory, Research, and Practice*, *7*(2), 122–134. https://doi.org/10.1037/1089-2699.7.2.122

Janwani, N. C., Daş, E., Touma, T., Wei, S. X., Molnar, T. G., & Burdick, J. W. (2024, May). A learning-based framework for safe human-robot collaboration with multiple backup control barrier functions. In *2024 IEEE International Conference on Robotics and Automation (ICRA)* (pp. 11676–11682). IEEE. https://doi.org/10.1109/ICRA57147.2024.10611634

Jenkins, P., Farag, A., Jenkins, J. S., Yao, H., Wang, S., & Li, Z. (2021, May). Neural utility functions. In *Proceedings of the AAAI conference on artificial intelligence* (Vol. 35, No. 9, pp. 7917–7925). https://doi.org/10.1609/aaai.v35i9.16966

Jonker, C. M., Van Riemsdijk, B., van de Kieft, I., & Gini, M. L. (2011, December). Towards Measuring Sharedness of Team Mental Models by Compositional Means. In *AAAI Fall Symposium: Robot-Human Teamwork in Dynamic Adverse Environment.*

Ju, H., & Aral, S. (2025). *Personality pairing improves Human-AI collaboration* (arXiv:2511.13979). arXiv. https://doi.org/10.48550/arXiv.2511.13979

Kaplan, S., LaPort, K., & Waller, M. J. (2013). The role of positive affectivity in team effectiveness during crises. *Journal of Organizational Behavior*, *34*(4), 473–491. https://doi.org/10.1002/job.1817

Koc, V., Verre, J., Blank, D., & Morgan, A. (2025). *Mind the metrics: Patterns for telemetry-aware in-IDE AI application development using the Model Context Protocol (MCP)* (arXiv:2506.11019). arXiv. https://doi.org/10.48550/arXiv.2506.11019

Kovbasiuk, A., Triantoro, T., Ciechanowski, L., Sowa, K., & Przegalinska, A. (2026). Is AI a good fit? The impact of personality on generative AI collaboration and enjoyment. *International Journal of Human-Computer Studies*, *210*, 103747. https://doi.org/10.1016/j.ijhcs.2026.103747

Krug, H., Geibel, H. V., & Otto, K. (2021). Identity leadership and well-being: Team identification and trust as underlying mechanisms. *Leadership & Organization Development Journal*, *42*(1), 17–31. https://doi.org/10.1108/LODJ-02-2020-0054

Landers, R. N., & Behrend, T. S. (2024). *Research methods for industrial and organizational psychology: Science and practice*. Routledge. https://doi.org/10.4324/9781315167473

Langley, C., Cirstea, B. I., Cuzzolin, F., & Sahakian, B. J. (2022). Theory of mind and preference learning at the interface of cognitive science, neuroscience, and AI: A review. *Frontiers in Artificial Intelligence*, *5*. https://doi.org/10.3389/frai.2022.778852

Larson, L., & DeChurch, L. A. (2020). Leading teams in the digital age: Four perspectives on technology and what they mean for leading teams. *The Leadership Quarterly*, *31*(1), 101377. https://doi.org/10.1016/j.leaqua.2019.101377

Lee, J. D., & See, K. A. (2004). Trust in automation: Designing for appropriate reliance. *Human Factors*, *46*(1), 50–80. https://doi.org/10.1518/hfes.46.1.50_30392

Lewis, K. (2003). Measuring transactive memory systems in the field: scale development and validation. *Journal of applied psychology*, *88*(4), 587. https://doi.org/10.1037/0021-9010.88.4.587

Li, H., Oguntola, I., Hughes, D., Lewis, M., & Sycara, K. (2022). Theory of mind modeling in search and rescue teams. *2022 31st IEEE International Conference on Robot and Human Interactive Communication (RO-MAN)* (pp. 483–489). https://doi.org/10.1109/RO-MAN53752.2022.9900572

Liao, M., & Sundar, S. S. (2021). How should AI systems talk to users when collecting their personal information? Effects of role framing and self-referencing on Human-AI interaction. *Proceedings of the 2021 CHI Conference on Human Factors in Computing Systems, CHI '21* (pp. 1–14). https://doi.org/10.1145/3411764.3445415

Lim, B., & Klein, K. J. (2006). Team mental models and team performance: A field study of the effects of team mental model similarity and accuracy. *Journal of Organizational Behavior*, *27*(4), 403–418. https://doi.org/10.1002/job.387

Ma, H., Vo, T. V., & Leong, T. Y. (2023). Hierarchical Reinforcement Learning with Human-AI Collaborative Sub-Goals Optimization. In *Proceedings of the 2023 international conference on autonomous agents and multiagent systems* (pp. 2310–2312). https://doi.org/10.1609/aaaiss.v1i1.27481

Marks, M. A., Mathieu, J. E., & Zaccaro, S. J. (2001). A temporally based framework and taxonomy of team processes. *The Academy of Management Review*, *26*(3), 356–376. https://doi.org/10.2307/259182

Marlow, S. L., Lacerenza, C. N., Paoletti, J., Burke, C. S., & Salas, E. (2018). Does team communication represent a one-size-fits-all approach?: A meta-analysis of team communication and performance. *Organizational Behavior and Human Decision Processes*, *144*, 145–170. https://doi.org/10.1016/j.obhdp.2017.08.001

Marques-Quinteiro, P., Ramos-Villagrasa, P. J., Passos, A. M., & Curral, L. (2015). Measuring adaptive performance in individuals and teams. *Team Performance Management*, *21*(7–8), 339–360. https://doi.org/10.1108/TPM-03-2015-0014

Mathieu, J. E., Heffner, T. S., Goodwin, G. F., Salas, E., & Cannon-Bowers, J. A. (2000). The influence of shared mental models on team process and performance. *Journal of Applied Psychology*, *85*(2), 273–283. https://doi.org/10.1037/0021-9010.85.2.273

Mathieu, J. E., Wolfson, M. A., Park, S., Luciano, M. M., Bedwell-Torres, W. L., Ramsay, P. S., Klock, E. A., & Tannenbaum, S. I. (2022). Indexing dynamic collective constructs using computer-aided text analysis: Construct validity evidence and illustrations featuring team processes. *Journal of Applied Psychology*, *107*(4), 533–559. https://doi.org/10.1037/apl0000856

Maynard, M. T., Conroy, S., Lacerenza, C. N., & Barnes, L. Y. (2021). Teams in the wild are not extinct, but challenging to research: A guide for conducting impactful team field research with 10 recommendations and 10 best practices. *Organizational Psychology Review*, *11*(3), 274–318. https://doi.org/10.1177/2041386620986597

McAllister, D. J. (1995). Affect-and cognition-based trust as foundations for interpersonal cooperation in organizations. *Academy of management journal*, *38*(1), 24–59. https://doi.org/10.5465/256727

Mehta, A., Feild, H., Armenakis, A., & Mehta, N. (2009). Team goal orientation and team performance: The mediating role of team planning. *Journal of Management*, *35*(4), 1026–1046. https://doi.org/10.1177/0149206308326773

Meneghel, I., Salanova, M., & Martínez, I. M. (2016). Feeling good makes us stronger: How team resilience mediates the effect of positive emotions on team performance. *Journal of Happiness Studies: An Interdisciplinary Forum on Subjective Well-Being*, *17*(1), 239–255. https://doi.org/10.1007/s10902-014-9592-6

Mohammed, S., Rico, R., & Alipour, K. K. (2021). Team cognition at a crossroad: Toward conceptual integration and network configurations. *Academy of Management Annals*, *15*(2), 455–501. https://doi.org/10.5465/annals.2018.0159

Müller, R., & Antoni, C. H. (2022). Effects of ICT shared mental models on team processes and outcomes. *Small Group Research*, *53*(2), 307–335. https://doi.org/10.1177/1046496421997889

Neerincx, M. A., van der Waa, J., Kaptein, F., & van Diggelen, J. (2018). Using perceptual and cognitive explanations for enhanced Human-Agent team performance. In D. Harris (Ed.), *Engineering psychology and cognitive ergonomics* (pp. 204–214). Springer International Publishing. https://doi.org/10.1007/978-3-319-91122-9_18

Neville, V., Dayan, P., Gilchrist, I. D., Paul, E. S., & Mendl, M. (2021). Dissecting the links between reward and loss, decision-making, and self-reported affect using a computational approach. *PLOS Computational Biology*, *17*(1), e1008555. https://doi.org/10.1371/journal.pcbi.1008555

Niederhoffer, K., Kellerman, G. R., Lee, A., Liebscher, A., Rapuano, K., & Hancock, J. T. (2025, September 22). AI-Generated "Workslop" is destroying productivity. *Harvard Business Review*. https://hbr.org/2025/09/ai-generated-workslop-is-destroying-productivity

O'Neill, T., McNeese, N., Barron, A., & Schelble, B. (2022). Human–Autonomy teaming: A review and analysis of the empirical literature. *Human Factors*, *64*(5), 904–938. https://doi.org/10.1177/0018720820960865

Pincus, J. D. (2025). Team emergent states as emotional needs. *Small Group Research*, 10464964251317474. https://doi.org/10.1177/10464964251317474

Podsakoff, P. M., Podsakoff, N. P., Williams, L. J., Huang, C., & Yang, J. (2024). Common method bias: It's bad, It's complex, It's widespread, and It's not easy to fix. *Annual Review of Organizational Psychology and Organizational Behavior*, *11*, 17–61. https://doi.org/10.1146/annurev-orgpsych-110721-040030

Premack, D., & Woodruff, G. (1978). Does the chimpanzee have a theory of mind? *Behavioral and Brain Sciences*, *1*(4), 515–526. https://doi.org/10.1017/S0140525X00076512

Rabinowitz, N., Perbet, F., Song, F., Zhang, C., Eslami, S. M. A., & Botvinick, M. (2018). Machine theory of mind. *Proceedings of the 35th International Conference on Machine Learning* (pp. 4218–4227). https://proceedings.mlr.press/v80/rabinowitz18a.html

Raetze, S., Staedter, C., & Hüllmann, J. A. (2025). Taking aim at research on esports teams: A systematic literature review and cross-disciplinary future agenda. *Team Performance Management: An International Journal*, *31*(5–6), 364–389. https://doi.org/10.1108/TPM-09-2024-0109

Ramzaninezhad, R., Keshtan, M. H., Shahamat, M. D., & Kordshooli, S. (2009). The relationship between collective efficacy, group cohesion and team performance in professional volleyball teams. *Brazilian Journal of Biomotricity*, *3*(1), 31–39.

Rapp, T., Maynard, T., Domingo, M., & Klock, E. (2021). Team emergent states: What has emerged in the literature over 20 years. *Small Group Research*, *52*(1), 68–102. https://doi.org/10.1177/1046496420956715

Reitman, J. G., Anderson-Coto, M. J., Wu, M., Lee, J. S., & Steinkuehler, C. (2020). Esports research: A literature review. *Games and Culture*, *15*(1), 32–50. https://doi.org/10.1177/1555412019840892

Rico, R., Gibson, C. B., Sánchez-Manzanares, M., & Clark, M. A. (2019). Building team effectiveness through adaptation: Team knowledge and implicit and explicit coordination. *Organizational Psychology Review*, *9*(2–3), 71–98. https://doi.org/10.1177/2041386619869972

Rico, R., Hinsz, V. B., Davison, R. B., & Salas, E. (2018). Structural influences upon coordination and performance in multiteam systems. *Human Resource Management Review*, *28*(4), 332–346. https://doi.org/10.1016/j.hrmr.2017.02.001

Rico-González, M., Pino-Ortega, J., Castellano, J., M. Oliva-Lozano, J., & Los Arcos, A. (2022). Reference values for collective tactical behaviours based on positional data in professional football matches: A systematic review. *Biology of Sport*, *39*(1), 101–114. https://doi.org/10.5114/biolsport.2021.102921

Salas, E., Cooke, N. J., & Rosen, M. A. (2008). On teams, teamwork, and team performance: Discoveries and developments. *Human Factors*, *50*(3), 540–547. https://doi.org/10.1518/001872008X288457

Salas, E., Grossman, R., Hughes, A. M., & Coultas, C. W. (2015). Measuring team cohesion: Observations from the science. *Human Factors*, *57*(3), 365–374. https://doi.org/10.1177/0018720815578267

Sawant, S., Mallick, R., McNeese, N., & Chalil Madathil, K. (2022). Mutually beneficial decision making in Human-AI teams: Understanding soldier's perception and expectations from AI teammates in human-AI teams. *Proceedings of the Human Factors and Ergonomics Society Annual Meeting*, *66*(1), 287–289. https://doi.org/10.1177/1071181322661355

Schelble, B. G., Filer, G., Flathmann, C., & Chalil Madathil, K. (2025). Examining the role of AI information-sharing on trust in Human-AI teams. *Proceedings of the Human Factors and Ergonomics Society Annual Meeting*, *69*(1), 1837–1842. https://doi.org/10.1177/10711813251367726

Seeber, I., Bittner, E., Briggs, R. O., De Vreede, T., De Vreede, G.-J., Elkins, A., Maier, R., Merz, A. B., Oeste-Reiß, S., Randrup, N., Schwabe, G., & Söllner, M. (2020). Machines as teammates: A research agenda on AI in team collaboration. *Information & Management*, *57*(2), 103174. https://doi.org/10.1016/j.im.2019.103174

Shen, Y. Y., Xing, Q. J., & Shen, Y. F. (2024). Markerless vision-based functional movement screening movements evaluation with deep neural networks. *Iscience*, *27*(1). https://doi.org/10.1016/j.isci.2023.108705

Shivashankar, V. (2015). *Hierarchical goal networks: Formalisms and algorithms for planning and acting* (Doctoral dissertation, University of Maryland, College Park). University of Maryland Digital Repository. https://doi.org/10.13016/M2TS69

Shteynberg, G., Hirsh, J. B., Wolf, W., Bargh, J. A., Boothby, E. J., Colman, A. M., Echterhoff, G., & Rossignac-Milon, M. (2023). Theory of collective mind. *Trends in Cognitive Sciences*, *27*(11), 1019–1031. https://doi.org/10.1016/j.tics.2023.06.009

Tarafdar, M., Page, X., & Marabelli, M. (2023). Algorithms as co-workers: Human algorithm role interactions in algorithmic work. *Information Systems Journal*, *33*(2), 232–267. https://doi.org/10.1111/isj.12389

Valls, V., González-Romá, V., & Tomás, I. (2016). Linking educational diversity and team performance: Team communication quality and innovation team climate matter. *Journal of Occupational and Organizational Psychology*, *89*(4), 751–771. https://doi.org/10.1111/joop.12152

Vinanzi, S., Patacchiola, M., Chella, A., & Cangelosi, A. (2019). Would a robot trust you? Developmental robotics model of trust and theory of mind. *Philosophical Transactions of the Royal Society B: Biological Sciences*, *374*(1771), 20180032. https://doi.org/10.1098/rstb.2018.0032

Wagner, M. G. (2006). *On the scientific relevance of eSports*. In H. R. Arabnia (Ed.), *Proceedings of the International Conference on Internet Computing (ICOMP 2006)* (pp. 437–442). CSREA Press.

Wang, X., He, C., Jiang, W., Wang, W., & Liu, X. (2025). Generative AI Based Dependency-Aware Task Offloading and Resource Allocation for UAV-Assisted IoV. *IEEE Open Journal of the Communications Society*. https://doi.org/10.1109/OJCOMS.2025.3562720

Weingart, L. R. (1992). Impact of group goals, task component complexity, effort, and planning on group performance. *Journal of Applied Psychology*, *77*(5), 682–693. https://doi.org/10.1037/0021-9010.77.5.682

Wespi, R., Birrenbach, T., Schauber, S. K., Manser, T., Sauter, T. C., & Kämmer, J. E. (2023). Exploring objective measures for assessing team performance in healthcare: An interview study. *Frontiers in Psychology*, *14*. https://doi.org/10.3389/fpsyg.2023.1232628

Wiernik, B. M., Raghavan, M., Caretta, T. R., & Coovert, M. D. (2022). Developing and validating a serious game-based assessment for cyber occupations in the US Air Force. *International Journal of Selection and Assessment*, *30*(1), 27–47. https://doi.org/10.1111/ijsa.12378

Wildman, J. L., Nguyen, D., Thayer, A. L., Robbins-Roth, V. T., Carroll, M., Carmody, K., Ficke, C., Akib, M., & Addis, A. (2024). Trust in Human-Agent teams: A multilevel perspective and future research agenda. *Organizational Psychology Review*, *14*(3), 373–402. https://doi.org/10.1177/20413866241253278

Wisakanto, A. K., Rogero, J., Casheekar, A. M., & Mallah, R. (2025). Adapting probabilistic risk assessment for AI. arXiv preprint arXiv: 2504.18536. https://doi.org/10.48550/arXiv.2504.18536

Wu, Y., Liang, S., Zhang, C., Wang, Y., Zhang, Y., Guo, H., Tang, R., & Liu, Y. (2025). *From human memory to AI memory: A survey on memory mechanisms in the era of LLMs* (arXiv:2504.15965). arXiv. https://doi.org/10.48550/arXiv.2504.15965

Wynne, K. T., & Lyons, J. B. (2018). An integrative model of autonomous agent teammate-likeness. *Theoretical Issues in Ergonomics Science*, *19*(3), 353–374. https://doi.org/10.1080/1463922X.2016.1260181

Zercher, D., Jussupow, E., Benke, I., & Heinzl, A. (2025). How can teams benefit from ai team members? Exploring the effect of generative ai on decision-making processes and decision quality in team–ai collaboration. *Journal of Organizational Behavior*. https://doi.org/10.1002/job.2898

Zhang, G., Raina, A., Cagan, J., & McComb, C. (2021). A cautionary tale about the impact of AI on human design teams. *Design Studies*, *72*, 100990. https://doi.org/10.1016/j.destud.2021.100990

Zhang, S., Wang, X., Zhang, W., Chen, Y., Gao, L., Wang, D., Zhang, W., Wang, X., & Wen, Y. (2024). *Mutual theory of mind in Human-AI collaboration: An empirical study with LLM-driven AI agents in a real-time shared workspace task* (arXiv:2409.08811). arXiv. https://doi.org/10.48550/arXiv.2409.08811

Zhang, Y., Zhao, J., Jiang, J., Zhu, Y., Wang, L., & Xiang, Y. (2023). Recording behaviors of Artificial Intelligence in blockchains. *IEEE Transactions on Artificial Intelligence*, *4*(6), 1437–1448. https://doi.org/10.1109/TAI.2022.3213531

Zhao, G., Li, Y., & Xu, Q. (2022). From emotion AI to cognitive AI. *International Journal of Network Dynamics and Intelligence*, 65–72. https://doi.org/10.53941/ijndi0101006

Zhong, W., Guo, L., Gao, Q., Ye, H., & Wang, Y. (2024). Memorybank: Enhancing large language models with long-term memory. *Proceedings of the AAAI Conference on Artificial Intelligence*, *38*(17), 19724–19731. https://ojs.aaai.org/index.php/AAAI/article/view/29946

14 Developing Autonomous System Training Models through Hierarchical Reasoning Representations

Mary L. Cummings

14.1 INTRODUCTION

Artificial intelligence (AI) is a key enabler for all robotic technologies, including military ground robots like bomb disposal robots and civilian robots like self-driving cars. In the aviation domain, the success of military unmanned aerial system (UAS) operations has been driven by advances in autonomy and AI. In addition to UAS platforms like medium- and high-altitude UASs (e.g., the Reaper Medium-Altitude Long-Endurance (MALE) UAS and the Global Hawk High-Altitude Long-Endurance (HALE) UAS), AI advances have led to the rapid rise in the use of small UASs. Indeed, the ability of the Ukrainian military to use small drones to penetrate deep inside Russian territory has shown just how tactically and strategically relevant such systems can be. Civilian applications of UASs have also been sharply growing, including agriculture, filming, delivery, and bridge inspection (Sharma, Kulkarni, & Pathak, 2022).

Given the widespread deployment of UAS and their complexities in human-AI teaming, companies and agencies need to know whether their training programs are efficient and correctly designed.[1] The US General Accounting Office (GAO) noted that the US Army and Air Force fall short in predicting UAS pilot shortages (GAO, 2017), as well as identifying training inadequacies (GAO, 2015). The GAO also similarly found that the Federal Aviation Administration is not doing enough to develop principled training programs for drone (small UAS) operations (GAO, 2023).

The aim of this research is to develop an unmanned system operator training model that links human operator reasoning in the operation of an autonomous system to training objectives, in order to better understand the design and effectiveness of a training program. To that end, this effort will outline the development of a system dynamic model based on a historical UAS training program that can be adapted to other programs. This quantitative modeling approach links operator behaviors to training resources and requirements and will be shown to demonstrate

DOI: 10.1201/9781003655923-17

robustness across meaningful variations in input. Finally, both theoretical and practical implications are discussed, as well as how such models will be able to support the introduction of new kinds of AI.

14.2 TRAINING FOR AUTONOMOUS AND AI-EMBEDDED UNMANNED SYSTEMS

The addition of autonomous capabilities and AI subsystems like path planners to unmanned systems has been the key enabler to the growth of UAS operations in both military and civilian operations. Despite the removal of the human from cockpits, it is essential that an operator understands the interfaces, functions, and limits of the system in order to ensure that missions are proficiently executed without the loss of one or more UASs. This proficiency is attained through a formal training process that prepares the trainee for operating the system within often dynamic constraints. Insufficient UAS training has led to numerous human-machine error-induced accidents (Tvaryanas, Thompson, & Constable, 2006; GAO, 2015), which potentially could have been avoided if training requirements aligned with requirements for proper human-machine collaboration.

It has been long recognized that increasing automation/autonomy/AI in a system often changes and possibly adds to human tasking instead of reducing it (Bainbridge, 1983; Sheridan, 2002). Despite the increasing use of autonomy and AI in unmanned systems, for which there is substantial research, there is substantially less research on how training programs should be adapted or reconfigured to account for new functionalities, including AI. Indeed, the bulk of AI research related to UAS training in the modern day includes how AI can be used to personalize training and reduce costs, with unclear effects on safety (e.g., Coombs, 2024; Wojciechowski, Wojtowicz, & Błaszczyk, 2023).

In the early days of human-autonomy interaction research, there was disagreement between whether the implementation of autonomy increases (Sarter, Woods, & Billings, 1997; Wiener & Curry, 1980) or decreases (Amalberti, 1999) the training time and requirements for operators. Recent research has shown that if skill-based training is reduced or eliminated because of the insertion of autonomy, trainee performance can suffer (Cummings, Huang, Zhu, Finkelstein, & Wei, 2019). Moreover, training programs for systems with embedded AI that include instruction for both skill and higher-level autonomy in the same program (to reduce costs, for example) can result in inferior training than just focusing on a single architecture (Zhou, Zhu, Kim, & Cummings, 2019). Lastly, AI is often used to reduce operator workload, but it is unclear how to determine the length of a training program to gain this benefit but not be too short such that operator mental models are incomplete (Li, Cummings, & Welton, 2022).

Given the complex interplay between the degree of AI and autonomy and human performance, any training model must consider the kinds of human reasoning needed in the presence of autonomy, in addition to other typical training variables like length, number of trainees, and number of instructors. Such a model requires a systems-theoretic approach that can include this mix of variables, which will take the form of a system dynamics model, described in the following section.

14.2.1 Embodying Human Reasoning in a Training Model

To develop a model of a UAS operator training pipeline, data is needed to build and validate such a model. To this end, a previous training program for military MALE and HALE UASs was selected, since this program had extensive documentation with well-structured objectives and training syllabi. In addition, these UASs are very mature, with more than 20 years of training outcomes. The objectives of this MALE system were to conduct intelligence, surveillance, and reconnaissance (ISR) missions with a crew of two, called the aircraft operator (AO) and payload operator (PO). The role of the AO was to set mission waypoints and monitor the trajectory of the aircraft, while the PO's role was to monitor and control the aircraft payload. The HALE, a larger UAS, also had a similar staffing structure and mission, but included instrument flying, more complicated ground operations, and the ability to launch weapons, which is why the training is more than twice as long as the MALE program.

Figure 14.1 represents this multi-phase UAS operator training program. Because the two UASs had similar missions, the training program included a common core of nine weeks with three modules that lasted just over two months. Phase 1 training occurred primarily in a classroom setting with lectures on flight safety, navigation, mission planning, aerodynamics, and aviation regulation. Upon graduation from Phase 1, trainees moved into Phase 2, where they received platform-specific training, which consisted of a 5-module, 3-month course for the MALE UAS, or a 7-module, 6-month program for the HALE UAS.

Given the different levels of autonomy for the various tasks and the need to represent human reasoning as previously discussed, a framework was needed to classify the module tasks and their associated operator behaviors. To this end, the Skills, Rules, and Knowledge (SRK) framework (Rasmussen, 1983) was leveraged. The SRK framework was created as a model for understanding how systems should be designed such that human perception and cognition information processing requirements are

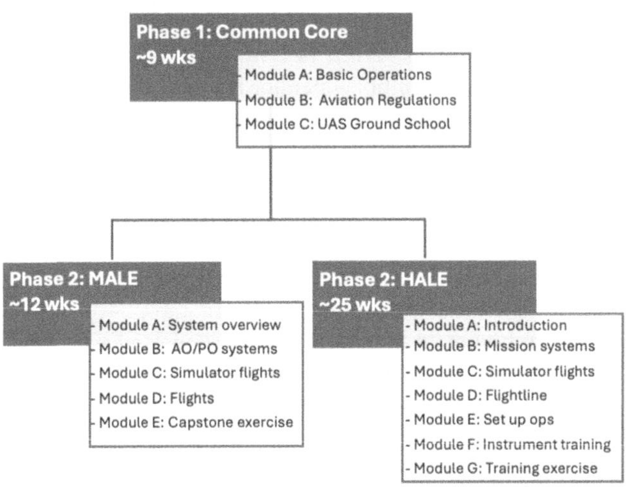

FIGURE 14.1 Historical UAS operator training program with modules per phase.

supported, and it was later updated to include specific considerations for AI and autonomy (Cummings, 2018, 2021). The resulting taxonomy is a three-tiered continuum that abstracts the ways in which human operators reason and interact with machines with embedded autonomy and AI. Other researchers have used the SRK framework to classify task complexity, including breaking air traffic control rerouting tasks into sub-tasks, depending on the complexity of the situation (Kilgore & St-Cyr, 2006).

At the lowest level, skill-based behaviors are defined as those that are highly automatic and lowest on the cognitive continuum, and thus serve as the best candidate behaviors for machines to perform (Cummings, 2014). An example of a Skill-Based Task (SBT) in UAS training is target tracking, which requires the PO to keep the UAS camera sensor locked onto a target using a trackball. This task is psychomotor in nature and highly automatic. Rule-based behaviors are the next highest on the cognitive continuum and require operators to follow prescribed procedures or checklists. Rule-Based Tasks (RBT) are common in military training operations, such as reviewing pre-flight checklists in a logical order before every flight, and are often candidates for AI.

Tasks such as UAS emergency procedures can require human judgment and reasoning, which embody knowledge-based behaviors, the highest on the SRK framework. For example, an engine can be accidentally shut off while in the middle of prosecuting a target (Carrigan, Long, Cummings, & Duffner, 2008), requiring reasoning beyond rote procedures. These behaviors are the most abstract and hardest to replace with AI since they require planning under the presence of uncertainty to reach a desired goal. Scenarios that are novel, uncertain, or have unpredictable exogenous variables influencing planning and decision-making, such as operations in bad weather, fall under the KBT classification (Cummings, 2014; Fleming & Pritchett, 2016). As unmanned systems increasingly rely on advanced forms of AI, it is essential that operator behaviors are linked to those tasks where autonomy may struggle, like KBT, and to what end this will affect overall training requirements.

In order to develop a systems-theoretic training model, each of the UAS operator training tasks in the modules in Figure 14.1 was categorized as SBT, RBT, and/or KBT. Some tasks can include multiple operator behaviors, such as reacting to emergency scenarios, which potentially requires SBT, RBT, and KBT. Table 14.1 depicts the breakdown of tasks per the SRK framework. For SBT, the task description must describe some physical interaction with the aircraft controls or equipment. For an RBT classification, which made up the majority of UAS operator tasks, the operator should have been required to execute procedures. Lastly, a KBT classification required complex reasoning in the presence of high levels of uncertainty. Tables linking all tasks to operator behaviors, such as those in Table 14.1, were completed for all training phases in Figure 14.1. In terms of AI-supported activities, the only one in Table 14.1 with any embedded AI was the Tactical UAS Mission Planning System task, which used path planning as an assistive tool.

Given the classification of all tasks for all the modules in Figure 14.1, Figure 14.2 shows the percentages of SBT, RBT, and KBT across each of the operator training phases for the representative program. The task times for the SBT, RBT, and KBT were determined by analyzing each of the tasks for lessons within the modules and

TABLE 14.1
Example Phase 2 Operator Behavior-Task Breakdown

Task	Operator Behavior
Perform UAS Mission Operations	R
Identify Data Collected by UAS	R,K
Perform Tactical UAS Launch Procedures	R
Perform Tactical UAS Recovery Procedures	R
Operate Tactical UAS Mission Planning System	S,R
Perform Tactical UAS Preflight Procedures	R
Perform Radio Communication Procedures	R
Track a Moving Target	S
Respond to Inadvertent Instrument Meteorological Condition	S,R,K
Perform Radio Failure Procedures	R,K

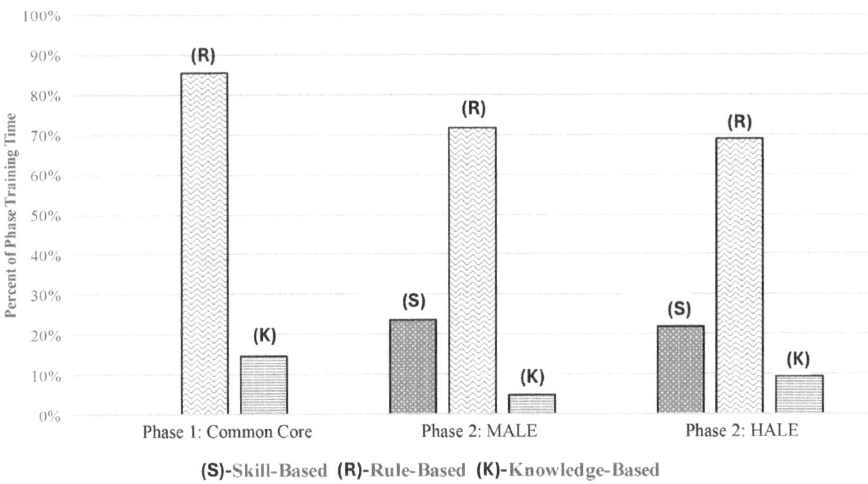

FIGURE 14.2 Percentages of overall training time for skill-, rule-, and knowledge-based tasks across the training phases.

assigning uniform times within Figure 14.2. For example, a lesson that takes 7.2 hours and has 10 tasks is assumed to divide evenly into 0.72 hours per task. This assumption of uniform distribution of task time is a limitation of this approach but is reflective of real-world constraints. The military did not document the time that it took instructors to teach a lesson or module but instead specified how long it *should* take.

Figure 14.2 illustrates that SBTs were understandably not taught until the platform-specific Phase 2, and the vast amount of training across both phases was spent on rule-based behaviors. In Phase 2 for the HALE platform, knowledge-based reasoning training was roughly twice that of the MALE, primarily because of the increased complexity of target identification and weapon release tasks, which occurred during simulation exercises in the later modules.

14.3 THE SYSTEM-THEORETIC TRAINING MODEL (STTM)

Given the assignment of tasks to different categories of behaviors via the SRK framework, the next step of model development was embedding the SRK tasks into a System Dynamic (SD) model, which is a modeling method for understanding nonlinear behavior of complex systems by taking advantage of causal feedback loops and stocks and flows (Sterman, 2000). System dynamic modeling was the selected technique for the STTM due to its ability to capture quantitative and qualitative parameters for human operators (Cummings & Clare, 2015; F. Gao, 2016; F. Gao, Clare, Macbeth, & Cummings, 2013), as well as its batching capability (Mikati, 2010). The ability to represent batching was critical since the military trains its personnel in batches, where a cohort of individuals track through the program together. While SD modeling is centered around modeling continuous, time-dependent behavior of a system (Sterman, 2000), since training occurs as a continuous flow of batches that respond to causal relationships, SD modeling is well suited to capture the rate of trainees moving throughout a training program. Indeed, SD models have been used to represent medical education (Bozikov, Relic, & Dezelic, 2018), physician training supply chains (Tomoaia-Cotisel et al., 2025), and education reform in public schools (Keith, Yadama, O'Neill, & Chung, 2024). However, this effort is unique in that it is the first to integrate different levels of human reasoning in the presence of AI and autonomy into a training pipeline SD model.

The goal of this effort is to model how the number and type of tasks (and their associated reasoning behaviors) and training time affect the number of UAS operator graduates and overall training times. Given that the overall training time for an individual is directly impacted by both the number of tasks and the requirements of the tasks, the STTM should capture the impact of alterations in a training program on its training time, including the attrition rates of trainees. The ability to connect the kinds of reasoning needed in training pipelines to training times, required resources (like instructors), and outputs would greatly assist managers of training programs in achieving their desired outcomes. This is especially important given the rise of AI in such systems.

To construct and validate this implementation of the STTM, two datasets were used. The first, UAS operator training data from year 1 (Y1) for the program in Figure 14.1, was used to initially design and parameterize the model, and the second, data for the same UAS operator training program from year 2 (Y2), was used for model validation. Most tasks between Y1 and Y2 were the same, but some tasks, such as dead reckoning, were added to the training requirements for Y2 due to Y1 trainees' struggles to navigate in GPS-denied environments. Despite some additions, the Phase 1 training program was reduced by ~1.5 weeks in Y2 and the MALE training program was reduced by ~2 weeks in Y2. Thus, an effective confidence builder for the STTM will be that it can accurately capture the change in the dataset, which is significantly different from the data used to build the model.

14.3.1 STTM STRUCTURE

The general structure of the STTM can be seen in Figure 14.3. The stocks (box variables) in the STTM represent the trainees that move through the modeled training program. Flows that connect the stocks are controlled by rate equations and are

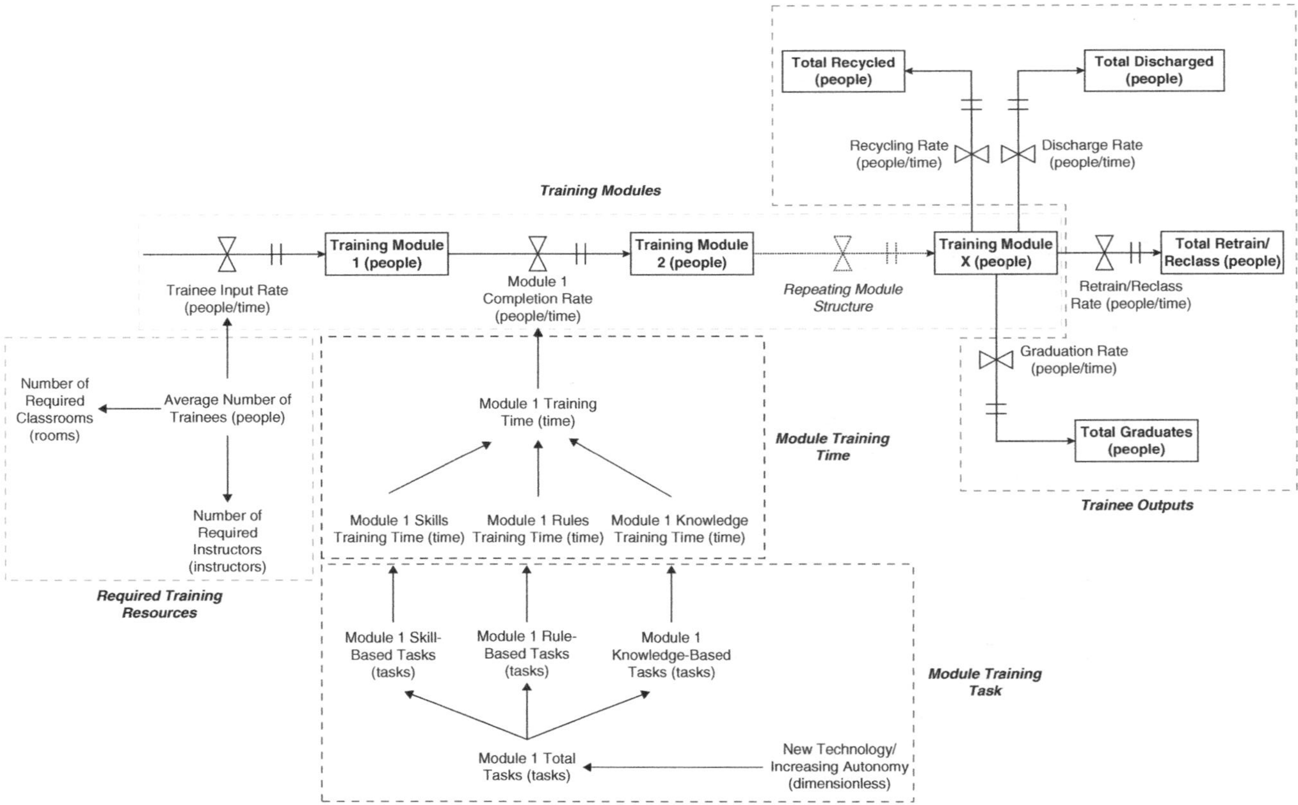

FIGURE 14.3 General structure of the STTM with units noted in parentheses. The Repeating Module Structure arrow and valve represent the repetition of the Module Training Time and Module Training Tasks sections for the respective number of modules within a training program.

derived from parameters that are linked with causal arrows. Trainees move from one stock to the next as designated by the direction the flow arrow is pointing and the value of the rate. The rate at which trainees flow between stocks is controlled by parameters that make up rate equations. SD models like STTM are time-dependent models that calculate new data points every time step, which is constant throughout the entire simulation. The STTM calculates the flow rates via the Euler integration method with 0.03-week time steps (0.2 days), which was selected by testing the sensitivity of the model to varying time steps. It was determined that time steps less than 0.03125 weeks (0.2 days) showed no variation in model outputs, and, thus, would serve as an accurate selection.

The structure of the STTM is made of five distinct sections (Figure 14.3): Training Modules, Required Training Resources, Module Training Tasks, Module Training Time, and Trainee Outputs. The Training Modules, Module Training Time, and Module Training Tasks sections are repeated for each of the modules that are in the training syllabus. Inputs to this UAS application of the STTM included the average number of trainees per group, frequency of new trainee group arrival, tasks per module, and module training times. The STTM outputs include the required number of resources (classrooms and instructors), the number of graduating trainees, the number of recycled trainees, the number of discharged trainees, the number of retrained/reclassified trainees, and the total training time.

The Training Modules section is made up of the trainees that are moving through the training program. The Required Training Resources section provides the user with the ability to plan and calculate the number of classrooms and instructors required to match the number of trainees. The Module Training Tasks section contains the tasks that the trainees must complete in each module. Those training tasks are converted into training time in the Module Training Time section. Finally, the Trainee Outputs section gives a breakdown of how many trainees successfully completed training as well as those that did not meet satisfactory requirements.

To predict the number of trainees that successfully graduate, the STTM estimates percentages for all possible trainee outcomes in each year, including recycled, discharged, and retrained/reclassified. Trainees that are recycled are those that are required to go back and redo training from the beginning of the phase with an incoming trainee group due to unsatisfactory performance on the tasks or missing too many days due to illness or personal issues. The discharged output represents trainees that are dismissed due to behavioral and legal issues or repetitive attempts at graduating without meeting satisfactory requirements. The retrained/reclassified output is reserved for trainees that are sent to a different training program. The graduating operators are the remaining trainees left after the recycled, discharged, or retrain/reclass outputs have been subtracted from the trainee group size. The training times for each of the modules are calculated by taking the total number of tasks per module (the Module Training Task section in Figure 14.3) and converting those tasks to training time (the Module Training Time section in Figure 14.3). Figure 14.4 shows a detailed sample from the actual STTM for Phase 1, with three of the five sections shown (Training Modules, Module Training Time, and Module Training Tasks). Figure 14.4 was created using the Vensim® SD modeling software, and these three sections made up the core of the STTM.

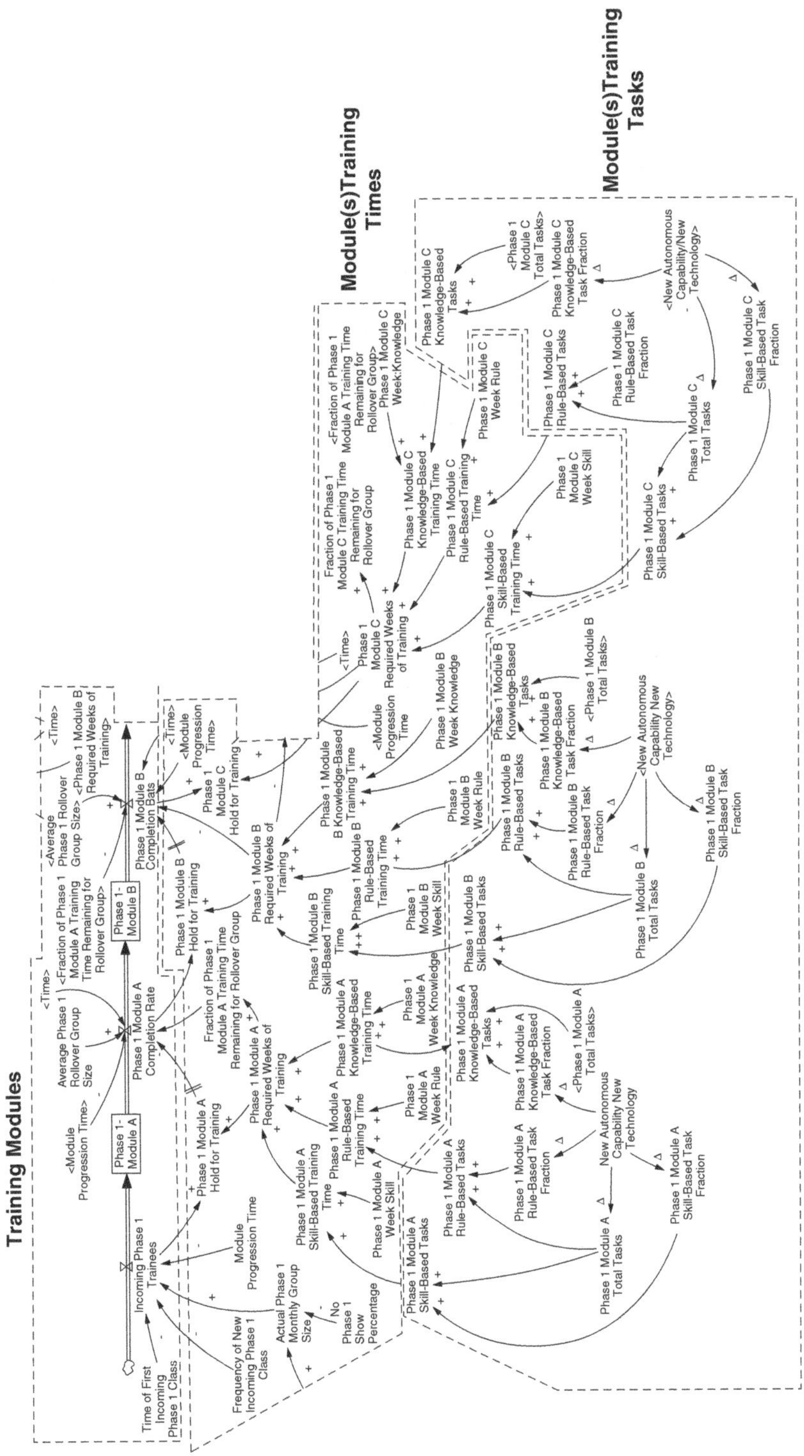

FIGURE 14.4 Sample of the STTM for Phase 1 in Y1. Shown are the Module(s) Training Tasks and Module(s) Training Times sections that control the rate functions between modules in the Training Modules section.

Initially, the total number of tasks that are taught in each training module is determined from the training syllabus. Then, the tasks are classified into SBT, RBT, or KBT in the Module Training Task section, using the method illustrated in Table 14.1. After the tasks are split into SBT, RBT, and KBT in the STTM, they are converted into training times. Each set of SBTs, RBTs, and KBTs has its total training time per module. This is calculated in the Module Training Time section of the STTM (Figure 14.4) by multiplying the number of SBTs, RBTs, or KBTs by the average time to train an SBT, RBT, or KBT. The resulting value is the total time to train all tasks in each module in terms of SBT, RBT, and KBT training time. Finally, the time to train the SBTs, RBTs, and KBTs is summed to give the total module training time. This total module training time parameter, *Phase 1 Module (A, B, or C) Required Weeks of Training* (Figure 14.4), should match the prescribed time from the training syllabus, assuming the equations and average SBT, RBT, and KBT training time per module are accurate. Recall that there are three modules in Y1 Phase 1 training, and, thus, this procedure occurs three times for each module.

The *Phase 1 Module (A, B, or C) Required Weeks of Training* parameter is used to control the rate functions of trainee groups moving from one module to the next. Since there are three modules in Y1 Phase 1, there are three *Phase 1 Module Required Weeks of Training* parameters, one for each module. When batches of trainee groups arrive at a module (stock), they will stay in that module for the duration of that module's required weeks of training. Once that time has expired, the trainee groups move into the next module, and the process begins again. This cycle continues until the trainees reach the Trainee Outputs section (Figure 14.3), where they will either graduate, be recycled, be discharged, or be retrained/reclassified depending on their performance.

14.3.2 Model Results

In order to determine that the STTM in Figure 14.3 accurately predicts training time and number of graduates, model validation was required, which meant comparing the model's performance against real-world data. For the Figure 14.1 training program, the STTM generated the number of graduates over one fiscal year in ~4-week intervals, which was equal to the frequency of new trainee group arrivals. A single fiscal year was chosen as the primary metric of interest because military UAS operator training operations are funded by Department of Defense budgets tied to the fiscal year cycle.

Figure 14.5, which shows the graduation rates for years 1 and 2 for both phases as well as the predicted graduation rates as determined by the STTM, demonstrates that the model accurately represents the numbers of graduates. The solid lines in Figure 14.5 represent the actual number of graduates, and the dotted lines are the predicted number of graduates. For Phase 1, the maximum deviation between the observed and expected Y1 data is 16.1% during weeks 12–14, and for the Y2 data, the maximum deviation is 23.4% in weeks 30–32. These deviations in model predictions are due to the fluctuating sizes of UAS operator trainee groups. Demand for UAS operators changes with operational needs, which can be seen in Figure 14.5 for the varying sizes of graduating UAS operator trainee groups. The STTM, however,

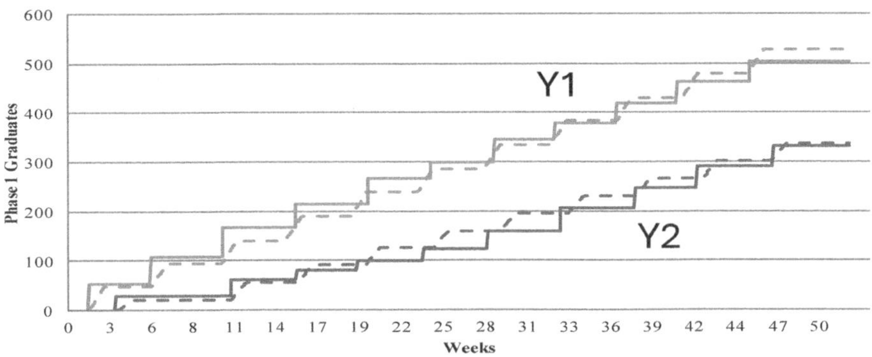

Phase 1 Graduates

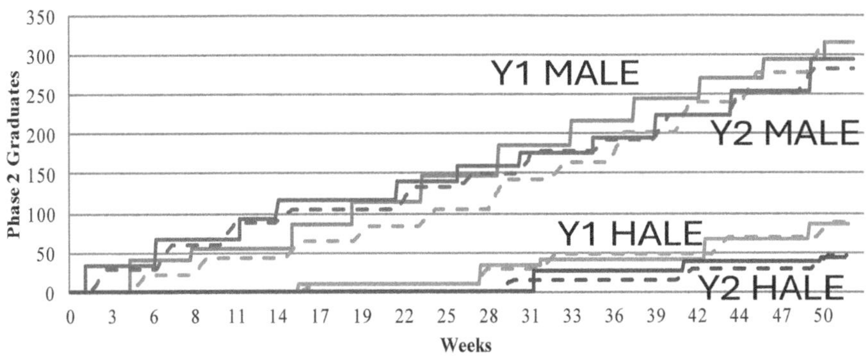

Phase 2 Graduates

FIGURE 14.5 STTM performance for Phase 1 graduates (upper) in year 1 (Y1) and year 2 (Y2) and for Phase 2 MALE and HALE graduates in Y1 and Y2 (lower). Solid lines are actual graduates and dotted lines are predicted graduates.

predicts the same number of graduating trainees per group due to using an average trainee group size. Thus, any deviations in the operational number of graduates from the average group size will result in discrepancies between the actual and STTM-predicted number of graduates. At the end of a year's cycle, the STTM error was 4.84% in Y1 and 2.02% in Y2.

For Phase 2, the STTM performed well, but not quite as well as for Phase 1 (Figure 14.5, lower). For the MALE operators, the STTM's average errors were 0.17% and 4.31% for Y1 and Y2, respectively. However, the maximum deviations were 29.5% and 10.4% for Y1 and Y2, respectively. The percent error for the Phase 2 HALE number of graduates was 1.30% for Y1 and 5.59% for Y2, with maximum deviations of 17.4% and 43.9%, respectively. The high maximum deviations in both the Phase 1 and Phase 2 graduates can be attributed to the difference in expected and observed training time. While the military specifies an expected time to complete the program, actual completion times often take longer than prescribed due to maintenance and budgeting issues. In addition, as previously mentioned, the size of

TABLE 14.2
Percent Error between Actual Vs. Model Outputs for UAS Operator Training for Y1 and Y2

STTM Output	Y1	Y2
Phase 1 – Common Core Recycled	0.63%	13.41%
Phase 1 – Common Core Discharged	6.25%	16.5%
Phase 1 – Common Core Retrain/Reclass	0.80%	13.22%
Phase 1 – Common Core Graduates	4.84%	2.02%
Phase 2 – MALE Recycled	1.56%	8.73%
Phase 2 – MALE Discharged	1.71%	8.75%
Phase 2 – MALE Retrain/Reclass	6.50%	9.00%
Phase 2 – MALE Graduates	0.17%	4.31%
Phase 2 – HALE Recycled	1.72%	2.88%
Phase 2 – HALE Discharged	0.50%	2.89%
Phase 2 – HALE Retrain/Reclass	5.14%	0.00%[a]
Phase 2 – HALE Graduates	1.30%	5.59%

[a] The 0% error reported for Phase 2 – HALE Retrain output in Y2 is due to no retrained/reclassified students that year.

trainee groups is constant in the STTM, as it uses the average group size. In reality, the size of the trainee groups fluctuates throughout the year due to demand for UAS operators.

While Figure 14.5 depicts the graduation rates, the STTM also produces outputs for recycled, discharged, and retrained/reclassified personnel. Table 14.2 shows an overview of the errors from each of these STTM outputs in Y1 and Y2. The errors reported in Table 14.2 can possibly be attributed to two factors, beyond the typical issues with SD models such as using averages. First, it is not clear that the recycling/retraining/reclassifying numbers are correctly reported, as there is significant room for confusion. One person who was recycled from a previous fiscal year may not be accurately counted in these numbers. Also, trainee data organization is not consistent. The datasets provide details for how many graduates there are per training group, but only give the total number of recycled, discharged, and retrained/reclassified trainees for the entire FY, which can contribute to the errors.

14.3.3 Sensitivity Analysis

For developers of training programs, it is important to understand which parameters contribute the most to the error in the STTM predicted number of graduates and training times. Identifying these numbers will aid in improving the model and allow organizations to understand which variables cause the greatest variability, thus providing critical objective feedback about possible training improvements. To this end, a sensitivity analysis is necessary for any model to understand where potential errors are present and where the boundaries exist for model adaptation. A sensitivity analysis addresses the question of how uncertainty in estimated parameters affects the final model output (Sterman, 2000). For the STTM, a numerical sensitivity analysis was chosen to measure model robustness under univariate perturbations of the exogenous parameters, i.e., when and how many students start and finish a program. To that end, two sensitivity analyses were conducted for the two primary outputs: the number of graduates and the training time.

A perturbation of +/–10% was selected after experimenting with the exogenous parameters, since 5% perturbations did not capture any variability in training time with the STTM discrete time steps at 0.03 weeks. Using +/–20% sensitivity resulted in identifying the same variables as those identified with the +/–10% test, so only the 10% perturbation results are reported here. Parameters such as trainee input, no-show percentages, recycling rates, discharge rates, and retraining/reclassified rates were expected to primarily influence how many trainees graduated and not the overall training duration. Similarly, parameters such as SBT, RBT, and KBT drive the training duration, and so would likely mainly influence how fast trainees progressed through the training program.

By changing the inputs at 10%, the percent change in the perturbed output as compared to the original output from the model was compared. Given this sensitivity analysis, there were no variables that when changed by 10%, changed outputs by 10%. For such sensitivity analyses, if inputs are changed by 10%, the variables that are the most concerning are those that change more than 10% (Sterman, 2000). Thus, this application of STTM is robust given that no outputs changed by 10% or greater.

Despite this robustness, it is still important to know which variables affected the model the most. The *Average Phase 1 Monthly Group Size* was found to be the most sensitive parameter (~ +/–8% of the predicted Y1 number of graduates). The likely reason the *Average Phase 1 Monthly Group Size* had the greatest impact on the number of graduates is that this parameter controls the number of people that enter training, i.e., the batch size. By altering this parameter, the size of the trainee groups will be altered accordingly. While not a particularly surprising result, it is also useful practical knowledge in that the military can be relatively sure that if they want more graduates, then the primary change they need to make is increasing their batch size. This result also implicitly suggests that the design of the training program in Figure 14.1 is relatively robust, and no one module is causing problems for trainees.

Only one input variable significantly impacted the time to graduation for Phase 1 when changed by 10%, which is the number of tasks in Module C (Ground School). This 10% change results in ~ +/–5% of the Y1 Phase 1 training time. During this module, all the lessons are made up of RBT. While a 5% change is excellent for this kind of model, it is still noteworthy that the most homogeneous module in terms of reasoning type was the only module that affected outputs when changed by 10%, albeit only ~5%. This is a surprising result because KBT are those that are the most open-ended, which in theory would have been the most susceptible to changes. However, ground school training contains a significant number of rules and facts that must be memorized by the trainee. It is possible that the quality in the actual instruction of this material is a driving factor given the volume of content. It is also possible that because of the significant memorization of material required in this module, trainees struggle to retain the material. This suggests it may be worthwhile to review the module to determine variability in instruction quality, as well as whether this module contains too much material and memorization. These are areas of future research.

14.4 STTM AND THE FUTURE OF AI

The previous sections lay out the mechanics of how to create a training model using skill-, rule-, and knowledge-based tasking for a historical unmanned aerial vehicle system, but what does this model tell us about either the past or future use of AI? First and foremost, skill- and rule-based reasoning tasks are excellent candidates for replacement by AI (Cummings, 2014). Indeed, UAS use is widespread in both military and commercial domains because the skill of flying and the rule-following of navigation were replaced with automation and AI. The one area of training that currently is not likely to be replaced by AI in the near or even distant future is knowledge-based reasoning, especially when such reasoning occurs under high uncertainty. This area is the best candidate for human-interaction tools.

Figure 14.2 illustrates that across all training elements, only roughly 10% of training focused on knowledge-based reasoning, like how to deal with unexpected weather. Given that uncertainty is a core element of the fog of war, driven by both unanticipated enemy behaviors and inevitable system failures, it is surprising that this training program focused more on low-level elements like target tracking—which is well within the domain of AI—than on the knowledge-based tasks. Perhaps the biggest lesson that should be learned from the STTM model is that not enough training time is spent on those knowledge-based activities that are critical for effective warfighting, including emergency responses.

To this end, given the core STTM model for these two UAS systems, it can now be used to determine how to use AI across all tasks to improve training. For example, AI can be used to effectively replace target tracking in Table 14.1 (Cummings et al., 2019), freeing up precious operator cognitive resources to tackle other knowledge-based tasks like coordination with other assets to respond to enemy movement. Indeed, new AI-based tools for UAS operators are rapidly expanding across both US and NATO forces (Erwin, 2025) to help with such tasks. However, such tools are fielded with little understanding of how training should be adapted to accommodate them and whether the tools adequately support human-AI teaming or even effective basic user interactions (Stone, 2025).

Using STTM could help both the military and defense contractors understand which skill-, rule-, and knowledge-based tasks are supported by new software, whether training for these new programs would cause longer training times than expected, and how much this training contributes to recycling or dropout rates. Indeed, such data would be critical in highlighting ineffective human-AI interaction and could signal the need for an in-depth audit of the problems caused by rapidly deployed AI decision support. Moreover, there are increasing calls for AI to be used in training programs to replace human instructors, especially given the rise of large language models. STTM could help to determine whether such AI-based approaches are appropriate and help to pinpoint exactly where outcomes may not meet requirements. STTM is an example of model-based engineering, and as such, it provides a roadmap for how to model the impacts of AI on the human-autonomy training environment.

14.5 STTM LIMITATIONS

As with all models, the quality of the data used to construct the model will directly impact the quality of the output. One limitation to the data used to construct the UAS application of STTM is that only the time required to teach each module was recorded, and not for the tasks that occur within the lessons. Thus, any temporal variable estimations may be too coarse to capture real-world variations. Training programs with finer detail for task training time within lessons would improve model outputs.

In addition, the UAS STTM application led to discrepancies in the number of graduates from each of the STTM phases. This can be attributed to multiple factors, with the first being the introduction of error into the model due to the use of average parameters. The nature of SD modeling requires averaging over datasets, which inherently introduces error into the model. For the STTM, the number of operators entering training was an annual average split across each of the months. In addition, the fraction of trainees recycled, discharged, and retrained/reclassified were average fractions across all the trainee groups within the STTM model. Finally, the model relies on prescriptive training times, not actual training times. This planned training time is not always met due to delays and holidays that occur throughout the fiscal year.

Lastly, SD models are generally poor at managing parameter variability because they are deterministic in nature and must take into account averages of parameters (Brailsford, Desai, & Viana, 2010). This makes them most applicable for a high-level understanding of the system under investigation, as illustrated in this UAS training application. There is ongoing research to address these limitations (Kobayashi, Takayasu, Havlin, & Takayasu, 2022; Kulhavý, 2007), which replaces traditional system dynamics models with discrete-time stochastic dynamic models where state variables are indirectly measured. Such models require significant amounts of data to estimate parameters, which may not be practical. Bespoke training programs will likely be the case for systems where autonomy and AI are used, so future research is needed to determine how to capture the non-determinism present in such systems in order to build more accurate system dynamics models.

14.6 CONCLUSION

This effort proposed a novel adaptation of SD models for training applications that explicitly represents increasingly complex levels of human reasoning, which is critical if such models are applied to training programs that include autonomy and AI. This model, the STTM, was validated for UAS operator training operations and was capable of predicting numbers of graduates and training times with changes in the numbers and types of tasks, number of incoming trainees, and fractions of attrition, retraining, and graduating graduates. This model is adaptable and has the potential to allow training designers to proactively determine how changes in resources like classroom size, number of instructors, and attrition rates affect overall training projections.

The unique element of the STTM is how it links operator behaviors to training through the skill-, rule-, and knowledge-based reasoning framework. Classifying task types in this way can aid developers of operator training programs to better understand the types of tasks that could and should be trained and their influence on training throughput and resources, especially as new capabilities are introduced. For example, if a new capability were introduced into a UAS, such as a new ground navigation system which effectively taxied the aircraft with no human intervention, then the training skill- and rule-based tasks would likely be reduced. However, the knowledge-based tasks would likely significantly increase, requiring new training modules.

Future work with the STTM should include the linking of different error classifications (both operational and during training) into the retraining and recycling rates as a feedback loop, particularly in light of the SRK taxonomy. With this improvement, the STTM could be used to analyze those common errors to determine their effect on the overall training process as well as the effectiveness of human-AI interaction designs, which is especially important given the rapid deployment of AI across the military. Additionally, the STTM could be extended to other robotic platforms with similar training programs, such as the da Vinci Surgical System, a tele-operated robot used for many surgical procedures, as well as many manufacturing robots that require some degree of human collaboration. Such STTM extensions will extend generalizability to other domains with varying degrees of autonomy and AI that will inevitably find their way into human-technology interactions.

ACKNOWLEDGEMENTS

This work was supported by the Army Research Laboratories under Grant W911NF-14-1-0632. Thanks to Andrew Hutchins, Michael Clamann, and Jasper Jeffers for assisting with an earlier version of this manuscript. The assistance from UAS trainers and operators at Fort Rucker, AL, Fort Huachuca, AZ, and Fort Bragg, NC was instrumental in understanding UAS training operations. The anonymous reviewers also helped in improving the manuscript.

NOTE

1 The need for improved training for ground, space and maritime robots is also relevant, but these systems are not as mature as UASs, so the remainder of this effort will focus on robots in aviation.

REFERENCES

Amalberti, R. (1999). Automation in aviation: A human factors perspective. In D. Garland, J. Wise, & V. Hopkin (Eds.), *Handbook of aviation human factors* (pp. 173–192). Mahwah, NJ: Erlbaum.

Bainbridge, L. (1983). Ironies of automation. *Automatica*, *19*(6), 775–779. https://doi.org/10.1016/0005-1098(83)90046-8

Bozikov, J., Relic, D., & Dezelic, G. (2018). Use of system dynamics modeling in medical education and research projects. *Student Health Technolology Information*, *247*, 830–834.

Brailsford, S. C., Desai, S. M., & Viana, J. (2010). Towards the Holy Grail: Combining system dynamics and discrete-event simulation in healthcare. *Paper Presented at the Proceedings of the 2010 Winter Simulation Conference (WSC),* Baltimore, MD.

Carrigan, G., Long, D., Cummings, M. L., & Duffner, J. (2008). Human factors analysis of Predator B crash. *Paper presented at the AUVSI 2008: Unmanned Systems North America*, San Diego, CA.

Coombs, R.A. (2024, May 2024). AI integration for scenario development training the whole-of-force. *Military Review Online.* Retrieved from https://www.armyupress.army.mil/Journals/Military-Review/Online-Exclusive/2024-OLE/AI-Integration-for-Scenario-Development/

Cummings, M. L. (2014). Man vs. machine or man + machine? *IEEE Intelligent Systems*, *29*(5), 62–69.

Cummings, M. L. (2018). Informing autonomous system design through the lens of skill-, rule, and knowledge-based behaviors. *Journal of Cognitive Engineering and Decision Making*, *12*(1), 58–61.

Cummings, M. L. (2021). Rethinking the maturity of artificial intelligence in safety-critical settings. *Artificial Intelligence Magazine*, *42*(1), 6–15.

Cummings, M. L., & Clare, A. S. (2015). Holistic modelling for human-autonomous system interaction. *Theoretical Issues in Ergonomics Science*, *16*(3), 214–231.

Cummings, M. L., Huang, L., Zhu, H., Finkelstein, D., & Wei, R. (2019). The impact of increasing autonomy on training requirements in a UAV supervisory control task. *Journal of Cognitive Engineering and Decision Making*, *13*(4), 295–309.

Erwin, S. (2025). Pentagon boosts budget for Palantir's AI software in major expansion of Project Maven. Retrieved from https://spacenews.com/pentagon-boosts-budget-for-palantirs-ai-software-in-major-expansion-of-project-maven/

Fleming, E., & Pritchett, A. (2016). Cognition, Technology & Work. *SRK as a framework for the development of training for effective interaction with multi-level automation*, *18*(3), 511–528.

GAO. (2015). *Unmanned aerial systems: Actions needed to improved DOD pilot training* (GAO-15-461). Washington, DC. Retrieved from https://www.gao.gov/products/gao-15-461

GAO. (2017). *Unmanned aerial systems: Air Force and Army should improve strategic human capital planning for pilot workforces* (GAO-17-53). Washington, DC.

GAO. (2023). *DRONES: FAA should improve its approach to integrating drones into the national airspace system* (GAO-23-105189). Washington, DC. Retrieved from https://www.gao.gov/products/gao-23-105189.

Gao, F. (2016). *Modeling human attention and performance in automated environments with low task loading* (PhD). Massachusetts Institute of Technology, Cambridge, MA.

Gao, F., Clare, A. S., Macbeth, J. C., & Cummings, M. L. (2013). Modeling the impact of operator trust on performance in multiple robot control. *Paper Presented at the AAAI Spring Symposium: Trust and Autonomous Systems*, Stanford, CA.

Keith, D. R., Yadama, A., O'Neill, E., & Chung, S. (2024). Anticipating the side effects of educational reform using system dynamics modeling. *Review of Research in Education*, *48*(1), 1–27. https://doi.org/10.3102/0091732X241260012

Kilgore, R., & St-Cyr, O. (2006). The SRK inventory: A tool for structuring and capturing a worker competencies analysis. *Paper Presented at the Human Factors and Ergonomics Society Annual Meeting*, San Francisco, CA.

Kobayashi, Y., Takayasu, H., Havlin, S., & Takayasu, M. (2022). Data-driven stochastic simulation leading to the allometric scaling laws in complex systems. *Physical Review E*, *106*(6), 064304.

Kulhavý, R. (2007). Bayesian analysis of stochastic system dynamics. *Paper Presented at the 25th International Conference of the System Dynamics Society*, Boston.

Li, S., Cummings, M. L., & Welton, B. (2022). Assessing the impact of autonomy and overconfidence in UAV first-person view training. *Applied Ergonomics*, *98*, 103580.

Mikati, N. (2010). Dependence of lead time on batch size studied by a system dynamics model. *International Journal of Production Research, 48*(18), 5523–5532.

Rasmussen, J. (1983). Skills, rules, and knowledge: Signals, signs, and symbols, and other distinctions in human performance models. *IEEE Transactions on Systems, Man, and Cybernetics, 13*(3), 257–266.

Sarter, N. B., Woods, D. D., & Billings, C. E. (1997). Automation surprises. In G. Salvendy (Ed.), *Handbook of human factors and ergonomics* (pp. 1926–1943). New York: Wiley.

Sharma, S., Kulkarni, P., & Pathak, P. (2022). Applications of Unmanned Aerial Vehicles (UAVs) for improved business management. *Paper presented at the International Interdisciplinary Humanitarian Conference for Sustainability (IIHC)*, Bengaluru, India.

Sheridan, T. B. (2002). *Humans and automation: System design and research issues*, Santa Monica, CA: Wiley-Interscience.

Sterman, J. D. (2000). *Business dynamics: Systems thinking and modeling for a complex world.* New York: Irwin/McGraw-Hill.

Stone, M. (2025). Anduril and Palantir battlefield communication system 'very high risk,' US Army memo says. Retrieved from https://www.reuters.com/business/aerospace-defense/anduril-palantir-battlefield-communication-system-has-deep-flaws-army-memo-says-2025-10-03/

Tomoaia-Cotisel, A., Allen, S. D., Cave, S., Mendel, P., Farr, W. W., Gordon, D. E., … Martineau, M. (2025). *Modeling physician workforce supply, demand, and need using system dynamics: Model structure and assumptions.* Santa Monica, CA. Retrieved from https://www.rand.org/pubs/research_reports/RRA2616-1.html.

Tvaryanas, A. P., Thompson, W. T., & Constable, S. H. (2006). Human factors in remotely piloted aircraft operations: HFACS analysis of 221 mishaps over 10 years. *Aviation, Space, and Environmental Medicine, 77*(7), 724–732.

Wiener, E. L., & Curry, R. E. (1980). Flight-Deck automation: Promises and problems. *Ergonomics, 23*(10), 995–1011.

Wojciechowski, P., Wojtowicz, K., & Błaszczyk, J. (2023). AI-driven method for UAV Pilot training process optimization in a virtual environment. *Paper presented at the IEEE International Workshop on Technologies for Defense and Security (TechDefense)*, Rome, Italy.

Zhou, J., Zhu, H., Kim, M., & Cummings, M. L. (2019). The impact of different levels of autonomy and training on operators' drone control strategies. *Journal of Human-Robot Interaction, 8*(4). https://doi.org/10.1145/3344276

15 Ambient Intelligent Command Center Testbed for the Navy

Jeffrey Bennett, Brian Lu, and Jaquelin Tsui

15.1 INTRODUCTION AND MOTIVATION

As the use of artificial intelligence (AI)-enabled technologies continues to proliferate in the commercial space, there is a renewed interest in developing and testing these technologies within the realistic environments where researchers have envisioned their paradigm-shifting usage. For the U.S. Navy, these environments include the watchfloors and operations centers that are home to the increasingly complex decision spaces that Naval commanders and their staff must navigate. We believe that successfully navigating these complexities and achieving competitive advantage will require a shift in the human-machine interaction paradigm, and that ambient intelligent (AmI) environments and their related features provide a viable option for this future. Here we outline our ongoing work in developing an Ambient Intelligent Command Center Prototype, which serves as a research and development space for prototyping new systems and workflows, as well as a testbed where human-AI teaming can be studied within the context of realistic scenarios to help guide further development and prepare for experimentation in the deployment environments these technologies will ultimately need to function within.

15.1.1 Background and Naval Operational Context

Naval Information Warfare Center (NIWC) Pacific has a rich history of developing vision videos to highlight the possibility of employing theoretical technologies into the battlespaces of the future. The most recent iterations of these vision videos imagine the possibility of Naval command staff interacting with an ambient intelligence to better understand their operational situation, and work through possible courses of action (COAs) in missions with challenging constraints. A subset of one of these videos is available publicly here.[1]

In developing the tools aligned to this vision, there are many technological challenges, but also a set of challenges unique to the Naval environment and use case. These include sensitivities with the data and systems such an implementation would require, including low network bandwidth, limited compute resource environments, and challenging visual and acoustic processing spaces, among others. To address these challenges and enable studies into how humans would interact with, and benefit

DOI: 10.1201/9781003655923-18

from, such technologies, it is necessary to create a testbed that can either (1) represent the operational environment to the necessary degree, or (2) be brought into actual operational environments themselves. In this chapter, we will outline how we conceived and developed an Ambient Intelligent Command Center Testbed that enables both criteria specified above, and we hope will serve as an ongoing resource for studying human-AI teaming in this unique and highly specialized domain.

15.1.2 Ambient Intelligence Concept and Need

Ambient intelligence first gained traction in the public domain as interest in smaller, multifunctional devices grew in popularity. Advances in technology ushered in a new range of devices that are situated within the environment and intended to be unnoticed by the user until needed. The AmI environment is further distinguished from the devices in the consumer realm by its persistent operation and seamless integration into the workspace. Rather than being a separate application or device that users need to turn on and off, the ambient intelligence is embedded within its environment—intuitive and responsive, yet imperceptible to the user (Garcia et al., 2024). When combined with the automation possible using large language models (LLMs), these AmI environments can be more adaptive and immersive than ever before (Dakshit & Prabhakaran, 2025).

The landscape of Naval warfare is constantly changing; however, due to long acquisition timelines and the complexities of developing and deploying military systems, the technology available to operators and Warfighters often reflects the needs of earlier eras, sometimes by multiple decades. An optimal environment from a human-system interaction perspective will draw from tools that Warfighters are familiar with in their daily lives, such that in situations demanding rapid deployment, minimal training is required to successfully use the system. The AmI space embodies this principle by providing an unobtrusive assistant that can utilize the verbal and visual cues humans employ when communicating with one another to perform tasks, relay information, and summarize data in the background.

The need for this push into modernization was precipitated by the influx of information that Warfighters must handle. The current state of self-defense and warfare necessitates the efficient and accurate tracking of a vast amount of data. Not only must Warfighters be apprised of the information flooding in, but they must also be able to synthesize this data to differentiate critical sources from noise.

Presently, our AmI environment serves as the vehicle for several other projects within the Naval Research and Development Enterprise. The creation of this centralized testbed allows teams to independently develop and test programs and subsequently integrate their technologies into the broader AmI space once they have achieved sufficient technical maturity and readiness.

15.1.3 Chapter Overview and Contributions

For this chapter, we begin with a discussion of the ambient intelligence Framework in Section 15.2, including the necessary features of these environments, and the types of human-AI teaming paradigms, hardware, and software considerations that result.

Section 15.3 provides an overview of our implementation of the Ambient Intelligent Command Center Testbed, and Section 15.4 highlights how this has been used for various presentations and experimentations to date. We conclude with a discussion on some of the current limitations and future directions for utilizing this testbed in Section 15.5.

15.2 AMBIENT INTELLIGENCE FRAMEWORK

To better understand the benefits and limitations of the ambient intelligence framework, we have identified a need for an ambient intelligence testbed. The testbed approach is commonly utilized and has been employed to conduct studies in related areas, such as simulating high-pressure environments for human-AI teaming in maritime defense Command and Control scenarios, and can be customized to replicate realistically degraded conditions (Schmidt et al., 2025). Outside of the military domain, other testbeds have offered the opportunity to reimagine the possibilities within the workplace, transforming meeting rooms from static spaces to dynamic, AI-enabled environments (Fender et al., 2017). Within the public sector, testbeds have been used to demonstrate the potential benefits that ambient intelligence can bring to education, healthcare, and smart homes (Ahmed et al., 2019). Utilizing a similar approach, our testbed was designed to support the necessary components of an AmI environment while also maintaining a core set of features necessary for conducting meaningful experimentation.

15.2.1 Core Features

We have identified a set of three features we believe are necessary for providing a useful and extensible testbed for our research purposes. These include (1) modular testbed deployments, (2) data logging and archival, and (3) third-party system hooks. With these features, the testbed is capable of being utilized in the appropriate environments, implementing relevant and accurate scenarios, and instantiating user studies effectively, given they are appropriately designed and approved by the necessary authorization entities (including Institutional Review Boards (IRBs), Facility Managers, Security Teams, etc.).

15.2.1.1 Modular Testbed Deployments

The testbed itself should be designed in such a manner that it can support a persistent deployment in one physical location if needed, but also have the flexibility for core components to be moved and temporarily set up in alternate environments. For our purpose, the persistent testbed is hosted in a converted conference room designated as the Command Center of the Future (CCoF), which is discussed in further detail in Section 15.3, *The CCoF as a Physical Implementation of Testbed*. However, it is also necessary to support human-AI teaming studies in other environments where we want to better understand their performance and potential implications. From a Naval perspective, this includes areas such as classified spaces where certain scenarios and data must reside, or shipboard locations which may need to be rapidly set up or disassembled in certain situations.

15.2.1.2 Data Logging and Archival

Any data within the testbed environment that we expect to inform the various systems and technologies being studied must also be recorded and archived in some manner. This gives researchers access to reproducible datasets for validating system performance and developing strategies for new capabilities and use cases. These data records can become large for long-duration, multi-user studies, as they require multi-modal sensor collection and additional system logging for all relevant components. Additionally, these data artifacts necessarily include human-derived data sources and require adherence to human subjects research (HSR) procedures, which are discussed further in Section 15.4, *Experiment Considerations.*

15.2.1.3 Third-Party System Hooks

Lastly, these intelligent spaces will rely on integrations with external third-party systems that researchers may have limited ability to control or customize. These systems will still be necessary to capture current user workflows or serve as sources of information to feed into other components within the environment. The details of these systems will be heavily dependent on the use case under study. For the Naval testbed, these have included interfacing with audio/visual (A/V) systems, enterprise systems (document management servers, email clients, chat systems, etc.), and mission-specific systems (planning tools, Tactical Decision Aids, Common Operational Picture providers, and various warfare-domain-specific systems).

15.2.2 Principles of Ambient Intelligent Environments

Persisting from space to space, AmI interacts with the user without the need for wearable devices (Verbeek, 2009). In the Naval warfare scenario, events can unfold at a rapid and often taxing pace for the Warfighter. Decision-making is complicated by the need to disseminate information quickly and accurately under pressure. The AmI environment envisioned for this project aims to consolidate the layers of communication while reducing cognitive load. With the system offering precise recommendations for specific operator tasks, Warfighters can make more informed decisions and reallocate limited staff resources efficiently.

15.2.3 Human-AI Teaming Paradigms

Human-AI teaming (Berretta et al., 2023) advocates for a future in which complex workflows remain efficient and sustainable even in complex and time-restricted environments. For the military in particular, the flow of information by the minute, hour, and day can be so substantial that even small oversights can lead to disastrous consequences. Chhetri et al. (2024), for example, describe the challenges of alert fatigue. In Security Operations Centers, operators are beleaguered by the number of alerts received, leaving them unable to ascertain which are critical and which can be safely postponed.

What distinguishes human-AI teaming from human-AmI teaming is both the utility of the system and its presence. In human-AI teaming, the intelligent machine agent is viewed as a tool with specific roles, engaged by the user as needed

(Berretta et al., 2023). Human-AmI teaming expands upon this definition by encompassing a wider range of abilities. It can capture, analyze, and present data to multiple users at a time, and its inherent nature of being embedded and automatically activated within the environment makes it more widely accessible. Most notably, the AmI environment is dynamic and evolves with the needs of its users (Constantine et al., 2021). Our goal is to elevate the efficacy of our AmI environment such that humans can interact with it naturalistically and seamlessly.

The current AmI testbed provides a space where Warfighters can develop trust in the system (Love et al., 2024), learn effective teaming (Vats et al., 2024), and better allocate their own knowledge and skillsets (Bosch & Barrué, 2009). Additionally, by incorporating the Warfighter in the development of this AmI environment, we are returning agency to the ultimate end user of the product. Warfighters participating in human-AI teaming within the testbed will be able to provide feedback that is limited to these studies, without fear of retribution in relation to their ongoing duties. By bringing the end user into the development process, we aim to build trust and investment in the products that are developed and to increase the chances of successful adoption in the Fleet.

15.2.4 Hardware Considerations

To implement the testbed, a few hardware aspects need to be considered. First, the testbed utilizes multiple LLMs as part of its services and requires robust graphics processing unit (GPU) resources to host them within the testbed. We have identified a threshold of GPUs with at least 16 GB VRAM to provide enough resources for hosting quantized open-source LLMs, such as Mistral[2] or Llama[3], while still reserving storage for any graphical interfaces displayed to the users. The next hardware consideration is network access, which largely depends on the deployment environment of the system. In unclassified environments, the restrictions on network access are much more flexible. However, in Controlled Unclassified Information (CUI) spaces, each system must meet the strict requirements of the U.S. Navy's Security Technical Implementation Guide to connect to the secure network. The information contained in these spaces, while not classified, still must adhere to dissemination restrictions.

Depending on how many services are deployed, the storage requirements for the testbed vary. Having more storage resources is beneficial; logs and volume sizes grow quickly as the testbed is persistently deployed and utilized. For mobile implementations of this testbed, the minimum requirements are visual displays, speakers, and microphones that are set up to properly route data to and from the testbed machine(s) (Figure 15.1).

15.2.5 Software Considerations

To streamline the development process while also conforming to the challenges of working in a secure testbed environment that is shared among developers, the development and deployment environments are divided and separately maintained. Before pushing to the testbed, all software must pass established acceptance criteria within a Continuous Integration/Continuous Deployment construct to minimize deployment risk.

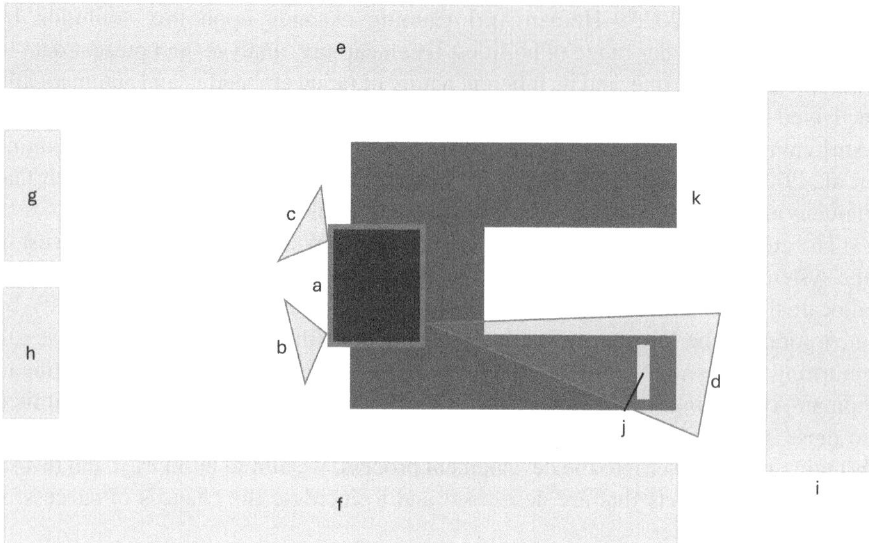

FIGURE 15.1 Diagram of physical testbed layout: The current iteration of the testbed allows users to enter the space and experience the AmI environment. Label "a" is where the microphone array is located, and labels "b," "c," and "d" are the microphone beams projected from the array. The beams are pointed toward the front and ends of the desks labeled "k." Beam "d" is where the presenter would stand to present the demo. Labels "e," "f," and "" are screens for the users and other staff to see different visualizations like the dashboard or the mission tool. Label "j" is an additional screen for the presenter to use. Labels "g" and "h" are additional seating for staff members not directly interacting with the environment or experiment.

The routes are flexible in their implementation. The only key requirement is that each tool will need to implement an integration service that can interact with a natural language speech or text interface to augment a typical visual user interface. This service would process the language data and route it to the proper API calls in the implemented tool. For example, when integrating a speech interface with a planning tool, we provide a separate service that ingests the audio transcript and converts it to a structured list of actionable items for use within the planning tool. Once that information is extracted, the service then processes it into the correct format for the API call and routes it to the planning tool. This is an example of integrating with a third-party service, and a case where the testbed's centralized message broker is not directly utilized as the data integration mechanism. The planning service API requires that certain actions are completed within the tool prior to receiving the data payload, which requires specialized integration paths to avoid race conditions. This highlights an example of third-party system hooks that are required by the testbed.

Commercially available LLMs often struggle when given highly specialized, Naval-specific questions. To address questions like those that Naval watchstanders may have when on the command floor, a retrieval-augmented generation (RAG) system is used. The RAG system offers responses that are grounded in doctrine, creating a more reliable and trustworthy experience. When a question is posed to the system, it is processed through the RAG system, which searches through an index of prepared documents. To get more accurate responses, the system runs two passes. First, an initial similarity search is conducted, using a low-compute resource algorithm that retrieves any related documents. This subset of documents is reranked using a higher-compute method to find the most similar documents. These top documents are returned to the LLM along with the initial query as context. Responses from the LLM can now be traced directly back to the document from which they were sourced. The RAG system offers the ability to process and retain a variety of document types, with examples for various use cases provided in subsequent sections.

15.3 IMPLEMENTATION

15.3.1 The CCoF as a Physical Implementation of Testbed

The testbed for this project is comprised of a framework and a collection of software services with many possible implementations. The CCoF serves as the physical prototype in this current work and emulates a modernized version of the Tactical Flag Command Center (TFCC) found onboard U.S. Navy aircraft carriers. A space like TFCC provides a rich and compelling use case for the application of ambient intelligence, human-AI, and human-AmI teaming studies because it is home to multiple users (upwards of ten staff members) working to aggregate information and data across separate domains, while still working towards achieving a common set of goals. Introducing new systems, workflows, or interactions to a space like this can be a complex undertaking, which is why introducing a surrogate environment that can slowly build towards the real-world complexity of a TFCC is ideal. Utilizing the CCoF as a controlled laboratory space with significant technical capabilities and specific development configurations, but that can also support compliant data processing within a controlled network zone, satisfies the needs for our persistent testbed. The space can be utilized by different projects to showcase their roster of tools, and it also provides the ability to test and train without disrupting the operations of an actual command center.

15.3.2 Hardware

The current implementation of the testbed uses three Quadro RTX 6000 GPUs, providing a total of 72 GB VRAM. A challenge of utilizing containerized deployments of services and linking them to multiple parallel GPUs is that, in many cases, only one GPU is regularly recognized, which will cause issues with GPU load balancing. For our purposes, this necessitates hosting a more robust resource allocation service to enable the orchestration of multiple resource-intensive services simultaneously, while still providing near real-time feedback to users within the space. Our current testbed utilizes 128 GB of hard drive space. Ideally, 500 GB of hard drive space would be the minimum recommendation to facilitate the addition of other features. This is especially true for those workflows and tasks that involve modeling and simulation, which generates asset libraries requiring more storage capacity.

Our testbed is deployed in an environment secured for up to CUI, but this creates restrictions on how the system can interface with other subsystems, such as the A/V equipment. These constraints need to be considered and planned for when creating new testbeds. The testbed currently requires a complex A/V solution that routes the graphics and audio to various screens and speakers. Using four networked Microsoft Azure Kinect depth cameras, this space can capture images up to a resolution of 3,840 x 2,160 at 30 FPS (color camera) or gather data up to 5.5 m at 30 FPS (Time-of-Flight depth camera). This feature allows the camera system to detect participants and track their movement around the room. The audio collection equipment, consisting of a Shure MXA910 ceiling microphone array, is connected to a Q-Sys Core 110f audio processor and has eight independently controlled audio beams. These beams can be positioned to specific coordinates or dynamically adjust to widths ranging from 35° to 55°. Using digital signal processing, the microphone adjusts volume and filters out background noise. Each beam independently captures, adjusts, and feeds audio back to the main source. From there, the data is sent to the appropriate microservice for processing the audio. This creates a unified and consistent stream of audio regardless of the subject's location in the room or their natural speaking volume.

15.3.3 Software

To integrate into speech or text language interfaces, software that interacts with the language component requires an API access point. Our systems utilize message brokers, like Kafka and MQTT, and HTTP requests to manipulate software tools and deliver information artifacts. Services connected to a message broker can generate and receive data across various topics defined within the testbed via a publish-subscribe messaging pattern. This enables asynchronous communication between the services and helps ensure that all services are only loosely coupled to support a more seamless addition or removal of services as the testbed is configured for different use cases and purposes. Once a topic is initialized, any service deployed on the same network can be configured to subscribe to that topic and receive information. An example of this usage can be found in the transcription service of the current testbed implementation. The "transcript" topic was established early in the project's development and has not since been modified. Other services can be quickly configured to consume from the transcript to receive updates as they are generated. Using this paradigm, services can readily communicate and synchronize across different machines on a secure network (Figure 15.2).

First, the intent classification agent, which is connected to multiple other services through the message broker or direct API integrations, processes and interprets the incoming dialogue, and then handles the orchestration of the processed and structured information to another layer of implemented agents. This feature monitors the transcript continuously and acts when necessary. Information processing is offloaded from the first agent to the next, and by doing this, the second agent in the workflow will always be able to extract information based on specified context without having the additional overhead of processing intent. The second agent will be aware of the most relevant services and can take action to assist the user using the appropriate tools and API calls. Currently, the testbed has an associated agent defined for each service, and once a more robust agent orchestration framework is implemented, the testbed will feature an agentic-deployment option for future testing. Most of the agents function similarly in that they are prompted with the current task and are directed to use the available tools to complete a specified task. As the tasks are routed to the specific agents, each generates guard-railed outputs to ensure the appropriate API formats and types are produced (Table 15.1).

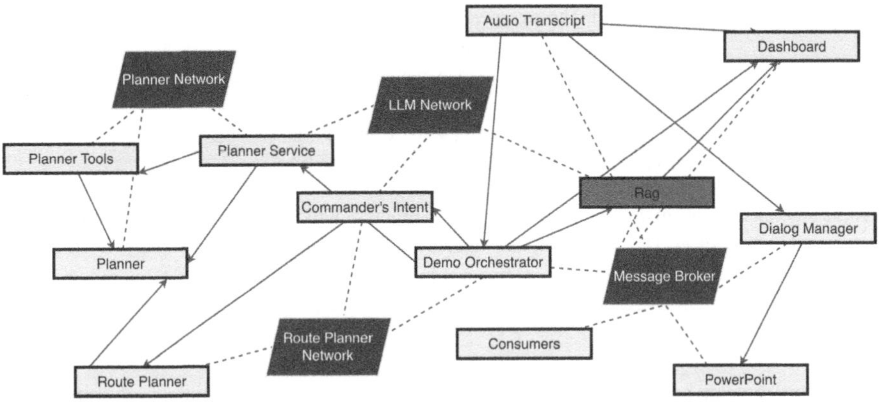

FIGURE 15.2 Software architecture diagram example: Our testbed involves various connections between different pieces of software across multiple managed networks. The light gray boxes are examples of services we have in the testbed. The dark gray parallelograms represent the networks connecting the various services. The dotted lines represent network connections for the services, while the solid arrows represent the flow of information from one service to another (enabled by the network connections). Shown here are the connections between services and the flow of information, starting from the audio transcript of a user and continuing all the way to the mission-specific tool input, in this case, a planning tool.

TABLE 15.1
Selected Examples of User Intents

Intent	Role
mission_planning	This routes the message to the planning_service, which will be a middleman service that additionally classifies what action in within the planning tool user wants to take.
do_demo_reset	This intent resets all the services to run through the demo. This involves resetting UI's and flags, so that they can be triggered again. This intent is used at the end of demos to be ready for the next one, or in the middle of one in case there is a failure at any point.
complete_reminder	This intent completes a reminder list in the dashboard. Reminders are handled by the system or agents and enable a mechanism for guiding user actions.
add_task	Adds a new user defined task to the task list in the dashboard. Associated intents for modifying, removing, and completing tasks also exist.
set_time	This intent sets the timer in the dashboard to the designated time and begins the countdown. Associated intents for modifying the timers and utilizing stopwatches are also present.

(Continued)

TABLE 15.1 (*Continued*)
Selected Examples of User Intents

Intent	Role
presenter_num	This intent changes the PowerPoint slide to the numbered slide in the current slide deck being presented.
presenter_tag	This intent changes the PowerPoint slide to the tagged slide in the current slide deck being presented. The tags are currently manually set in a dictionary. The tag is extracted from the user message and matched with the dictionary. In the future, multimodal LLM's will be used to summarize each slide and that will be used to better identify the slide that a user is referring to.
semantic_query	This intent sends a message to the RAG system for the purpose of routing synthesized answers and grounded information to the dashboard.

These user intents listed in the "Intent" column give an example of the types of categories that are currently supported in the testbed. The "Role" column outlines actions triggered by these events and represents the types of roles assumed by a given agent instantiated by the triggering of the event.

Visualization of relevant processes within the testbed is contained within the modular dashboard, which acts as a unified hub for the events of a presentation or experimentation session. The dashboard is the primary front-end for the custom digital assistant agents developed by the team, providing similar capabilities to that of traditional commercial smart assistant technologies. Through this, users interact with what we refer to as "AISI" (the AmI Speech Interface) in a familiar way. The current capabilities displayed on the dashboard include speech transcripts, basic tools, and system responses, each contained within separate widgets. The dashboard displays the current transcript, showing users their spoken dialogue in real-time. This provides a real-time view of the transcript quality and gives visual feedback on interactions with the system, allowing users to more easily correct course if commands are misinterpreted. The dashboard displays a task list widget, a timer widget, and a reminder list widget. The task list provides a list of actions for the user to undertake, while the reminder list offers a place to queue incomplete, system-generated tasks for later. The dashboard also displays the response to the user, replete with links to relevant documents, when a question is answered by the RAG system. A separate widget for system-generated responses is also included as a dashboard model and is where the query results are displayed. The dashboard, built in Vue.js, requires additional socket configuration to communicate with other systems in the testbed environment. Updates to the front-end are a continual source of development focus and aim to achieve modern designs and better displays as new services are integrated.

For the current use case, a collection of Naval doctrine used in planning operations, which were initially available as PDFs, were processed and chunked to be indexed in a vector database using open-source libraries. RAGatouille[4] is leveraged for the current RAG service's indexing and search mechanisms to easily enable the use of the ColBert model, which has "been shown to generalize better to new or complex domains than dense embeddings" (Clavié, 2025). We utilize Colbert V2 in our implementation of RAGatouille and PyPDFLoader from Langchain[5] to handle PDF document processing. During the indexing stage, the documents are also tagged with various types of metadata which can be utilized to further refine retrieval or associate source documents with specific user tasks or phases of an experiment.

15.3.4 Deployment Options

The services hosted in the testbed can be deployed using three methods: fully onboard, room-integrated, and server-based. The fully onboard option utilizes a laptop or desktop computer to host all the services needed for the testbed, with a microphone directly connected to the machine to enable voice interactions. This setup is sufficient for basic testing and small-scale experimentations. In a room-integrated deployment, the local network links all the available services. It is also possible to route to other services on different machines and to connect to the audiovisual equipment in the room for greater control over the demo. Finally, for a server-based deployment, services can be hosted on a server, or a collection of connected servers, and run persistently in a hybrid compute configuration. If any services fail, health-check monitoring helps identify issues and automatically re-deploys the offending services.

An additional benefit to having the testbed machine connected to the onsite network is the ability to connect to the High-Performance Computing Center (HPCC). Locally, due to the real-time constraints of such a system, single-GPU machines are currently limited to hosting roughly 7–14B parameter models, but through the HPCC, real-time access is available to 120B parameter models with significantly improved results.

15.3.5 Key Subsystems and Integrations

A unique ecosystem of tools, both commercial and specialized, sustains the current testbed. In designing this testbed, we considered the needs and use cases within the command center. Features commonplace to Warfighters, such as an instant messaging interface or integration with commercially available software, are incorporated to build familiarity and comfort with the testbed. Conversely, we also implement applications that are completely new to the Warfighter, such as planning tools or intent classification through machine learning (ML), to showcase how improved technology can optimize their tasking.

15.3.5.1 Voice Exploitation

The voice exploitation tool in the testbed environment transcribes audio and allows for speaker identification using the microphones in the room. By allowing the system to recognize the speaker, it can categorize and rank points of focus. To illustrate, when Naval watchstanders in the room are identified, the system can steer the microphone beams toward those individuals to prioritize their audio inputs. In addition to this hardware guidance, it also provides software guidance by prioritizing responses to designated individuals within a command hierarchy. The significance of utilizing voice exploitation in the testbed is the ability to showcase the suite of tools being tested without users needing to know specific implementations or interactions with the software, allowing them to use natural language to interact with systems. Voice exploitation is tightly coupled to AmI environments as it provides an immediate mechanism for user interactions in the space via ambient audio collection and processing such that no additional microphones or equipment need to be worn by users.

15.3.5.2 Planning Tools

To further showcase the capabilities of the testbed, we include planning tools that are used by Warfighters to better simulate wartime environments. An intermediary service routes API requests to the planning tool, offering the ability to easily control Naval plan development using voice commands. These planning tools significantly update the current suite of applications used for planning, which often relied significantly on manual input and had gaps and opportunities for human error. Voice activation is naturalistic and easy for new users to learn. At the same time, the interface design is familiar to Warfighters and accurately represents the command environment. The software manages the visualization and presentation of the routes planned by the Warfighter, presenting a consistent and structured result. This optimizes the planning process; there is no longer a need to expend valuable time and resources on learning formatting in cumbersome, outdated applications. Other tools and services can also easily be incorporated into the testbed without needing to worry about how data or actions are presented.

15.3.5.3 Understanding Intent

For all tasks and workflows, the system must understand the user's intents and goals. The testbed incorporates tools that can parse out intent from a speaker and route that data to specific actions that the system can take. While the concept seems simplistic, its usage is critical for the system. Being able to successfully route user requests to the appropriate application allows for the user to interact with the system seamlessly. The user does not need to spend time searching for the correct application or checking that it is available prior to issuing a command. The intent recognition system runs continuously in the background so that user commands are always recognized and routed accurately. It allows users to test the tools in the testbed without additional knowledge of their implementations. Our default speech interface implements one version of this capability, but there is additional, more specialized software that would be possible to implement in the future to detect intent under more specific domain conditions.

15.3.5.4 Common Operational Pictures

This route planning service is integrated with the Commander's Intent analysis to construct routes based on any constraints created by the Commander's Intent. Once the route is built, the data artifact is sent to the Planning tool to visually display the newly created route to the user. The benefits this feature adds are twofold: not only does it automatically consider Commander's Intent in generating a specific route, but it also produces an accurate and easily navigable map, allowing for visual representations within the planning tool.

15.3.5.5 Chat Messages

Chat messages are displayed on the screen to introduce events that occur in a natural manner and allow for a more immersive environment by providing messages that may appear in an operational environment. This feature mirrors the realistic reliance on instant messaging applications in Naval domains. The inclusion of a chat feature shows that future implementations would allow external interaction with this system. Rather than being closed off or limited in use, the testbed creates a hub for collaboration. It also allows some realism within the testbed, allowing for a better simulation of the existing operational environment.

15.3.5.6 Integration with A/V Equipment

Integrating with A/V equipment is one of the core attributes of the testbed system. The audio system routes the acoustic inputs to an automated speech recognition service to be transcribed and exposed to subsequent system services. Cameras and computer vision services track the movement of registered speakers and are later used to direct the microphone beam towards prioritized users to optimize the quality of the audio stream. The ability to track participants and roles ensures that all feedback is heard, while still prioritizing the most urgent tasks. These services allow for more robust use of the testbed and increase the number of participants such that multi-user human-AI teaming studies can be effectively supported.

15.3.5.7 Integration with Commercial Software

As an example of a third-party commercial software integration, the current testbed can work directly with the Microsoft PowerPoint[6] application. While this is one of the simpler services deployed in our testbed, it excels at showcasing the integration of LLMs as a service. With the LLMs acting as a passive listener, it allows the user to control slide progression using only voice activation. The purpose of these integrations is to provide easier use cases for the audience to understand the usage of the system within the scope of applications they already find familiar. This application is one example of such an implementation and serves as a blueprint for incorporating other commercially available software in the future. Users can see that the testbed is not limited in its possibilities to customized software systems, but rather that nearly any secure software or product that they may leverage in their existing workflows can be successfully incorporated into the testbed.

15.4 DEMONSTRATION AND EVALUATIONS

15.4.1 Testbed Design

The design process for this testbed requires the integration of hardware and software capabilities that are also aligned with user needs. If the product has interest and relevance to the broader cohort, we can then examine where it would best fit into the existing scenario. Some products are easily added to the baseline testbed, while others require a separate, controlled implementation to fully showcase their capabilities (Figure 15.3).

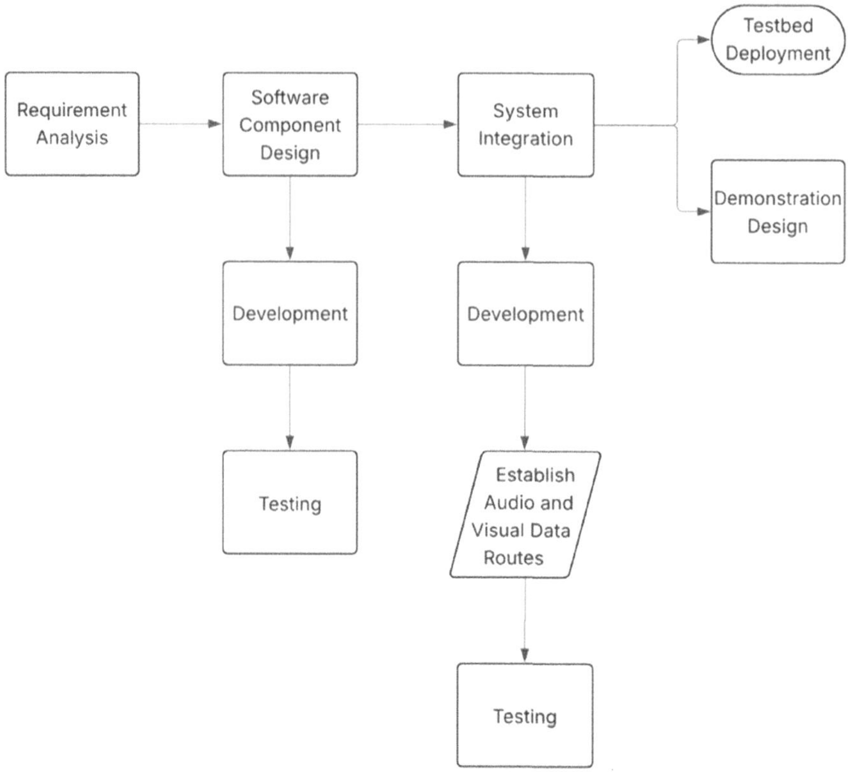

FIGURE 15.3 Testbed design workflow: When adding services to the testbed, an analysis of the requirements for the service is first done to see how it fits into the existing testbed processes. Then, the necessary software components are identified and can be developed and tested. Once initial development is complete, these components are integrated with any necessary A/V routes. Then, an analysis and review of the new process flow is done to fit the service into the testbed and allow the developers to either highlight or conduct experimentation with the updated component.

As part of the integration process, the software development teams need to collaborate to identify the types of data needed for a given feature, as well as how the information will be collected. Some data are captured through speech, while others require data artifacts to be created and transferred to specific services or tools. Throughout this process, we must assess the security requirements involved in transferring documents and data streams into the testbed. When adding software to the testbed, we need to identify the user inputs and external data sources it will need, then establish a route to ensure both can be sent to the tool. For most teams, the typical process for integrating additional capabilities involves the containerization of applications and software services before they are added to the primary testbed compute environment. Some projects have capabilities that only run on standalone machines, and these can be included as part of the networked testbed configuration. When the testbed is deployed on a single machine, it eliminates the need for complicated networking; however, this introduces compatibility issues when services have differing package or library requirements and cannot be deployed within a container orchestration service. To circumvent this issue, we require all services to support some level of containerized deployment, allowing them to be independently developed, managed, and hosted, even from within the same physical computer.

Beyond this technical aspect, the narrative component is critical to show how any tool will be incorporated into a simulated scenario to support testbed experimentation. We need to determine how each tool fits into the narrative flow and how it contributes to the user's mission and workflows. Lastly, we need to determine whether the service we are incorporating should exist as a baseline capability of the testbed or if it is better suited as an add-on feature specific to a certain configuration, deployment, or experiment.

15.4.2 Experiment Considerations

To realize the benefits of the AmI Command Center testbed, it is necessary to record how users work in the space and analyze these interactions as new technologies are integrated. These human-AI teaming studies are conducted in accordance with approved IRB protocols tied to the testbed. As the scope and emphasis of these studies change, active protocols are amended as needed, and new protocols are constantly being developed to support the necessary HSR activities. In lieu of the specifics of any individual protocol, we highlight here the most critical features of the various protocols that have leveraged or will leverage the testbed.

15.4.3 Data Collection Mediums

To effectively characterize the environment under study as users complete various tasks and workflows, we have found it necessary to collect four distinct data types (audio, video, system, and survey). Each of these data types has the potential to encode personally identifiable information (PII), so it is imperative to outline how each will be collected, handled, and utilized as part of any testbed-related IRB protocol. The current implementation of the testbed relies heavily on voice interactions, which places a large emphasis on the audio collection medium. Recordings of users' voices are considered PII since leakage has the potential to be traced back to a specific individual. For participants to engage with the system authentically, they need to trust that their personal data will be anonymized to the greatest extent possible and that their feedback will be used solely for the research purposes outlined in the informed consent process they complete.

Stored data is encrypted, and raw forms containing PII are accessible only to members explicitly included in the IRB research protocol. As an initial data processing step, the data is extensively deidentified for use in the various studies. As an example, for the audio data, attempts are made to transcribe the content and further remove any references to specific individuals or other identifying information. The resulting transcript would contain only key information about the way users interact with the system rather than focusing on the performance or specific identifiers of any individual.

15.4.3.1 Data Usage

As part of the IRB research protocol, the informed consent process completed by study participants must outline methods for how the collected data may be used. We have found it beneficial to provide multiple data usage options to enable increased utilization of the space while maintaining the ability for researchers to collect cohesive sets of data from a subset of users. For any subjects consenting to participate in studies within the testbed, we suggest options for utilizing the space with no data archival, utilizing the space with temporary data storage that enables presentations, or utilizing data storage options that will record the users' interactions and be used by researchers in accordance with the procedures outlined in the research protocol.

15.4.3.2 Data Handling

The means by which the data will be handled must also be outlined in detail. Active-duty service members are considered a protected class under IRB protocol and are covered under DoD Directive 3216.02, with specific guidelines that may limit or otherwise prohibit their participation (Office of the Under Secretary of Defense for Research and Engineering, 2025). Special care must be taken to ensure that participation in the study poses no risk of physical injury to the service members, and civilian and military supervisors need to be vigilant that they are not inadvertently influencing a service member to participate. Service member participation in these studies is often at the discretion of their chain of command, and Warfighters need to feel confident that their responses will be handled responsibly and securely to participate meaningfully and provide constructive feedback (Rhon et al., 2021). These restrictions on data handling are not only important from an HSR perspective but must also adhere to any additional considerations regarding data sensitivities and classification in various environments. Ensuring that data is stored in access-controlled systems and encrypted in accordance with security procedures are examples of how this data handling must be implemented in practice.

15.4.3.3 Participant Recruitment

How participants are recruited will depend on the implementation of the testbed. For the persistent testbed example of the CCoF, participant recruitment will be defined by the protocol to ensure the desired study population is sampled. For a mobile testbed deployment, it is necessary to determine an event or location that will host participants from the desired population and then take steps to ensure the testbed can be brought into that environment.

15.4.3.4 Study Design

It is imperative that each protocol outlines the specific research objectives and study design to ensure all the previously discussed considerations are properly addressed. We believe having a testbed with modular components that can be independently controlled and quickly configured to meet study needs is imperative for providing a useful analytic framework for exploring human-AI teaming, which has led to many of our environment design decisions. An example of a current experimental design is discussed in the following sections to showcase how the current testbed is being utilized by our research team.

15.4.4 Scenario Design

See Figure 15.4

15.4.4.1 Experiment Design

The testbed in its current iteration consists of two sections: a presentation on system features and an interactive scenario. The NIWC Pacific technical vision film serves as the backdrop for our scenario. With these objectives as the inspiration, we then survey our existing capabilities, as well as those in development, and incorporate them into the design. Our testbed serves as the focal point for different groups to collaborate on presentations to interested stakeholders and, ultimately, user experiments.

In designing the testbed, we not only consider the software and hardware requirements discussed earlier, but we also script the entire scenario to provide a

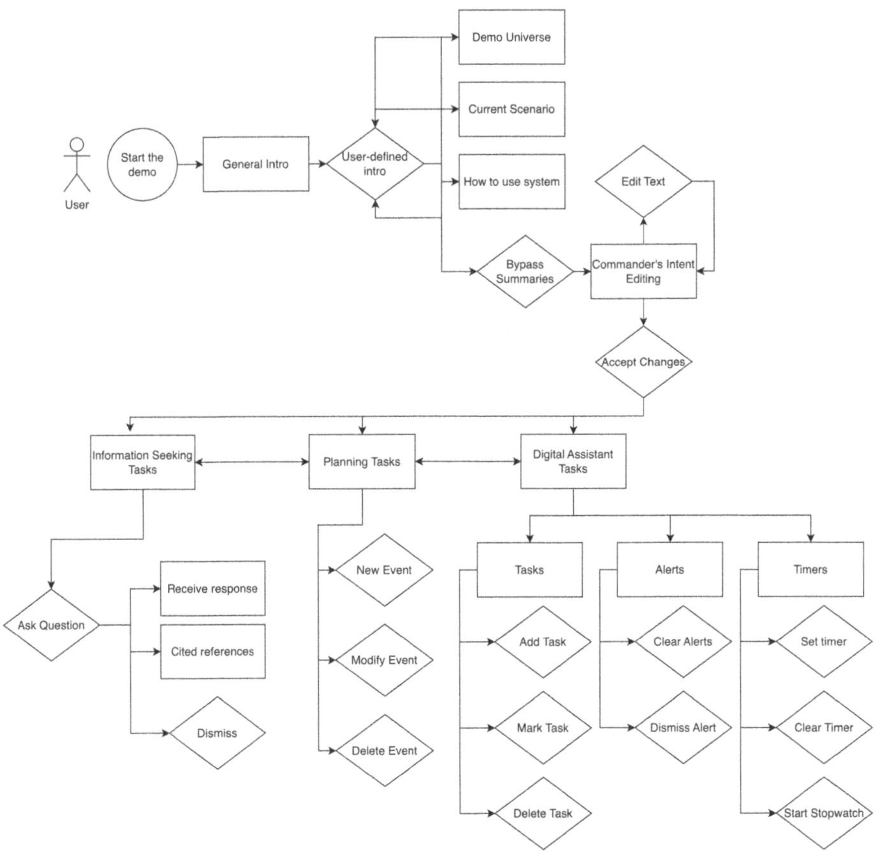

FIGURE 15.4 Workflow diagram of initial AmI experimentation: Diagram of the flow of the AmI testbed as events occur and participants take actions. The rectangles are discrete events within the scenario, while the diamonds represent user interactions and potential scenario branching points. This diagram shows a basic introductory flow that introduces a user to the AmI testbed and provides a set of example decisions.

well-defined transition from one tool to another. Pacing is a critical aspect of this planning process. We must ensure that the limited time we have for a presentation or experiment can encompass all the tools that need to be showcased or studied. For programs that take longer to finish processing, we preload certain features to run on command in the background.

The interactive aspect of the testbed trains users on the speech interface and provides an immersive experience. Through a process of trial and error, our team learned that many of the potential subjects who were accustomed to digital assistants tended to believe the AmI environment could respond to any verbal instruction rather than the more limited set we expose during our experiments and demonstrations. Alternatively, others had limited experience with digital assistants and assumed that they needed to speak at an elevated volume for their voices to be picked up, which strained the highly sensitive audio system. These challenges demonstrated a need to refine the testbed introduction in a way that engaged the audience in the learning process.

The tasks users complete in our initial demonstration and experimentation environment have been carefully curated to introduce the system in a naturalistic manner and to guide the subjects on the speaking volume and format needed for the system to process directed interactions. After the scenario is complete, the research teams present a debrief on the features and receive feedback for future iterations.

15.4.4.2 Scenario-Based Problem Solving

When participants enter the CCoF, they are introduced to the space that simulates the TFCC room familiar to Naval Carrier Strike Group command staff. After an overview of the system's capabilities, the participants receive a briefing on a fictional scenario. The participants are then given opportunities to prompt the system using different voice commands to show how the tools facilitate and expedite the planning process.

The participants interact with the testbed through tasks progressing in complexity, starting with low-stakes system interactions, such as changing slides or using a digital assistant, and leading into scenario-driven interactions that convey a sense of urgency. In the scenario crafted for this purpose, the audience is embedded in a universe in which Naval forces are engaged in active conflict and tasked with evacuating civilians and military personnel from a flooded island, simulating a Humanitarian Assistance and Disaster Relief mission. Complicating this mission is the presence of adversarial forces surrounding the area and the geographic limitation of a restricted operating zone. Acting as the commanding officer and staff within the TFCC, the participants must swiftly and safely conduct a rescue mission while avoiding any escalation of the conflict. This realistic scenario offers multiple opportunities for participants to engage in various tasks, such as requesting summaries of documents, pulling up maps, and implementing planning operations.

15.4.5 Naval Use Case

The CCoF is a secure testing and development environment that mirrors its counterpart on a Naval ship. It is used internally by our research team to show external

TABLE 15.2
Comparison of Naval and Public Use Cases

Component	Naval Use Case	Public Use Case
Location	Tactical Flag Command Center (TFCC)	Emergency Operations Center (EOC)
Scenario	Civilian evacuation	Wildfire response
Operational picture	AMPT	Open map server
Documents	Naval Doctrine/Commander's Intent	Emergency Operations Reference

This table presents the current implementation for naval use cases and the proposed use cases for public domain.

collaborators the current capabilities under development and being prioritized for further study and deployment. It additionally serves as a space for both Warfighters to contribute to HSR studies and engineers to test new software and hardware. In a typical event, the Commander uses the TFCC space to work with their staff to execute a mission in a specific warfare domain. The system ingests Naval doctrine and Commander's Intent and provides guidance regarding decisions on how to allocate units for the safety of both civilians and military personnel. Using a specialized Naval planning tool designated as AMPT, Warfighters can see where forces are and speak to the system, updating results in real time (Table 15.2).

15.4.6 General Public Use Case

When examining wider use cases of our AmI environment, one compelling application would be in training courses for disaster response and situations that require the establishment of emergency operation centers (EOCs). Existing tools have been developed that use AI in wildfire response to create route planning (Zhao et al., 2020) and predict wildfire spread (Giannakidou et al., 2024). Other studies have already found success in implementing elements of AI in disaster response, such as the use of human and UAV teaming tools (Ramchurn et al., 2016; Sun et al., 2020). Our AmI environment could integrate all these tools into an immersive testbed and be used to create a realistic scenario for students training to become first responders.

During a natural disaster, local authorities must rapidly set up an EOC (US EPA, 2024) to manage the crisis. In our testbed, students could replicate all the typical stages of emergency response while also experimenting with the latest tools. Our current capabilities would easily allow for the incorporation of an open map server to instantaneously display areas of urgency. The system would be able to ingest emergency operations reference documents and advise the students on the best COA. Students would be able to practice allocating resources appropriately using built-in planning tools. The testbed would offer students the opportunity to rehearse unexpected scenarios and balance a rapid and efficient response with a limited roster of personnel and supplies.

15.5 LIMITATIONS AND FUTURE DIRECTIONS

15.5.1 Security Considerations

Some unique challenges of working with LLMs for military use cases are the consideration of wartime scenarios and the delicate handling of controlled unclassified and classified information. The CCoF operates at an Impact Level 5 (IL5) (Microsoft, 2023), meaning it is certified to hold CUI. In a Denied and Degraded Environment, during which there is a significant decrease in critical utilities ranging from a reduction in performance to a near-total loss of capabilities, systems designed for military use must be able to function seamlessly without interruption. This eliminates the possibility of employing any LLMs that are only accessible through third-party cloud services, lack the necessary security qualifications, or are not open source. Additionally, the U.S. Navy's Research, Development, Test, and Evaluation network adheres to strict requirements to secure assets. GPU capacity is limited on standard-issue devices, with additional restrictions on Naval ships due to increased security requirements. With all these constraints in mind, our current work is limited to open-source, locally available LLMs that can function within a fixed GPU capacity.

15.5.2 Ongoing Work

The current iteration of the testbed utilizes traditional ML techniques to implement intent classification. However, a similar approach utilizing a multi-agent framework is also under development and can classify speech or text inputs into the same intent categories, such as question-asking, confirmation/verification, or editing. The testbed is currently able to use LLMs to extract keywords that signify actions needed for specific services and can route the requests as needed. For future work, more thorough information extraction to encompass all available services is planned. Additional plans include more advanced orchestration mechanisms to support the more complex user workflows and tool interactions that are expected in upcoming human-AI teaming studies utilizing the testbed.

Initial experimentation within the testbed has begun but is in its nascent stages. To date, there has been one completed study, which focused primarily on attempts to reduce the transcription error rate to functional levels within the environment. Current studies are exploring the workflows that the system enables and how these interactions can impact efficacy and efficiency for individuals and teams. These early studies are beginning with more of an individual focus, but the testbed is designed to support team-based studies as well. There are plans to integrate more robust planning tools within the system before further studies are conducted. Due to the complexity and scheduling challenges of recruiting active-duty participants, it is our goal to produce a high-functioning and sustainable prototype before launching large-scale studies within the testbed.

15.6 CONCLUSION

The CCoF provides a modernized testbed for human-AI teaming studies and seeks to demonstrate what is possible within AmI environments. We have shown the ability to utilize automated speech recognition within complex planning scenarios by integrating it into the A/V systems of advanced demonstration spaces, defining workflows, and showing proof of concepts for how these systems work and how to easily integrate additional systems into the workspace. Looking ahead, the team has plans for additional integration with other tools. We would also like to develop a framework that allows the team to easily compare metrics such as system performance and user interaction efficacy between testbed versions. This will lead to a more effective experimentation environment that accounts for past failures and successes and efficiently introduces users to the necessary concepts.

We envision future implementations of this AmI environment in public-facing emergency response settings and aim to further improve our processes so they are portable and readily deployable in a variety of settings, both within the Navy and beyond.

NOTES

1 https://www.niwcpacific.navy.mil/VISION/
2 https://mistral.ai/news/ministraux
3 https://www.llama.com/models/llama-3/
4 https://github.com/AnswerDotAI/RAGatouille
5 https://www.langchain.com/
6 https://www.microsoft.com/en-us/microsoft-365/powerpoint

REFERENCES

Ahmed, S., Ilyas, M., & Raja, M. Y. A. (2019). Smart living: Ubiquitous services powered by ambient intelligence (AmI). *2019 IEEE 16th International Conference on Smart Cities: Improving Quality of Life Using ICT & IoT and AI (HONET*–ICT*)*. https://doi.org/10.1109/honet.2019.8908054

Berretta, S., Tausch, A., Ontrup, G., Gilles, B., Peifer, C., & Kluge, A. (2023). Defining human-AI teaming the human-centered way: A scoping review and network analysis. *Frontiers in Artificial Intelligence*, *6*. https://doi.org/10.3389/frai.2023.1250725

Bosch, G., & Barrué, C. (2009). Managing ambient intelligence sensor network systems, an agent based approach. In *Lecture Notes in Computer Science* (pp. 1121–1128). https://doi.org/10.1007/978-3-642-02478-8_140

Chhetri, M. B., Tariq, S., Singh, R., Jalalvand, F., Paris, C., & Nepal, S. (2024). Towards human-AI teaming to mitigate alert fatigue in security operations centres. *ACM Transactions on Internet Technology*, *24*(3), 1–22. https://doi.org/10.1145/3670009

Clavié, B. (2025). RAGatouille (0.0.9). Github. https://github.com/AnswerDotAI/RAGatouille

Constantine, S., Antona, M., & Ntoa, S. (2021). Chapter 41: Human factors in ambient intelligence environments. In G. Salvendy, & W. Karwowski, Eds., *Handbook of human factors and ergonomics*. Wiley. https://doi.org/10.1002/9781119636113

Dakshit, S., & Prabhakaran, B. (2025). Ambient intelligence in immersive realities. *IEEE MultiMedia*, *32*(2), 4–6. https://doi.org/10.1109/mmul.2025.3590111

Emergency Operations Center | US EPA. (2024, October 29). US EPA. https://www.epa.gov/emergency-response/emergency-operations-center

Fender, A. R., Benko, H., & Wilson, A. (2017). Meetalive: Room-scale omni-directional display system for multi-user content and control sharing. *Proceedings of the 2017 ACM International Conference on Interactive Surfaces and Spaces* (pp. 106–115).

Garcia, K., Vontobel, J., & Mayer, S. (2024). A digital companion architecture for ambient intelligence. *Proceedings of the ACM on Interactive, Mobile, Wearable and Ubiquitous Technologies*, *8*(2), 1–26. https://doi.org/10.1145/3659610

Giannakidou, S., Radoglou-Grammatikis, P., Lagkas, T., Argyriou, V., Goudos, S., Markakis, E. K., & Sarigiannidis, P. (2024). Leveraging the power of Internet of Things and artificial intelligence in forest fire prevention, detection, and restoration: A comprehensive survey. *Internet of Things*, *26*, 101171. https://doi.org/10.1016/j.iot.2024.101171

Love, J., Gronau, Q. F., Palmer, G., Eidels, A., & Brown, S. D. (2024). In human–machine trust, humans rely on a simple averaging strategy. *Cognitive Research Principles and Implications*, *9*(1). https://doi.org/10.1186/s41235-024-00583-5

Microsoft. (2023, December 7). *Department of Defense (DOD) Impact Level 5 (IL5) - Microsoft Compliance*. Microsoft Learn. https://learn.microsoft.com/en-us/compliance/regulatory/offering-dod-il5

Office of the Under Secretary of Defense for Research and Engineering. (2025, July 25). *DOD Instruction 3216.02 Protection of Human Subjects and Adherence to Ethical Standards in DoD-Conducted and -Supported Research.*

Ramchurn, S. D., Huynh, T. D., Wu, F., Ikuno, Y., Flann, J., Moreau, L., Fischer, J. E., Jiang, W., Rodden, T., Simpson, E., Reece, S., Roberts, S., & Jennings, N. R. (2016). A Disaster response system based on human-agent collectives. *Journal of Artificial Intelligence Research*, *57*, 661–708. https://doi.org/10.1613/jair.5098

Rhon, D., Oh, R., & Teyhen, D. (2021). Challenges with engaging military stakeholders for clinical research at the point of care in the U.S. military health system. *Military Medicine*, *187*(7–8), 209–214. https://doi.org/10.1093/milmed/usab494

Schmidt, A., Markowitz, J., Ashcraft, C., & Ramsden, D. (2025). Testbed for human-autonomy teaming (Testbed4HAT). *Artificial Intelligence and Machine Learning for Multi-Domain Operations Applications VII*, 11. https://doi.org/10.1117/12.3053992

Sun, W., Bocchini, P., & Davison, B. D. (2020). Applications of artificial intelligence for disaster management. *Natural Hazards*, *103*(3), 2631–2689. https://doi.org/10.1007/s11069-020-04124-3

Vats, T., Kumar, S., Singh, S. K., Madan, U., Preet, M., Arya, V., Bansal, R., & Almomani, A. (2024). Navigating the landscape: Safeguarding privacy and security in the era of ambient intelligence within healthcare settings. *Cyber Security and Applications*, *2*, 100046. https://doi.org/10.1016/j.csa.2024.100046

Verbeek, P. (2009). Ambient intelligence and persuasive technology: The blurring boundaries between human and technology. *NanoEthics*, *3*(3), 231–242. https://doi.org/10.1007/s11569-009-0077-8

Zhao, X., Lovreglio, R., Kuligowski, E., & Nilsson, D. (2020). Using artificial intelligence for safe and effective wildfire evacuations. *Fire Technology*, *57*(2), 483–485. https://doi.org/10.1007/s10694-020-00979-x

Index

For Product Safety Concerns and Information please contact our EU representative GPSR@taylorandfrancis.com Taylor & Francis Verlag GmbH, Kaufingerstraße 24, 80331 München, Germany